Müllen, Feng (Eds.)
Chemistry of Carbon Nanostructures

Also of interest

Organic Materials.
An Introduction
Ye, Xu, Neo (Eds), 2018
ISBN 978-3-11-047940-9, e-ISBN 978-3-11-047941-6

Nanocarbon-Inorganic Hybrids.
Next Generation Composites for Sustainable Energy Applications
Eder, Schlögl (Eds.), 2014
ISBN 978-3-11-026971-0, e-ISBN 978-3-11-026986-4

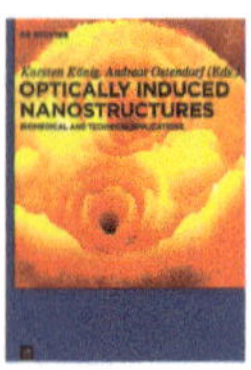

Optically Induced Nanostructures.
Biomedical and Technical Applications
König, Ostendorf (Eds.), 2015
ISBN 978-3-11-033718-1, e-ISBN 978-3-11-035432-4

Structure Analysis of Advanced Nanomaterials.
Nanoworld by High-Resolution Electron Microscopy
Oku, 2014
ISBN 978-3-11-030472-5, e-ISBN 978-3-11-030501-2

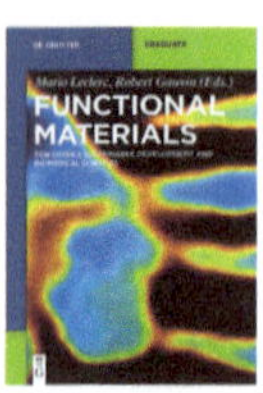

Functional Materials.
For Energy, Sustainable Development and Biomedical Sciences
Leclerc, Gauvin, 2014
ISBN 978-3-11-030781-8, e-ISBN 978-3-11-030782-5

Grignard Reagents and Transition Metal Catalysts.
Formation of C-C Bonds by Cross-Coupling
Cossy (Ed.), 2016
ISBN 978-3-11-035266-5, e-ISBN 978-3-11-035272-6

Klaus Müllen, Xinliang Feng

Chemistry of Carbon Nanostructures

Edited by
Klaus Müllen, Xinliang Feng

DE GRUYTER

Editors
Prof. Dr. Klaus Müllen
Max Planck Institute for Polymer Research
Ackermannweg 10
55128 Mainz
Germany
muellen@mpip-mainz.mpg.de

Prof. Dr. Xinliang Feng
Chair for Molecular Functional Materials
Center for Advancing Electronics Dresden (cfaed) &
Department of Chemistry and Food Chemistry
Technische Universitaet Dresden
Mommsenstrasse 4, 01062 Dresden, Germany
xinliang.feng@tu-dresden.de

ISBN 978-3-11-028450-8
e-ISBN (PDF) 978-3-11-028464-5
e-ISBN (EPUB) 978-3-11-038162-7
Set-ISBN 978-3-11-028465-2

Library of Congress Cataloging-in-Publication Data
A CIP catalog record for this book has been applied for at the Library of Congress.

Bibliographic information published by the Deutsche Nationalbibliothek
The Deutsche Nationalbibliothek lists this publication in the Deutsche Nationalbibliografie; detailed bibliographic data are available on the Internet at http://dnb.dnb.de.

Cover image: RUSSELL KIGHTLEY/SCIENCE PHOTO LIBRARY/Agentur Focus
Typesetting: Integra Software Services Pvt. Ltd.
Printing and binding: CPI books GmbH, Leck
♾ Printed on acid-free paper
Printed in Germany

www.degruyter.com

Contents

List of Contributors

Chapter 1
Polina Angelova
CNM Technologies GmbH
Herforder Str 155a
33609 Bielefeld
Germany

Chapter 5
Alberto Bianco
CNRS
Institut de Biologie Moléculaire et Cellulaire
Laboratoire d'Immunopathologie et Chimie
Thérapeutique
67000 Strasbourg
France
E-mail: a.bianco@ibmc-cnrs.unistra.fr

Chapter 6
Dr. Artur Ciesielski
Nanochemistry Laboratory
ISIS & icFRC
Université de Strasbourg & CNRS
8 allée Gaspard Monge
67000 Strasbourg
France

Chapter 9
Soumyajit Das
Department of Chemistry
National University of Singapore
3 Science Drive 3
117543 Singapore

Chapter 2
Siegfried Eigler

Chapter 6
Matilde Eredia
Nanochemistry Laboratory
ISIS & icFRC
Université de Strasbourg & CNRS
8 allée Gaspard Monge
67000 Strasbourg
France

Chapter 10
Holger Frauenrath
Ecole Polytechnique Fédérale de Lausanne
(EPFL)
Institute of Materials
Laboratory of Macromolecular and Organic
Materials
EPFL – STI – IMX – LMOM
MXG 037, Station 12
1015 Lausanne
Switzerland
E-mail: holger.frauenrath@epfl.ch

Chapter 1
Armin Gölzhäuser
Universität Bielefeld
Physik supramolekularer Systeme und
Oberflächen
Universitätsstr. 25
33615 Bielefeld
Germany

Chapter 2
Andreas Hirsch

Chapter 3
Kenichiro Itami
JST, ERATO
Itami Molecular Nanocarbon Project
Chikusa, Nagoya 464-8602
Japan
and
Graduate School of Science
Nagoya University
Chikusa, Nagoya 464-8602
Japan
and
Institute of Transformative
Bio-molecules (WPI-ITbM)
Nagoya University
Chikusa, Nagoya 464-8602
Japan

Chapter 4
Hongtao Liu

Chapter 4
Yunqi Liu

Chapter 5
Cécilia Ménard-Moyon
CNRS
Institut de Biologie Moléculaire et Cellulaire
Laboratoire d'Immunopathologie et Chimie Thérapeutique
67000 Strasbourg
France

Chapter 6
Prof. Paolo Samorì
Nanochemistry Laboratory
ISIS & icFRC
Université de Strasbourg & CNRS
8 allée Gaspard Monge
67000 Strasbourg
France
E-mail: samori@unistra.fr

Chapter 10
Stephen Schrettl
Ecole Polytechnique Fédérale de Lausanne (EPFL)
Institute of Materials
Laboratory of Macromolecular and Organic Materials
EPFL – STI – IMX – LMOM
MXG 037, Station 12
1015 Lausanne
Switzerland

Chapter 10
Bjoern Schulte
Ecole Polytechnique Fédérale de Lausanne (EPFL)
Institute of Materials
Laboratory of Macromolecular and Organic Materials
EPFL – STI – IMX – LMOM
MXG 037, Station 12
1015 Lausanne
Switzerland

Chapter 3
Yasutomo Segawa
JST, ERATO
Itami Molecular Nanocarbon Project
Chikusa, Nagoya 464-8602
Japan
and
Graduate School of Science
Nagoya University
Chikusa, Nagoya 464-8602
Japan

Chapter 7
Mikhail Shekhirev
Department of Chemistry
University of Nebraska – Lincoln
Lincoln, NE 68588
USA

Chapter 7
Alexander Sinitskii
Department of Chemistry
University of Nebraska – Lincoln
Lincoln, NE 68588
USA

Chapter 5
Isabella Anna Vacchi
CNRS
Institut de Biologie Moléculaire et Cellulaire
Laboratoire d'Immunopathologie et Chimie Thérapeutique
67000 Strasbourg
France

Chapter 8
Tanja Weil
Max-Planck-Institute for Polymer Research
Ackermannweg 10
D-55128 Mainz
Germany
and
Institute of Organic Chemistry III
Ulm University
Albert-Einstein-Allee 11
89081 Ulm
Germany

Chapter 9
Jishan Wu
Department of Chemistry
National University of Singapore
3 Science Drive 3
117543 Singapore
and
Institute of Materials Research and Engineering
A*STAR
3 Research Link
117602 Singapore

Chapter 8
Yuzhou Wu
Max-Planck-Institute for Polymer Research
Ackermannweg 10
D-55128 Mainz
Germany
and
Department of Chemistry and Chemical Engineering
Huazhong University of Science and Technology
037 Luoyu Road
430074 Wuhan
China

Chapter 3
Akiko Yagi
Graduate School of Science
Nagoya University
Chikusa, Nagoya 464-8602
Japan

Polina Angelova and Armin Gölzhäuser

1 Carbon Nanomembranes

Abstract: This chapter describes the formation and properties of one nanometer thick carbon nanomembranes (CNMs), made by electron induced cross-linking of aromatic self-assembled monolayers (SAMs). The cross-linked SAMs are robust enough to be released from the surface and placed on solid support or over holes as free-standing membranes. Annealing at ~1000K transforms CNMs into graphene accompanied by a change of mechanical stiffness and electrical resistance. The developed fabrication approach is scalable and provides molecular level control over thickness and homogeneity of the produced CNMs. The mechanisms of electron-induced cross-linking process are discussed in details. A variety of polyaromatic thiols: oligophenyls as well as small and extended condensed polycyclic hydrocarbons have been successfully employed, demonstrating that the structural and functional properties of the resulting nanomembranes are strongly determined by the structure of molecular monolayers. The mechanical properties of CNMs (Young's modulus, tensile strength and prestress) are characterized by bulge testing. The interpretation of the bulge test data relates the Young's modulus to the properties of single molecules and to the structure of the pristine SAMs. The gas transport through the CNM is measured onto polydimethylsiloxane (PDMS) - thin film composite membrane. The established relationship of permeance and molecular size determines the molecular sieving mechanism of permeation through this ultrathin sheet.

Keywords: Carbon Nanomembrane (CNM), Self-Assembled Monolayer, radiation induced cross-linliking, Graphene, Helium Ion Microsocopy.

Carbon nanomembranes (CNMs) [1] are very thin (~1 nm), synthetic two-dimensional (2D) layers or sheets with tailored physical, chemical or biological function. With their two opposing surfaces they interface and link different environments by their distinct physical and chemical properties, which depend on the thickness, molecular composition, structure and environment on either side. Due to their minute nanometer thickness and 2D architecture, they can be regarded as "surfaces without bulk" separating regions with different gaseous, liquid or solid components and controlling any exchange of materials between them [2].

CNMs have an extremely large surface-to-volume ratio but at the same time are robust enough to stand freely in air and to be prepared as freestanding structures (cf. Figure 1.1). Their uniqueness comes from the combination of their molecular-scale thickness and their macroscopic dimensions, thus linking the properties of the macroscopic materials with those of individual molecules. CNMs are predicted to possess a superior performance [3] and withstand pressure of a few bar without rupturing. This allows their use in technical applications, such as force sensing, and separation of materials where CNMs with pores of controlled size allow a faster passage of the selected gas or liquid molecules than any conventional filter [4].

Although very young, the group of functional supported or freestanding 2D nanolayers is already large. Among those, graphene (Figure 1.2(c)) [5] that only consists

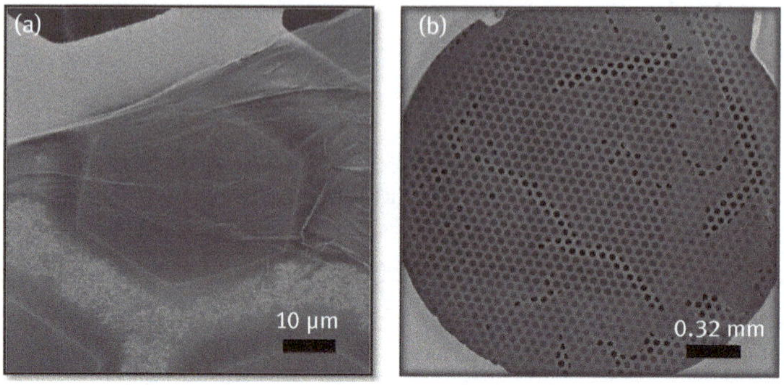

Figure 1.1: Helium ion microscopy image of a CNM, placed over a copper transmission electron microscopy hexagonal mesh grid: (a) field of view 70 μm, (b) field of view 2.25 mm.

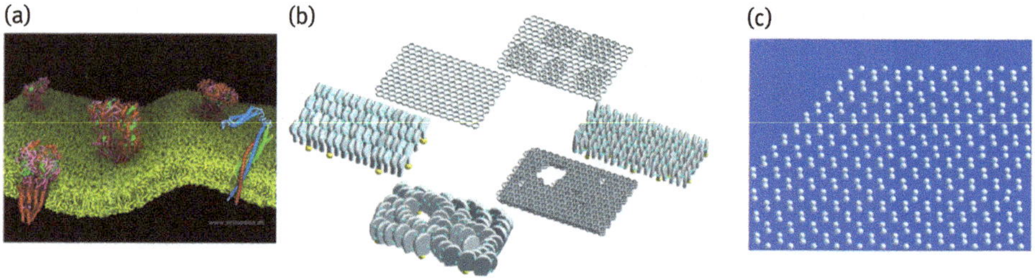

Figure 1.2: Schematic presentation of ultrathin carbon-rich nanomembranes: (a) biological membrane, (b) the route from self-assembled monolayer through CNM to graphene [6], (c) atomically thick graphene sheet.

of a single layer of carbon (only 3 Å) is by far the thinnest and structurally most simple one. On the other side, biomembranes (7–9 nm), consisting of lipid bilayers with embedded membrane proteins, are probably the most complex nanolayers that perform numerous tasks of living cells. Figure 1.2 shows a scheme with these "extreme" complex and simple nanolayers placed on the left and right, respectively. Other nanomembranes – whether of *artificial*, i.e., synthetic or engineered, or of *biological* origin – lay between these cases.

The structure and function of biological membranes are not in the focus of this chapter. However, one has to keep in mind that biological membranes are, by far, the most functional 2D "machines" on earth. Their applications range from encapsulation and mechanical protection of the cell content to molecular recognition, filtration and purification. Biomembranes can be considered as the "reference system" for functional 2D materials. They demonstrate the manifold of possible membrane application and they benchmark membrane performance.

In the following, we will focus on synthetic, man-made nanomembranes. These can be divided into inorganic, organic and hybrid ones. The inorganic membranes possess superior chemical and thermal stabilities and offer higher fluxes in contrast to polymer materials. One of the central positions among the inorganic membranes belongs to semiconducting silicon-based nanomembranes [7]. They combine thickness of 10–50 nm with extreme flexibility and robustness, thus allowing their incorporation in stretchable electronic devices. An intermediate class, which received deserved attention in the last decade, is the organic–inorganic hybrid membranes that include relatively new classes of porous materials like the organic–inorganic interpenetrated networks (IPNs) [8], metal–organic frameworks [9] and mixed matrix membranes [10]. They carry the macroscopic robustness of the ceramic materials and the flexibility, light weight and economical processability of polymers. Technologically important class of nanomembranes is the polymeric membranes [11]. Scaled up to industrial needs, they are mostly utilized in water and gas separation processes and biofiltration in medicine. But despite their easy preparation, low price and propensity to versatile chemical functionalizations, they lack chemical and mechanical stability. They also suffer from aging and fouling by time and to date their thickness cannot be reduced below 8 nm [12], thus limiting their permeance performance. There are two main processes of fabrication of ultrathin organic nanomembranes. Since the 1990s the layer-by-layer (LbL) technique is employed to form polymer membranes for corrosion protection, sensing and drug delivery [11a, 13]. In the LbL process, an electrically charged surface is sequentially dipped into positively and negatively charged polyelectrolytes, leading to the formation of polymeric membranes of well-defined molecular composition with thicknesses from ~15 nm to several hundreds of nanometers. An essential characteristic of this method is that it offers easy control over the thickness. Thin polymer membranes are also obtained by interfacial polymerization [12, 14], including air/liquid interface and liquid/liquid interface.

In the literature the term "nanomembrane" often denotes sheets with thicknesses below 100 nm. However, in this chapter we treat much thinner nanomembranes, namely those with a thickness of one molecule. The synthesis of freestanding nanomaterials with an atomic or molecular thickness is a challenging task and to date only a scarce number of chemical pathways have been developed. Atomically thin membranes can be made by exfoliating single sheets out of a layered material [15]. Graphene, consisting of a few (~1–5) layers of carbon atoms, was initially made by exfoliation [5] and is nowadays produced by a variety of techniques [16], allowing researchers to gain new insights into physics and chemistry in two dimensions. However, the surface of graphene is homogeneous and chemically inert, which is unfavorable for an efficient surface functionalization. On the contrary, covalent organic frameworks (COFs) [17] of arylalkynes [18] and related 2D systems [19] are nanomembranes with heterogeneous, chemically reactive surfaces. The first step in their creation is the synthesis of organic molecules of well-defined size and shape, with functionalities at defined positions. When brought into proximity, adjacent molecules form multiple covalent bonds and assemble in a 2D lattice. The second stage of nanomembrane fabrication is detaching the carbon nanolayer from its initial or sacrificial surface. A common feature of all these synthetic approaches is that they

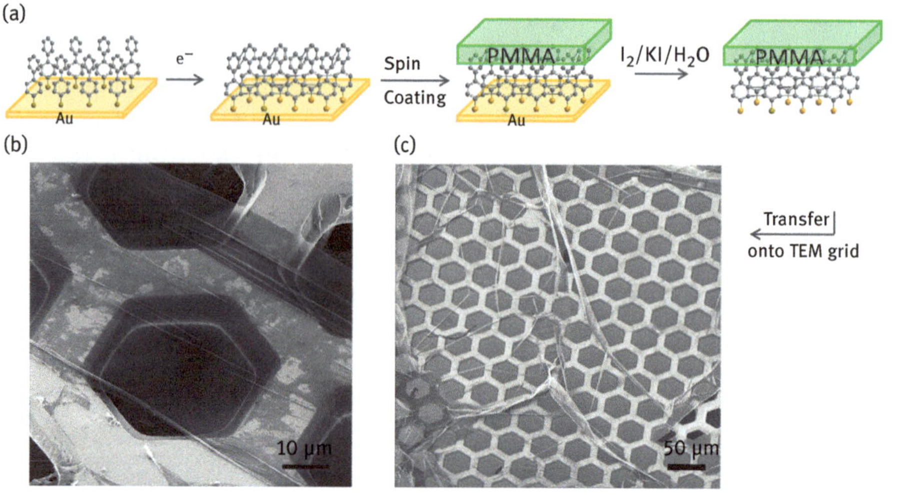

Figure 1.3: (a) Schematics of the structure and fabrication of a CNM, (b and c) HIM images of CNM, prepared from polyaromatic (HBC, precursors 3b and 3c, presented in Figure 1.9) molecules.

all are driven by spontaneous self-assembly. This quite simple and energy-saving approach, inspired by the biological membranes and mechanisms in living organisms, has gained wide applicability due to its superiority in tailoring organization with molecular-level precision. However, the necessity of specially designed molecular precursors requires high skills in organic synthesis and is time-consuming. Most of the described approaches are also restricted to micrometer lateral dimensions, expensive and sophisticated fabrication and low mechanical stability.

A molecular route toward the large-scale fabrication of a CNM that combines the thinness of graphene with the chemical functionality of a COF and the ease of fabrication of LbL films was described by Angelova et al. [6]. The fabrication route (Figure 1.3(a)) utilizes a sequence of (i) molecular assembly on a solid surface, (ii) radiation-induced 2D polymerization and (iii) a liftoff of the network of cross-linked molecules [20]. Figure 1.3(b) and (c) shows freestanding CNMs, formed from hexa-peri-benzocoronene (HBC) derivatives (Figure 1.9, precursor 3b and 3c), which, after etching of the initial gold film, are transferred [20] onto a perforated support (copper TEM grid) and imaged with a helium ion microscope (HIM) [21]. This novel charged particle microscopy combines high resolution (~4 Å) with high surface sensitivity and the possibility to image nonconducting ultrathin specimens [22].

Thereby the surface-bound self-assembled monolayers (SAMs) are converted into freestanding CNMs, which are structurally amorphous, chemically durable [23] and thermally [24] and mechanically robust [20] to be suspended over orifices of micrometer sizes. Furthermore, upon annealing they can be converted into well-conducting nanocrystalline graphene sheets (Figure 1.4) [25]. In addition, the freestanding CNM has two faces, the chemistry of which is determined by the surface-active and the end group of the molecular precursors. Thus a smart molecular design can easily tailor

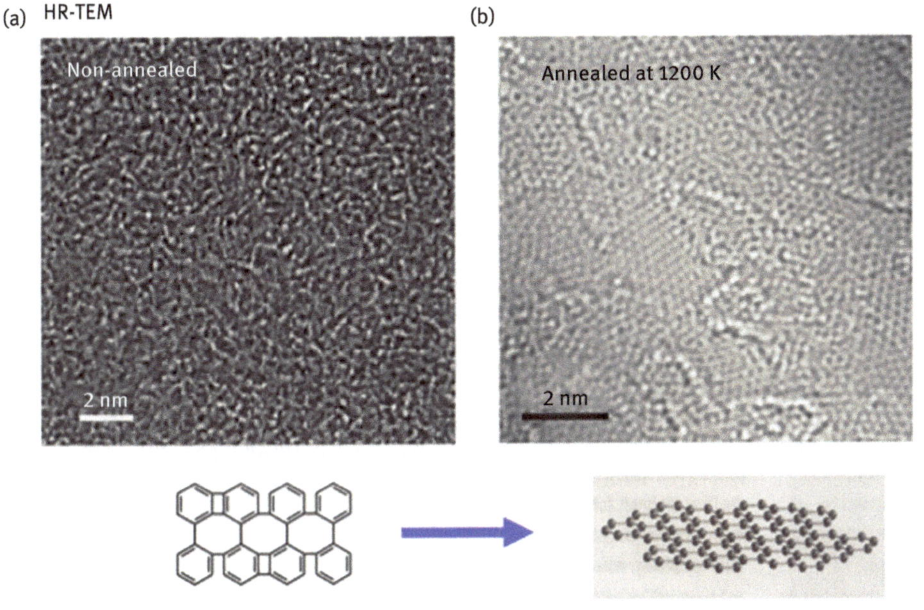

Figure 1.4: High-resolution TEM images of a CNM (a) before and (b) after annealing in vacuum at 1,200 K. Reprinted with permission from Ref. [25]. Copyright (2011) American Chemical Society.

the chemical functionalities on both faces of the membrane, and by means of a 2D synthesis a refashioning of its properties and functions is easy to achieve [26]. But the most remarkable aspect of this pathway is that the thickness, porosity and surface chemistry of the nanomembranes are determined by the molecular order of the monolayers, and structural and functional features are passed on from the molecules through their monolayers to the CNMs and finally on to the graphene.

1.1 Molecular Mechanisms of Electron-Induced Cross-Linking

It has been shown that electron, ion and photon beams induce specific chemical reactions, i.e., bond dissociation, oxidation/reduction or polymerization, in SAMs that depend strongly on their building blocks – the molecules. The response of a SAM to low-energy electron irradiation depends on the chemical structure of the ordered molecules. In general the aliphatic monolayers are heavily destroyed and thus broadly utilized as positive tone resists [27], but the phenyl-based SAMs are stabilized through the formation of intermolecular cross-links and can be used as negative resists. In aliphatic SAMs a cleavage of C–H bonds through resonant and nonresonant processes is induced, which leads to an orientational and conformational disorder of the chains, desorption of material and formation of C=C double bonds in the fragments, remaining on the surface, whereas in aromatic SAMs, it was found that upon irradiation with electrons, C–H cleavage occurs, which is then followed by cross-linking between neighboring phenyl units. During this process, cross-linked molecules maintain their preferred orientation and almost no material desorbs. Hence the electrons generate a molecularly dense monolayer that could be used as a negative tone electron resist [23, 27, 28].

In addition, electron irradiation also changes the surface chemistry of the nanomembrane. Several groups have investigated the chemical transformation of nitro-, cyano-, trifluoromethyl- *para*-substituents upon electron irradiation [29, 30]. While the underlying aromatic rings are partially dehydrogenated, the hydrogen atoms liberated from the aromatic cores locally reduce the nitro and cyano groups to amino and aminomethyl groups, which can be further chemically modified by electrophilic agents [26a, 26b, 31]. By employing a local electron exposure, one can define surface regions that can be used as a template for a site-selective molecular immobilization. This process is named chemical lithography, as the lithographic exposure directly affects the surface chemistry of the SAM [29a, 32].

Resolving the molecular mechanisms of the e-beam-induced polymerization in aromatic SAMs is a complicated task, since the long-range order in the cross-linked layers is distorted and a broad spectrum of molecular species can be formed. Turchanin et al. [33] employed complementary spectroscopic techniques such as X-ray and UV photoelectron spectroscopy (XPS and UPS), near-edge X-ray absorption fine structure spectroscopy (NEXAFS) and thermal desorption spectroscopy (TDS) to investigate the mechanisms of cross-linking in biphenyl-4-thiol (BPT) SAMs on gold (Figure 1.5). The experimental data are completed by quantum chemical calculations that allow the derivation of a model for the e-beam-induced chemical and structural transformations in aromatic thiol SAMs and agree well with the response to electron irradiation of other oligophenylene-based SAMs [29b, 29c, 34].

When a molecular layer is exposed to low-energy electrons, the electrons are effectively captured by the ordered molecular film, followed by a partial decomposition, desorption of hydrogen and molecular fragments, orientational and conformational disordering, and intermolecular cross-linking. For alkanethiol SAMs on gold [35], it was found that the electron energies, resonant with the electron orbitals of the aliphatic components, have a maximum at ~10 eV with a cross section of 10^{-16} cm^2. Therefore, dissociative electron attachment (DEA) is most likely the dominating process. DEA [36] is observed when a free electron interacts with an atom or molecule and is temporally captured to form a negative ion resonance (also called transition negative ion). This electronically excited anionic state, in which the attached electron resides on one of the unoccupied molecular orbitals, decays in a release of hydrogen anion and formation of carbon radical centers within the film or a release of hydrogen atom and formation of molecular anionic species. These reactive centers further react with the adjacent phenyl rings to form intermolecular C–C covalent bonds as the delocalization of π-electrons over the δ-framework of the aromatic ring retains its integrity through the irradiation process. Desorption of hydrogen molecules is generated through scattering of the formed hydride/hydrogen atom within the dense molecular layer.

For BPT SAMs ~650 primary electrons per molecule with energy of 50 eV are necessary to create the cross-linked molecular network [33]. It has been shown that secondary electrons (SEs) generated in the substrate [37] also contribute to the cross-linking

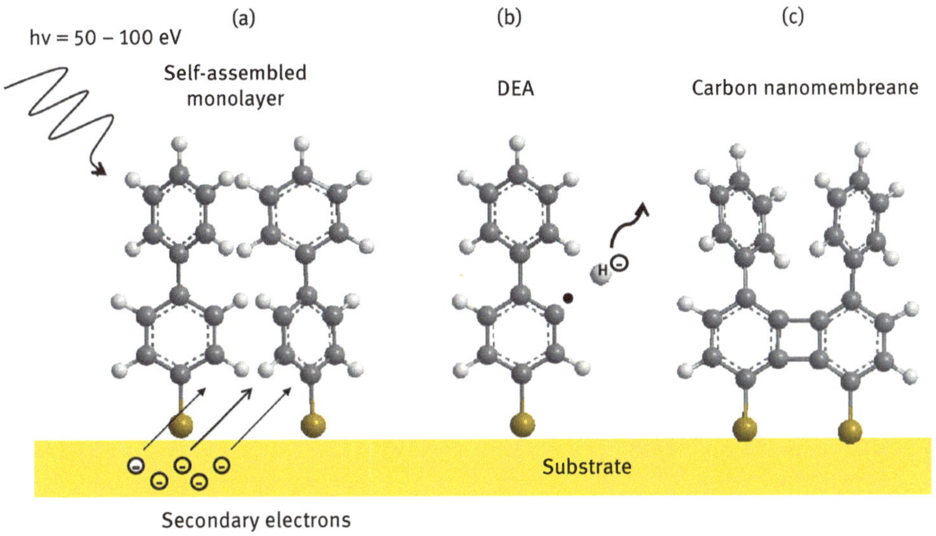

Figure 1.5: Schematic presentation of cross-linking of biphenylthiol-based monolayers: (a) low-energy electron irradiation and emission of secondary electrons, (b) dissociation of C–H bonds, (c) cross-linked molecules.

along with the primary e-beam (Figure 1.5) [38]. Taking into account the SE yield [39], the cross section for the cross-linking of BPT was estimated to be ~10^{-18} cm^2. This number is an average value without specifying an exact number of the involved chemical bonds and their chemical nature.

Electron-induced processes in polycyclic aromatic hydrocarbons were studied in the gas phase [40]. Resonances, characteristic for DEA, were observed at energies of 7–8 eV with effective cross sections of 10^{-17}–10^{-16} cm^2. As these cross sections are close in magnitude to the one found for biphenyl cross-linking, it is likely that DEA contributes to the C–H cleavage in BPT SAMs.

In Figure 1.6, XPS, UPS and NEXAFS data for a BPT SAM before and after irradiation with 50 eV electrons and a dose of 45 mC/cm^2 are presented. A comparison of these data reveals specific transformations upon cross-linking of the carbonaceous part of the monolayer and of the sulfur/gold interface. While the XP C1s signal (cf. Figure 1.6(a)) mostly preserves its initial intensity, demonstrating small reduction in the amount of carbon atoms after electron irradiation, the corresponding data of the C1s X-ray absorption show a reduction of the π^* resonance intensity while at the same time the δ^* resonances become more intense (cf. Figure 1.6(d)). The width of all resonances increases; however, their positions remain unchanged. These observations are attributed to a partial loss of aromaticity within the monolayer due to the interconnection of adjacent molecules via C–C bonds or even due to their partial decomposition. Measurements of angular dependencies of the resonance intensities of the pristine and e-beam-irradiated SAMs revealed an increase of the averaged molecular tilt within the monolayer upon cross-linking from $\phi = 31°$ to $41°$.

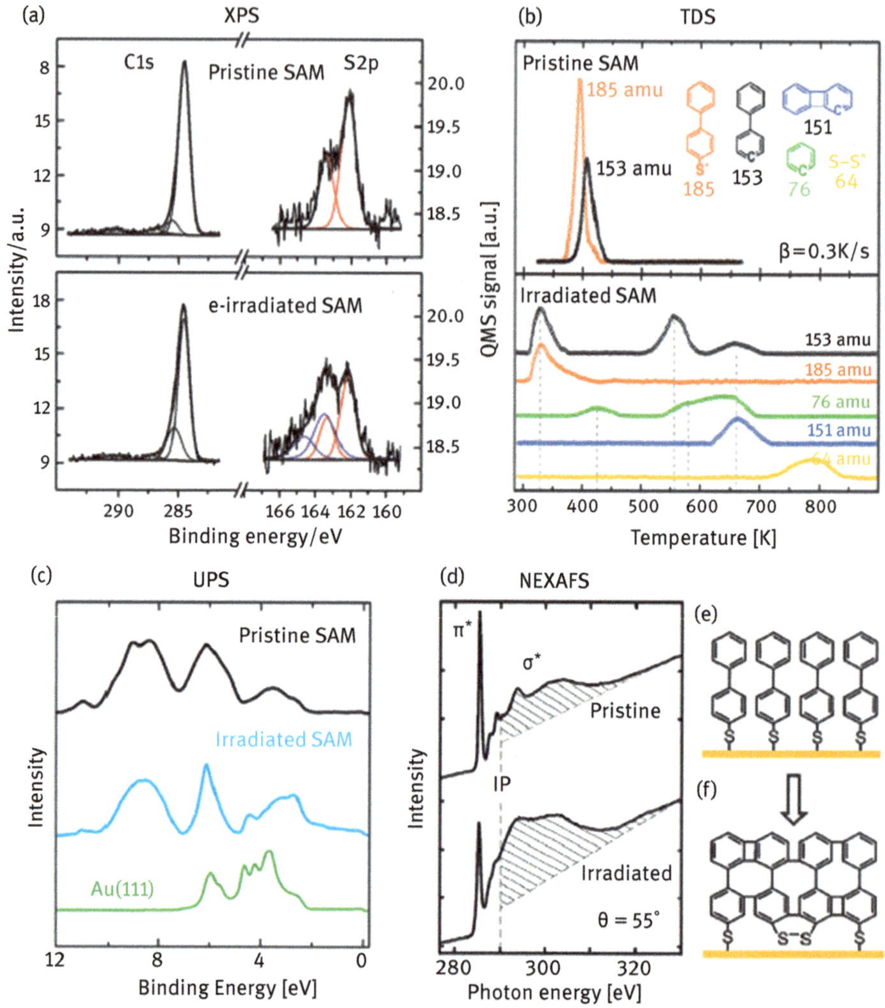

Figure 1.6: Spectroscopic characterization of the e-beam-induced cross-linking of a biphenyl-4-thiol SAM (e-beam energy of 50 eV, electron dose of 45 mC/cm²): (a) X-ray photoelectron spectroscopy (XPS), (monochromatic Al Ka, 1,486.7 eV), (b) temperature desorption spectroscopy (TDS), (c) ultraviolet photoelectron spectroscopy (UPS), (He I, 21.22 eV), (d) near-edge X-ray absorption fine structure spectroscopy (NEXAFS) (schematic representation of a pristine BPT SAM), (e) an e-beam cross-linked BPT SAM. Reprinted from Ref. [1]. Copyright (2012) with permission from Elsevier.

The UPS data show photoemission bands at 2.7 and 4.4 eV as well as a loss of fine structures around 7–10 eV upon cross-linking (Figure 1.6(c)). Based on quantum chemical calculations of the electronic structure, which were carried out for numerous possible cross-linked BPT dimers, the features in the UPS can be attributed to electronic excitations and thus allow the identification of individual molecular species such as BBDS, B1BPT, B4BPT, BBPDT and BBPDTn (Table 1.1). For the analysis of the

Table 1.1: Molecular species whose formation upon cross-linking of a BPT SAM was considered and for which UPS and HOMO–LUMO gaps were calculated.

Molecule	Abbr.	Structure	Stoichiometry	E_{gap}[eV]
Biphenylthiol	BPT	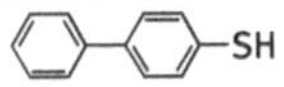	$C_{12}H_{12}S$	4.77
Bis-1,1-biphenylthiol	B1BPT	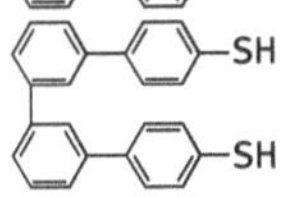	$C_{24}H_{18}S_2$	4.64
Bis-2,2-biphenylthiol	B2BPT	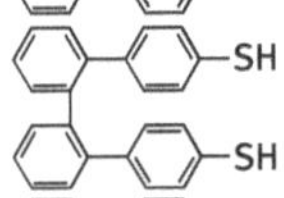	$C_{24}H_{18}S_2$	4.75
Bis-3,3-biphenylthiol	B3BPT	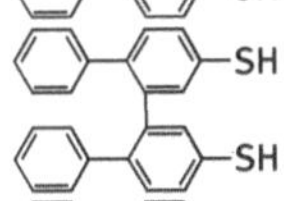	$C_{24}H_{18}S_2$	4.76
Bis-4,4-biphenylthiol	B4BPT	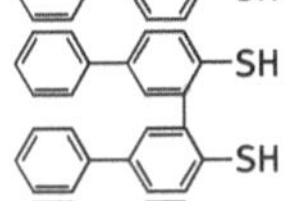	$C_{24}H_{18}S_2$	4.59
Diphenylthiol-biphenytene	DPTBP	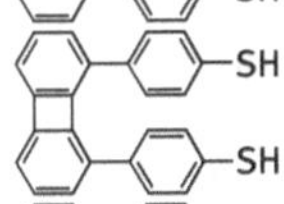	$C_{24}H_{16}S_2$	3.94
Diphenyl-biphenylenedithiol	DPBPDT	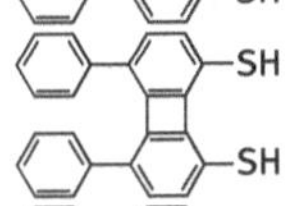	$C_{24}H_{16}S_2$	3.88
Bisbiphenylene-dithiol	BBPDT	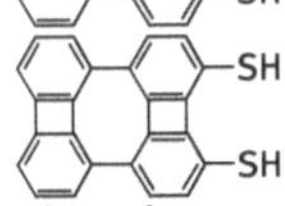	$C_{24}H_{12}S_2$	3.20
Biphenylene-dithiol	BPDT	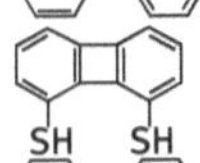	$C_{12}H_8S_2$	3.91
Bisbiphenyl-disulfanyl	BBDS	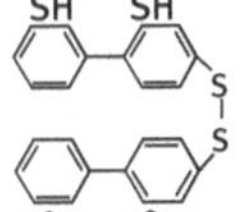	$C_{12}H_{18}S_2$	4.59
Biphenylene-dithian	BPDTn	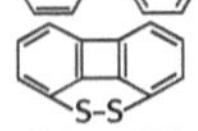	$C_{12}H_6S_2$	3.11
Bisbiphenylene-dithian	BBPDTn	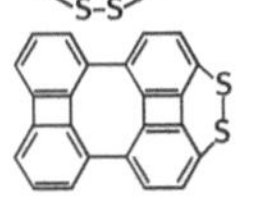	$C_{12}H_{10}S_2$	3.00

photoemission lines at first the molecular energy levels were calculated and then the allowed dipole transitions were considered [41]. The comparison of calculated and measured UPS suggests a prevailing formation of BBPDT and BBPDTn species upon e-beam irradiation.

This conclusion is consistent with the thermal desorption of biphenylene fragments (m/z = 151) in a partially cross-linked BPT SAM, which was observed by TDS (Figure 1.6(b)). The appearance of dimers is also consistent with a time of flight secondary ion mass spectrometry study of irradiated 4′-methyl-1,1′-biphenyl-4-thiol (MBPT) SAMs on gold by Cyganik et al. [34a].

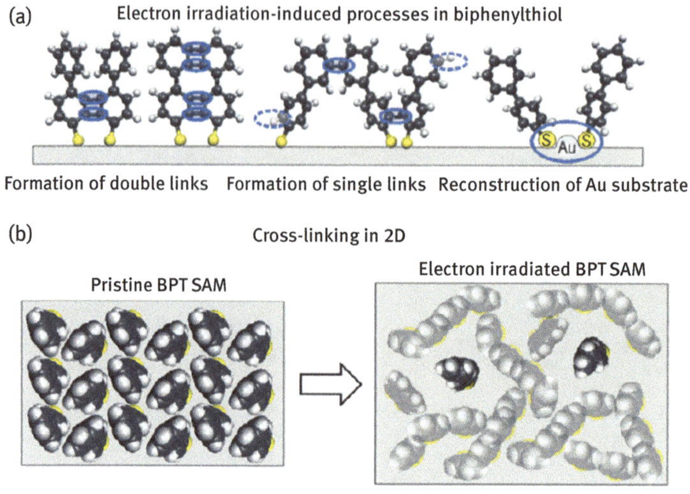

Figure 1.7: Schematic summary of the structural properties of pristine and cross-linked BPT SAMs: (a) molecular species formed upon e-beam irradiation, (b) conversion of a pristine BPT SAM into cross-linked BPT SAM. Reprinted with permission from Ref. [1]. Copyright (2012) with permission from Elsevier.

Distinct changes upon e-beam-induced cross-linking take place at the sulfur/gold interface of a BPT SAM. Figure 1.6(a) shows that besides thiolate SAM species [42] with a $S2p_{3/2}$ binding energy (BE) of 162.0 eV, new sulfur species with a BE of 163.5 eV form in a cross-linked SAM. Although its BE coincides with that of thioethers (R–S–R) or organodisulfides (R–S–S–R) [43], the presence of other species cannot be unequivocally excluded. It has been demonstrated that metal surfaces are not that rigid and actually reveal a substantial rearrangement upon formation of thiol-based SAMs [44]. In addition to vacancy depressions or ad-islands, which were attributed to a stress release [45], formation of thiolate-dimers (RS–Au–SR) also takes place [46, 47] and the measured S2p doublet at 162.0 eV is attributed to these sulfur species. It was shown that some molecules within thiol-SAMs form pairs, stabilized by an additional gold atom between neighboring sulfur atoms as depicted schematically in Figure 1.7(a). The density functional theory (DFT) - based geometry optimization (carried out for the related bisbiphenyl-disulfanyl species (cf. Table 1.1)) [33] revealed a substantial outward tilting of the aromatic backbones (see middle of Figure 1.7(a)) due to steric repulsion of the lower phenyl rings, which is in a qualitative agreement with the observed by NEXAFS increase in the tilt of the aromatic backbone upon cross-linking from $\phi = 31°$ to $41°$ [33]. Since cross-linked BPT molecules with multiple cross-links are less separated, they may provide sufficient space for formation of thiolate-Au-thiolate pairs with a higher downward tilt. These pairs can be stabilized by additional links at the upper phenyl rings as shown in Figure 1.7(a), middle.

To realize a cross-linking in two dimensions and the formation of CNMs, it is important to consider the molecular packing motifs adopted in the pristine BPT SAMs (Figure 1.7(b), left) [48, 49]. Isolated BPT molecules reveal a characteristic torsional angle between the upper and lower phenyl rings. Within the crystalline SAM both phenyl rings may not be coplanar but still have a substantial twisting. Theoretical studies indicate that this molecular degree of freedom is of key importance for the formation of ordered films of oligophenylene-based SAMs [50]. Moreover, in view of the herringbone arrangement, which is adopted in the crystalline phase of oligophenylenes [51], a lateral cross-linking occurring upon BBPDT (Table 1.1) formation comprises a molecular rearrangement, such as a rotation. Therefore, the formation of extended chains exhibiting the double Ph–Ph bonding motif (Figure 1.7(a), left) is sterically unfavorable and a 2D network can propagate via molecules that are linked either at the upper or lower ring as schematically shown in Figure 1.7(a) middle.

A further consequence of the cross-linking is a reduction of the spatial separation of neighboring BPT molecules. While the intermolecular distances within the pristine film are essentially given by the van der Waals dimensions of the molecules, the irradiation-induced additional carbon–carbon links also enable shorter distances. The density of the cross-linked films increases locally and may lead to a formation of "nano-voids," containing isolated and noncross-linked molecules (denoted as dark molecules in the scheme in Figure 1.7(b), right).

Due to the multiple cross-links the CNM exhibits a thermal stability up to 1,000 K [24]. Pristine biphenylthiol SAMs desorb above 400 K (~130°C) (cf. Figure 1.6(b)), whereas e-beam-irradiated SAMs become more stable with an increasing irradiation dose, i.e., with an increasing degree of cross-linking. This process saturates at a dose of ~50 mC/cm^2 for 50 eV electrons. For such samples only a slight (~10 %) reduction of their carbon XP signal is observed even upon annealing at ~1,000 K. But the data show that the sulfur atoms initially present in the monolayer, and in the CNM, continuously desorb upon heating through cleavage of C–S bonds until they completely vanish at temperatures above 800 K (cf. Figure 1.6(b)). The absence of any sulfur signal in the cross-linked monolayer after annealing at elevated temperatures demonstrates that the remaining carbonaceous film is not anchored by thiolate but is solely stabilized by covalent bonds within the aromatic network, which is directly coupled to the gold surface via van der Waals interactions. This conclusion is well corroborated by the temperature-dependent NEXAFS measurements [33], revealing a downward tilting of the aromatic rings and a substantial broadening of the π^* resonances at elevated temperatures. As demonstrated by quantum chemical calculations, the saturation of the cross-linking process at a dose of ~50 mC/cm^2 can be attributed to the electronic quenching of the anionic molecular states by means of tunneling or hopping (Figure 1.7).

Upon electron irradiation of an aromatic monolayer, a dense molecularly thick network is formed, which is robust enough to be further removed from its initial substrate (Au [20], Si_3N_4 [30], Cu [52]) and placed onto solid or perforated support (Figure 1.8) as the thickness of the resulting CNM is adopted by the thickness of its parenting

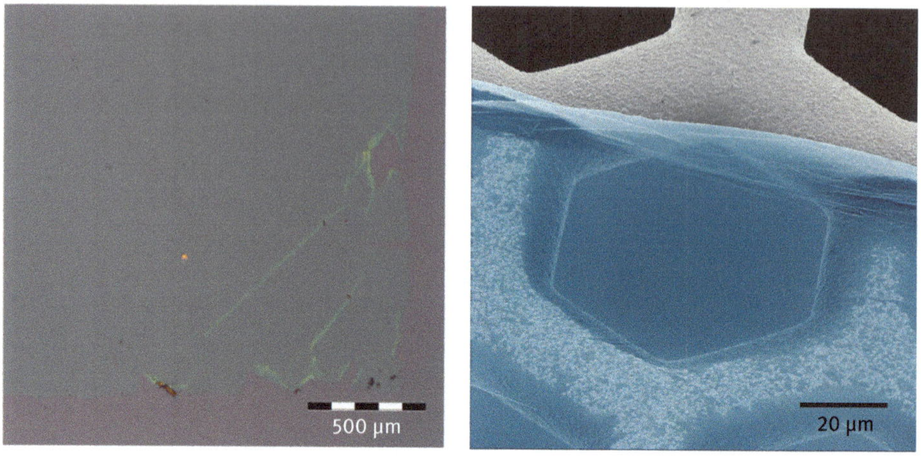

Figure 1.8: (a) Interference contrast of transferred CNM onto Si substrate with 300 nm silicon oxide layer, (b) helium ion microscopy image of a freestanding CNM (in blue color), transferred onto copper TEM hexagonal mesh grid.

SAMs. Motivated by these findings, we applied similar protocols to a variety of other polyaromatic molecules, aiming to examine the change in structural features and functionality of the membrane. Figure 1.9 shows schematic drawing of the molecular route, applied to different types of monolayers.

1.2 Tuning of CNM's Properties on a Molecular Level

Various types of thiol-based precursors were studied on Au (111) polycrystalline substrates: nonfused oligophenyl derivatives (Figure 1.9(d), structures 1a–c), which possess linear molecular backbones providing an improved structural ordering of the formed SAMs; condensed polycyclic precursors like naphthalene (NPTH), anthracene (ANTH) and pyrene (MP) derivatives (Figure 1.9(d), structures 2a–f), which are more rigid and should result in a higher stability and an increased carbon density of the monolayers; "bulky" molecules, like the noncondensed hexaphenylbenzene derivative with a propeller-like structure (structure 3a) and extended disk-type polycyclic aromatic hydrocarbons such as HBC derivatives (Figure 1.9(d), structures 3b–c) [53]. The former are equipped with long alkyl chains and a surface-active group, which is attached to the π-conjugated backbone through a flexible methylene linker. This molecular structure enables a control over the thickness and packing density of the SAMs by varying the conditions of preparation. In Figure 1.9(a) oligophenyls with a linear molecular backbone form well-ordered monolayers that can be cross-linked into homogeneous CNMs. After pyrolysis, the CNMs transform into graphene, whose thickness depends on the density of carbon atoms in the monolayers. In Figure 1.9(b), condensed polycyclic precursors also form monolayers that are cross-linked into

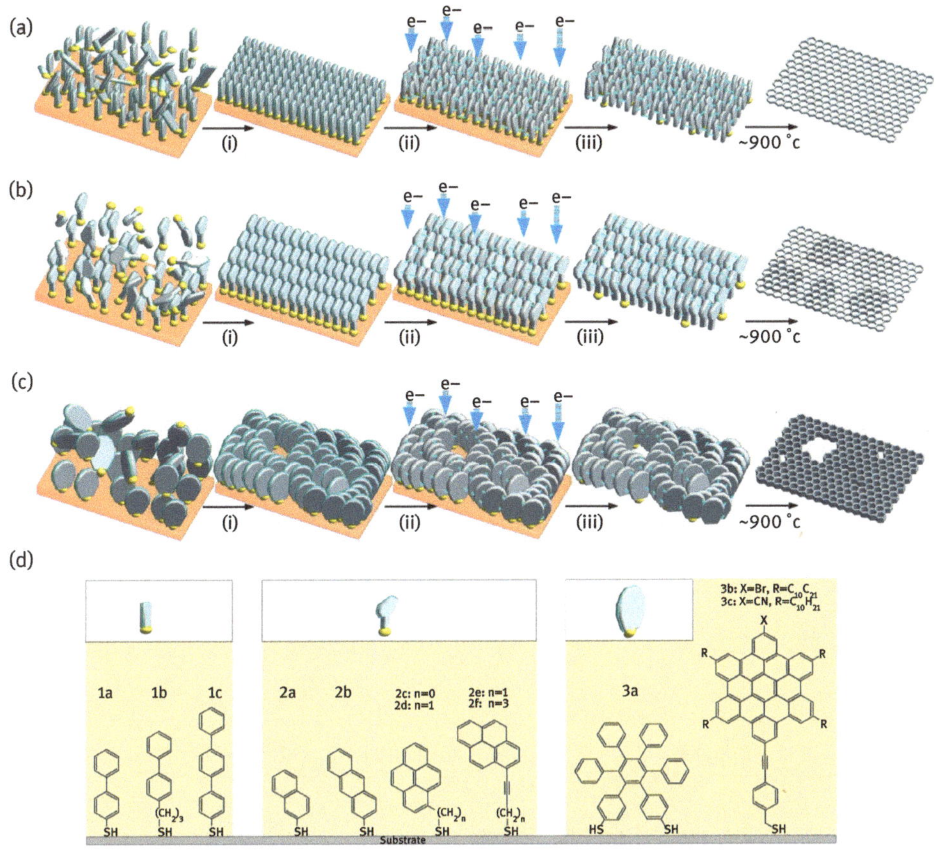

Figure 1.9: Schematic for the formation of CNMs and graphene from various molecular precursors: (a–c) schematic illustration of the fabrication route for CNMs and graphene. SAMs are prepared on a substrate (i), then cross-linked by electron irradiation to form CNM of a monomolecular thickness (ii). The CNM can be released from the underlying substrate (iii) and transferred onto a new substrate as a freestanding or supported material. Annealing to 900°C transforms CNM into graphene. (a) Fabrication of atomically thin CNMs and graphene from precursors 1a–c in (d), (b) fabrication of thicker CNMs and few layer graphene sheets from precursors 2a–f in (d), (c) fabrication of CNMs and graphene sheets with nanopores from precursors 3a–c in (d), (d) chemical structures of the molecular precursors used for preparation of CNMs and graphene. Reprinted with permission from Ref. [6]. Copyright (2013) American Chemical Society.

CNMs. After pyrolysis, the CNMs transform into graphene, whose thickness is higher than that in structure 1a, Figure 1.9(d), even if the carbon density is the same as in Figure 1.9(a). Figure 1.9(c) shows bulky aromatic hydrocarbons that assemble in a less-ordered monolayer and polymerize into CNMs with pores. After annealing, these nanomembranes transform into thicker graphene sheets with pores. Hence, the produced graphene adopts features from the preceding CNM, which itself adopts features from the monolayer, i.e., from molecules and surface.

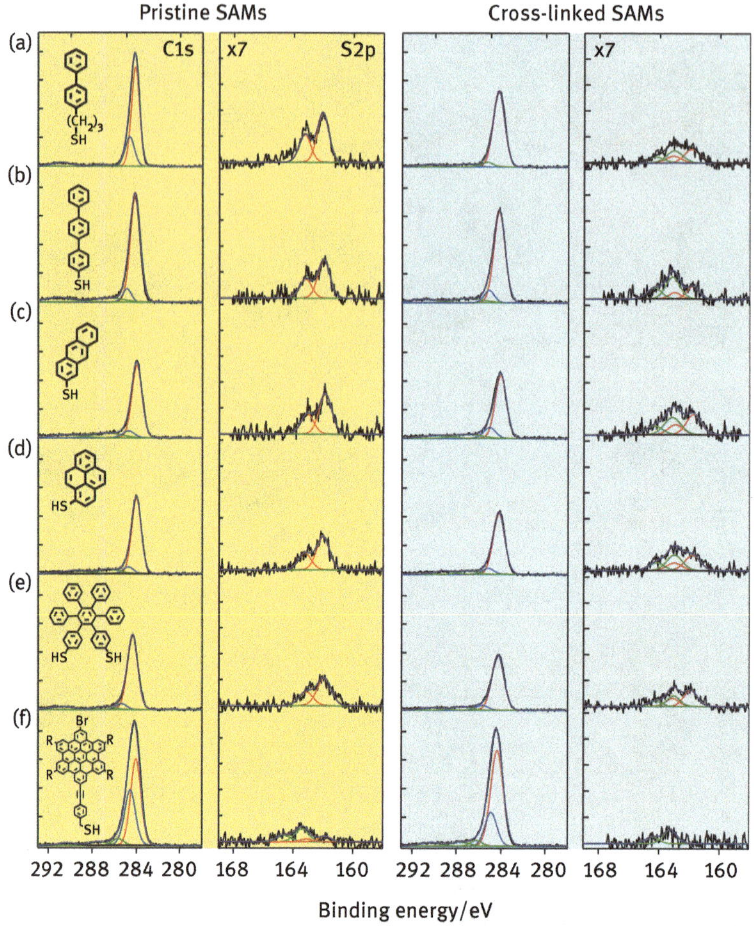

Figure 1.10: XPS data of pristine SAMs and CNMs: (a–f) XP spectra of C1s and S2p signals of the pristine (in left) and electron-irradiated monolayers (in right). In the insets are presented the chemical structures of the monolayer precursors. Reprinted with permission from Ref. [6]. Copyright (2013) American Chemical Society

SAMs form due to formation of strong bonds between the sulfur and the gold atoms, which is accompanied by van der Waals interactions between the carbon atoms. To obtain SAMs with a desired molecular packing, one can adjust parameters like immersion time, temperature, concentration and polarity of the solvents. XPS data of aromatic SAMs, representing diverse types of precursors, are shown in Figure 1.10 (left). For each SAM, its chemical composition and thickness can be derived from the BE and intensity of the C1s, S2p and Au4f photoelectron signals. The sulfur signal consists of a doublet with a $S2p_{3/2}$ BE of 162.0 eV, which unambiguously demonstrates the formation of sulfur–gold bonds [54].

Table 1.2: Effective thickness of pristine SAMs and CNMs; carbon reduction upon electron irradiation.

Sample	Thickness SAM [Å]	Thickness CNM [Å]	Reduction of C [%]
1a (BPT)	10	9	5
1b (BP3)	12	10	16
1c (TPT)	13	12	4
2a (NPHT)	6	6	9
2b (ATRH)	9	9	2
2c (1MP)	9	8	10
2d (MP1)	9	8	4
2g (MP3)	11	10	8
2f (MP5)	10	8	11
3a (HPB)	8	8	13
3b^1 (HBC-Br)	10	10	4
3b^2 (HBC-Br)	12	11	5
3b^3 (HBC-Br)	19	17	2
3c^1 (HBC-CN)	12	10	6
3c^2 (HBC-CN)	14	12	5
3c^3 (HBC-CN)	24	22	3

* Different conditions were applied for preparation of SAMs from **3b**$^{1-3}$ and **3c**$^{1-3}$.

Only for the HBC derivatives (structures 3b–c), the presence of a second sulfur species with the BE of the $S2p_{3/2}$ signal at 163.6 eV is observed. This signal originates from physisorbed molecules and/or of disulfides [54], which may be stabilized by π–π interactions between the large aromatic cores. Aromatic and aliphatic carbons contribute to the C1s signal at BEs of ~284.2 and ~285.0 eV, respectively [55]. Stoichiometry and thickness as obtained from XPS correspond to the composition of the precursor molecules and indicate the formation of SAMs with an "upright" molecular orientation. By varying the precursors, the thickness of the aromatic monolayers can be adjusted from ~6 Å for NPHT (Figure 1.9(d): structure 2a) to ~24 Å for HBC-CN (Figure 1.9(d): structure 3c), which directly correlates with their molecular lengths (Figure 1.10 and Table 1.2).

In addition, the temperature- and solvent-dependent intermolecular interactions of the HBC derivatives [56] allow one to tune the final SAM thickness by varying the preparation conditions (Table 1.2).

Scanning tunneling microscopy (STM) and low-energy electron diffraction (LEED) data of three molecular precursors, containing different numbers of carbon atoms per molecule – 2-anthracene (C14), 3-biphenylpropane (C15) and terphenyl (C18) thiols – are shown in Figure 1.11. The structure and surface density of the SAMs were investigated by STM and LEED. They showed that precursor molecules 1b, 1c, 2a, 2b, 2c and 2e (Figure 1.9(d)) form well-ordered SAMs with the densely packed ($\sqrt{3} \times \sqrt{3}$) [57] unit cell of the adsorption places and with the ($2\sqrt{3} \times \sqrt{3}$) superstructure of the molecular

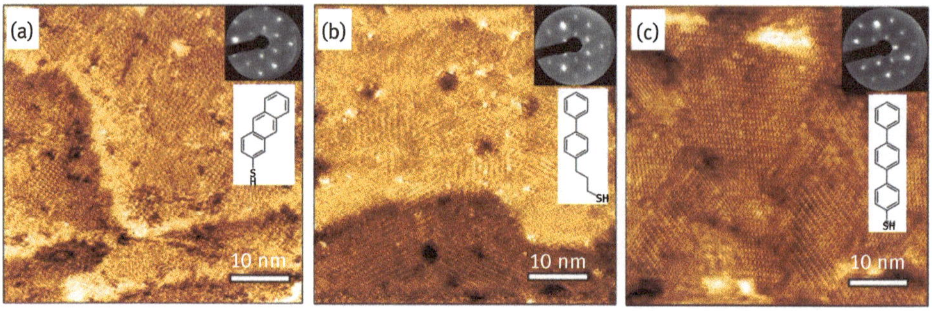

Figure 1.11: Structure of pristine SAMs. STM micrographs and LEED patterns (insets) of SAMs from molecular precursor: (a) ANTH, 2b; LEED pattern at 116 eV, (b) BP3, 1b; LEED pattern at 127 eV, (c) TPT, 1c; LEED pattern at 129 eV. For molecular structures see Figure 1.9(d). Reprinted with permission from Ref. [6]. Copyright (2013) American Chemical Society.

backbones. These structures correspond to a surface area of 21.6 Å^2 per molecule. Note that for these SAMs the surface density of carbon atoms in the monolayer can be precisely tuned by the carbon content of the respective molecular precursors. Short biphenylthiols (Figure 1.9(d), 1a) exhibit a (2×2) arrangement of the adsorption places, which corresponds to a less densely packed monolayer of 28.7 Å^2 per molecule [49a]. The formation of LEED patterns and well-ordered SAMs by STM for precursor molecules 2d and 2f and 3a–c (Figure 1.9(d)) was not verified. As XPS indicates a formation of sulfur–gold bonds for all precursors, we conclude that monolayers of the “bulky” and polycyclic molecules (structures 2d, 2f, 3a–c from Figure 1.9(d)) are less ordered and probably less densely packed than the monolayers of oligophenyls (1a–c) and the small fused-ring systems (2a–c, 2e).

The irradiation of these SAMs with low-energy electrons (50 or 100 eV), using typical doses of ~60 mC/cm^2, corresponding to ~3,500 electrons per 1 nm^2, leads to a loss of order, as observed in LEED and STM. Figure 1.10 shows XPS data before and after electron irradiation in the left and right parts, respectively. As seen from the intensities of the C1s and Au4f (not shown), for purely aromatic SAMs the irradiation reduces the carbon content and the monolayer thickness by 5–10 %; in the SAMs that also contain aliphatic chains (structures 1b and 2f from Figure 1.9(d)) the carbon loss is up to ~16 %.

The obtained freestanding (self-supported) CNMs from the above-mentioned molecular precursors can be seen in Figure 1.12. It shows HIM micrographs of CNMs from three different types of aromatic molecular precursors (Figure 1.9(d): molecular structures 1–3). Simply the fact that one can take these images demonstrates that the SAMs of all these molecules have been cross-linked into mechanically stable CNMs. Figure 1.12(a–c) shows the freestanding CNMs from precursors 1c, 2c and 2d. These HIM micrographs were acquired at different magnifications, demonstrating the successful fabrication of CNMs of various sizes. The field of view in Figure 1.12(a) is 15×15 µm^2, which allows to observe some folds in the freestanding 1.2-nm-thick CNM. In the lower left corner of Figure 1.12(b), the boundary between the freestanding or supported CNM and substrate can clearly be seen. Figure 1.12(c) shows the field of

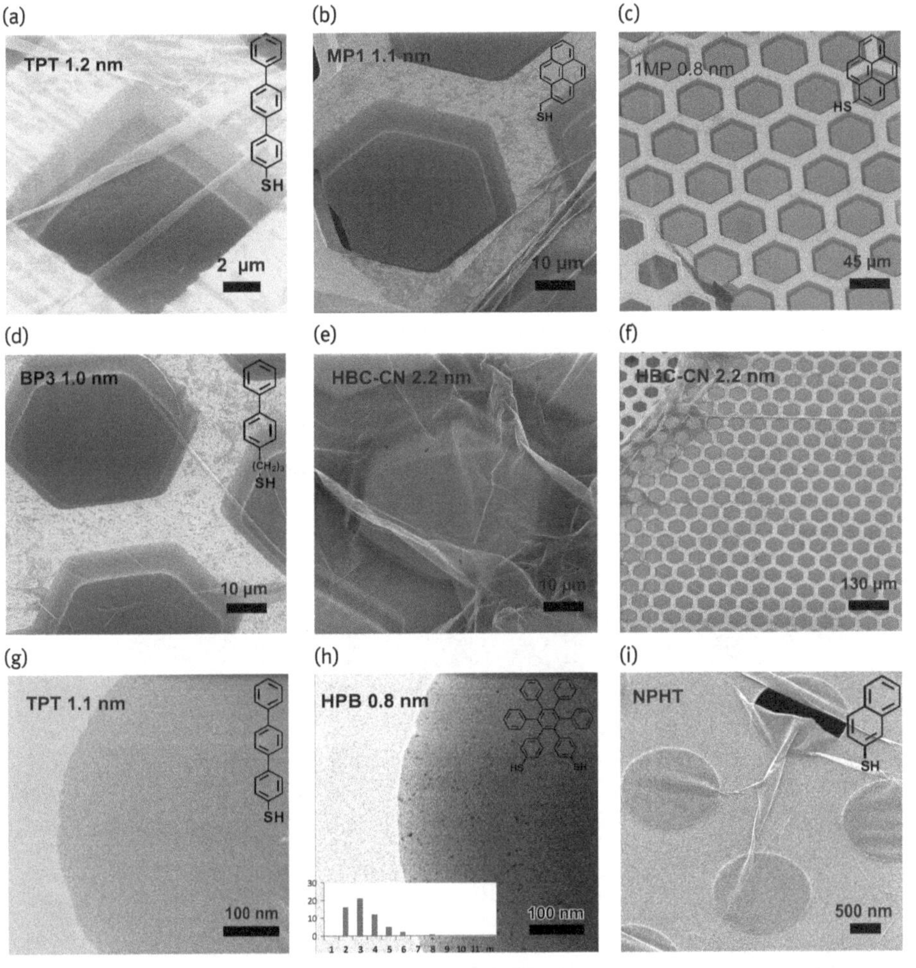

Figure 1.12: HIM micrographs of freestanding CNMs. After cross-linking the nanomembranes were transferred onto TEM grids. CNMs prepared from (a) TPT, structure 1c in Figure 1.9(d), suspended over gold TEM grid, (b) MP1, 2d, (c) 1MP, 2c, (d) BP3, 2b, (e) HBC-CN, 3c, (f), HBC-CN, 3c, (g) TPT, 1c, (h) HPB, 3a, in Figure 1.9(d) the inset shows the histogram of the pore size distribution over the displayed area, (i) freestanding nanocrystalline graphene after pyrolysis of NPHT (2a) CNM, (b–f) CNMs suspended over copper TEM grids, (g–i) CNMs and graphene are suspended over a grid with thin carbon film on Cu grid. Reprinted with permission from Ref. [6]. Copyright (2013) American Chemical Society.

view of 300 × 300 µm^2 with a large and homogeneous CNM of a thickness of 0.8 nm. Macroscopic defects in these nanomembranes are practically negligible on the length scale of these images.

Since the thickness of CNMs is determined by the precursor molecules and their packing density in SAMs, it can be controlled by tailoring these parameters. Figure 1.12(e) and (f) displays an example of the thickest CNM (2.2 nm) from precursor 3c, Figure 1.9(d) at two scales that has about four times higher thickness than the thinnest nanomembrane, prepared from precursor 2a, Figure 1.9(d) with a thickness of only 0.6 nm (Table 1.2). An image of an annealed NPHT CNM, transformed into nanocrystalline graphene sheet, can be seen in Figure 1.12(i). The opportunity to flexibly tune the thickness of CNMs opens broad avenues for the engineering of nanomembranes. A thorough investigation of the surface and structural features of different CNMs by HIM revealed the relation between properties of the precursor molecule, its SAMs and the appearance of the ensuing CNM. If the molecule forms a densely packed SAM (1a–c, 2a–c, 2e in Figure 1.9(d)), the following CNM is continuous and free of holes. Figure 1.12(g) shows a high magnification HIM image of a homogeneous CNM made from terphenylthiol. Conversely, CNMs made from HBC (3b–c in Figure 1.9(d)) or HPB (3a in Figure 1.9(d)) precursors, two molecules that possess larger sizes and form less well-ordered SAMs, exhibit pores (cf. the HIM images in Figures 1.12(h) and 1.13(a–f)). The dark spots in these images are pores that have a very small diameter and a narrow size distribution, as shown in the respective histograms (shown in the insets).

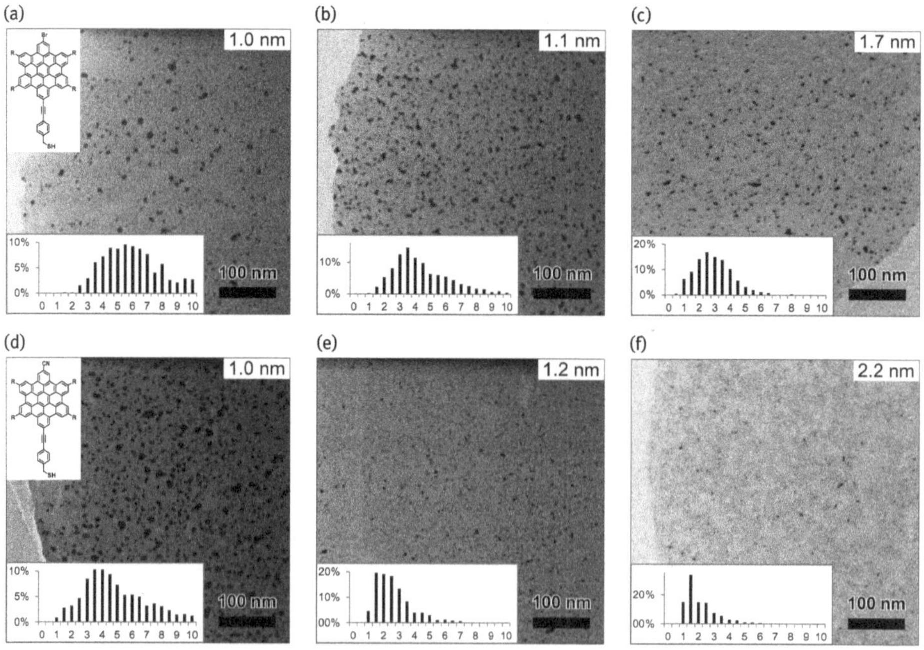

Figure 1.13: HIM images of freestanding CNMs: (a–c) prepared from HBC-Br (Figure 1.9(d), 3b) monolayers of different thickness, (d–f) prepared from HBC-CN (3c) monolayers of different thickness. The thickness of the membranes is written at the top right corner of each image. The histograms in the insets show the distribution of the pore diameter with a step of 0.5 nm. The membranes are placed onto Quantifoil TEM grids type Multi A. Reprinted with permission from Ref. [6]. Copyright (2013) American Chemical Society.

In case of the HBC precursor the mean size of the nanopores is ~6 nm with the surface density of 9.1×10^{14} pores/m^2; the more compact HPB precursor shows a size of ~2.4 nm with a surface density of 1.3×10^{15} pores/m^2. The formation of nanopores in these CNMs is thus attributed to the large van der Waals radii of HBC and HPB structures and in the case of HBCs to the propensity of the disk-like molecules for intermolecular stacking, which competes with the molecule–substrate interactions and lowers the homogeneous coverage in the respective SAMs. We also observe that the average pore diameter of the HBC-based CNMs decreases from 6.4 to 2.5 nm when the SAM thickness increases from 1 to 2 nm (Figure 1.13).

Further studies demonstrated that the mechanical stiffness and the resonance frequency of a monolayer membrane can also be finely tuned via varying the precursor molecules. To investigate the mechanics of such ultrathin 2D sheets, mechanical

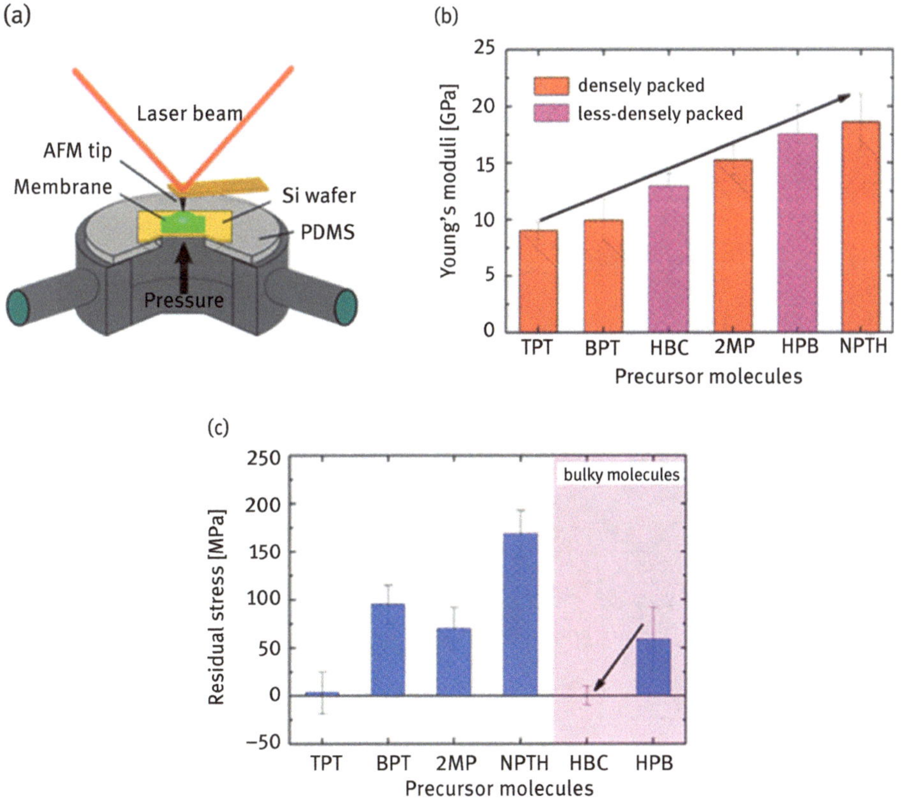

Figure 1.14: (a) Schematic diagram of the AFM bulge test setup where a gas pressure is applied with compressed nitrogen, and the corresponding membrane deflection is recorded with an AFM. (b) Plot of Young's moduli of CNMs prepared from different precursor molecules. (c) Residual stress of CNMs, where CNMs from bulky molecules are separately presented. Reprinted with permission from Ref. [59]. Copyright (2014) from American Chemical Society.

characterization methods must be adapted to the nanoscale by combining them with nanoanalytical tools. Bulge testing was used to characterize the mechanical properties of freestanding films. The technique involves clamping of a freestanding membrane over an orifice and application of an overpressure to one side. The deflection of this 1-nm-thin carbon nanosheet was monitored by atomic force microscopy (AFM). The Young's modulus and prestress were calculated from the obtained pressure–deflection relationship. Figure 1.14(a) shows a schematic diagram of the AFM bulge test setup. Freestanding CNMs with a thickness between 0.6 and 1.7 nm from diverse polyaromatic precursors were investigated by Zhang et al. [58, 59]. A correlation between the rigidity of the precursor molecules and the macroscopic mechanical stiffness of CNM was found. The data show that CNMs from rigid and condensed precursors like NPTH and pyrene thiols prove to exhibit higher Young's moduli of 15–19 GPa, while CNMs from nonfused oligophenyls possess lower Young's moduli of ~10 GPa. Figure 1.15 shows the elastic performance of CNM among other material classes. Materials with similar stiffness are porous ceramics and composites, whereas the class of polymers exhibits lower mechanical stability than the CNM.

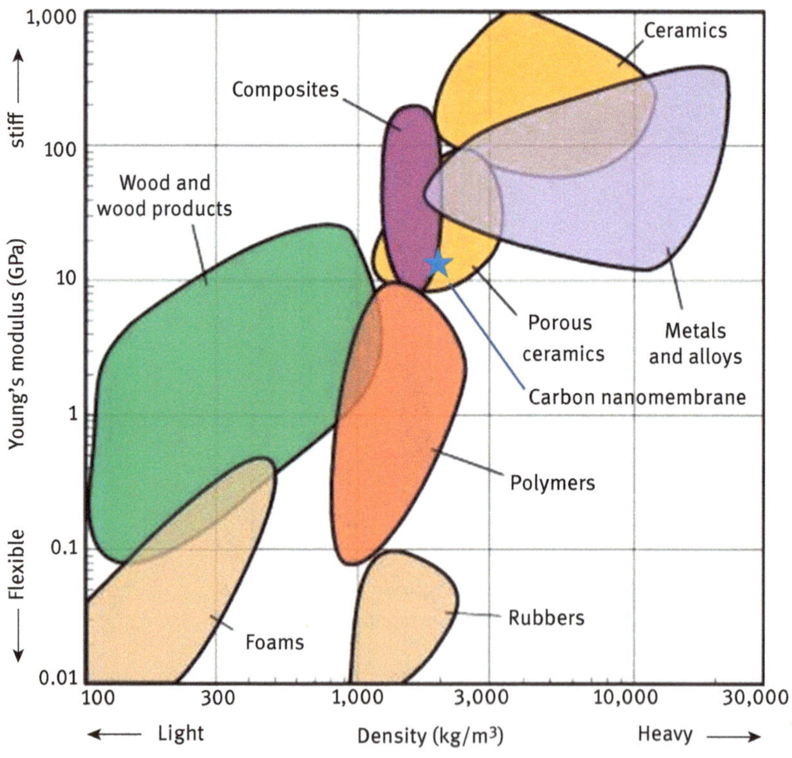

Figure 1.15: Young's modulus plot of different materials [60].

The interpretation of Zhang et al.'s data relates the Young's modulus to the properties of single molecules and to the structure of the pristine SAMs. The membranes with the highest stiffness are formed from molecules that exhibit higher molecular rigidity (less conformational freedom), whereas the terphenyl-based CNM is more mechanically stable than biphenylthiol-based CNM due to the denser packing of molecules within the SAM and to the larger number of cross-linking sites in the molecular skeleton. The degree of cross-linking of the SAM proved to be also an important parameter that has a strong impact on the membrane prestress or tensile stress. Although van der Waals interactions between the sheet and the substrate are the main factors that determine the constant tension of a self-supported monolayer, different prestresses for CNMs prepared from different precursor molecules, in spite of their similar thickness, were estimated. This implies that electron irradiation plays a key role, as a higher number of conformational degrees of freedom of the precursor molecules enable more ways to cross-link the molecules and thus achieve a lower prestress in the CNM, as in the case of TPT-CNMs. The ultimate tensile strength of CNM was also determined by means of bulge tests, measuring the ultimate pressure at which the membrane ruptures [58]. The tensile strength of biphenyl-based CNMs ranges from 400 to 700 MPa. In this range materials like graphite, iron and silicon nitride break. Of course, the tensile strength of CNM is orders of magnitude lower than that of graphene and carbon nanotubes, but compared to other nanomembranes, such as polymer membranes and the organic–inorganic IPNs, exhibiting a tensile strength of 105 MPa, the ultimate tensile strength of CNMs is five to six times higher.

Zhang et al. [61] also studied the mechanical motion of the CNM and showed that they can be applied as mechanical resonators. Their vibrational mode shapes were characterized and visualized by optical interferometry (Figure 1.16). The freestanding CNM was transferred onto Si substrates with square or rectangular orifices.

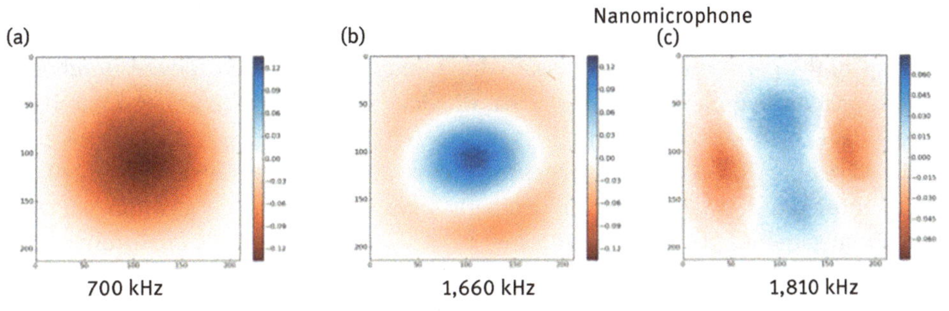

Figure 1.16: Several representative vibration modes of a 4′-nitro-1,1′-biphenylthiol CNM with dimensions of 50 × 50 μm^2 at (a) 700 kHz, (b) 1,660 kHz, (c) 1,810 kHz. Reprinted with permission from Ref. [61]. Copyright (2015) from AIP Publishing LLC.

A vibration of the membrane was actuated by applying a sinusoidal voltage to a piezoelectric disk on which the sample was glued. The dynamic behavior of the CNM's vibrational modes is described by linear response theory of a membrane with negligible bending rigidity. A phase velocity of 61.6 ± 10 m/s and 55.06 ± 9 m/s was obtained for 4′-nitro-1,1′-biphenyl-4-thiol (NBPT)-based CNM and TPT–CNM, respectively. By comparing the dispersion relation to an analytical model, the static stress of the membranes of ~5 MPa was determined and found to be caused by the fabrication process. The extremely low mass and bending stiffness ensures that CNM resonators are in the membrane regime where the pretension of the CNM dominates over its elastic stiffness.

Stimulated by the development of polymer and asymmetric hollow fiber membranes, membrane separation processes gained rapid progress not only in medical applications (artificial kidneys and lungs) and in the field of environmental protection (wastewater treatment, desalination, purification of flue gases), but even in energy renewable technologies. Since the end of the twentieth century, carbon materials, such as carbon molecular sieves, fullerenes, single and multiwall nanotubes, were extensively studied for their implementation in membrane technology. Two-dimensional nanomembranes with a thickness below ~5 nm and pores tuned to act as molecular sieves are predicted to be ideal separation membranes with many advantages over bulk membranes [62]. Especially, such extremely thin and dense carbon-based nanomembranes are considered to be effective and superior separation media for chemical and gas purification [63], offering the advantage of fast flow and lack of aging. Ai et al. [64] investigated the gas permeation performance of CNMs, built up from biphenylthiol (BPT) and p-nitrobiphenylthiol (NBPT) SAMs. Their permeance was measured onto a polydimethylsiloxane-supported membrane (PDMS-TFC, 150 nm PDMS/PAN/nonwoven) [65, 66] in order to enhance the mechanical stability and minimize roughness-induced strain. Considering the geometry and composition of the measured system – CNM/PDMS/PAN/nonwoven – it is intuitive to consider that the gas transport properties of PDMS determine and contribute significantly to the measured permeance values of CNM-PDMS. Therefore, it is necessary to consider the change of the permeance of bare PDMS due to the deposition of CNMs. A number of gases, such as hydrogen, helium, carbon dioxide, oxygen, nitrogen, argon, methane and ethane, have been tested and their corresponding permeance and ideal gas selectivity were obtained. The data are presented in relation to the penetrant kinetic diameters of the gases in Figure 1.17. They clearly show that the gas permeation of PDMS membranes reduces by the deposition of CNMs. This reduction is more pronounced for gas species with larger kinetic diameters. A simple model to extract the permeance of CNMs from the measured values of CNM-PDMS membranes revealed a different gas permeation behavior for single-layer and three-layer CNMs. In both cases, He and H_2 showed substantially higher permeance values in comparison to gases with larger kinetic diameters with the exception of CO_2, which displayed high permeances for single-layer CNMs and low values for three-layer CNMs. The established relationship

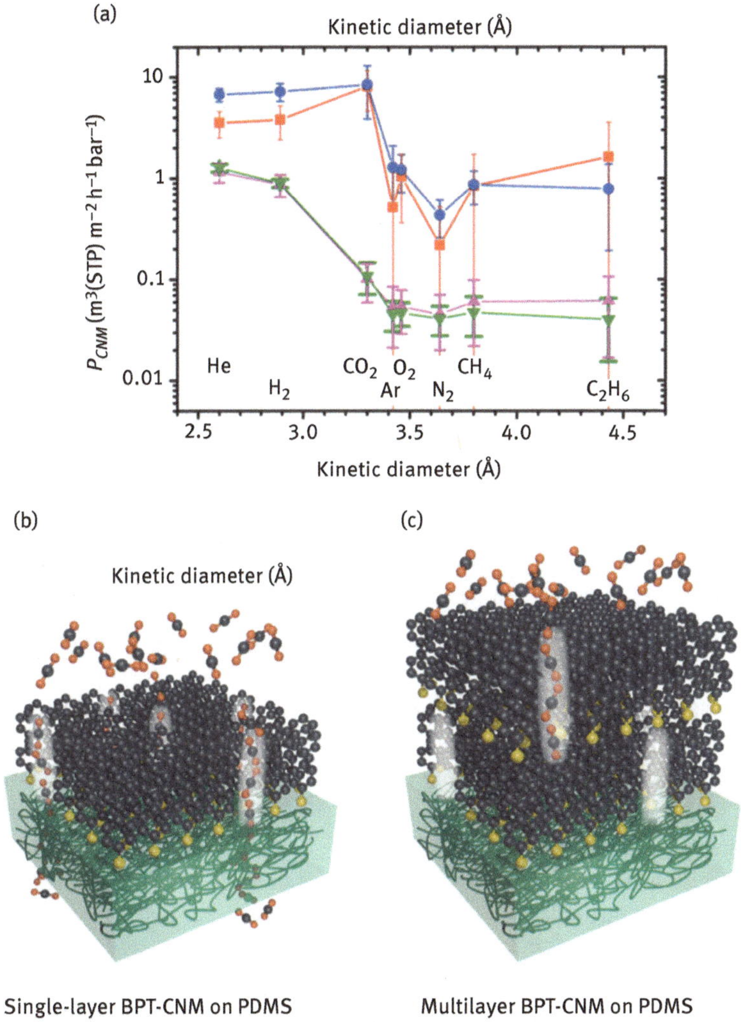

Figure 1.17: (a) Permeance of the CNM part of the composite membranes that accounts for the finite permeance of the PDMS support as well as a partial coverage of the PDMS surface by CNM. round symbol – single layer of NBPT-CNM, square symbol – single layer of BPT-CNM, triangle, pointing up – three-layer BPT-CNM, triangle, pointing down – three-layer NBPT-CNM. (b) Schematic depiction of the proposed gas transport mechanism in single-layer CNMs. Molecular-sized channels (highlighted by bright regions) favor the permeation of CO_2 and smaller gas molecules. (c) Schematic depiction of the proposed gas transport mechanism in multilayer CNMs. A low diffusion of CO_2 in between the individual CNMs hinders its permeation through multilayer CNMs. Reprinted with permission from Ref. [64]. Copyright (2014) from Wiley-VCH.

of permeance and molecular size determines the molecular sieving mechanism of permeation through this ultrathin sheet. It can be understood by assuming that molecular-sized channels in the CNM dominate the permeation properties. These molecular-sized channels may form during the cross-linking process due to its statistical nature, which is expected to result in a random variation of intermolecular distances. Thus, even in a defect-free CNM, a high density of small free volumes (or pores) may form at locations of relatively large intermolecular distances, which act as the proposed molecular-sized channels.

The transport through a three-layer CNM is consistent with lateral diffusion that is high for He and H_2 but hindered for the larger gases including CO_2. Lateral diffusion of gas molecules may be described by Knudsen-like diffusion or by condensation and surface flow of gas molecules in between the CNMs. Both possible mechanisms result in a permeance characteristic, which prefers the transport of smaller gases. The three-layer CNMs exhibit very small and constant values for all gases except the two gas species with the smallest kinetic diameters, which show more than one order of magnitude higher permeances. These higher permeance values cannot be explained by considering defects in the membrane as they would increase the permeance for all gas species that implies that this behavior is an intrinsic property of CNMs, resembling the permeation mechanism through glassy polymers with stiff polymer backbones, which also act like molecular sieves [67].

In addition, these experiments demonstrate that the gas permeation of PDMS membranes can be modified by the deposition of CNMs in a controlled way. The flexibility in the assembly of CNMs, by employing different precursor molecules as well as functionalizing CNMs, opens a path to molecular designed membranes also with respect to desired gas permeation characteristics.

This chapter describes a universal scheme to fabricate tailored CNM. CNMs of different size, thickness, elasticity and perforation have been generated. Their properties can be tailored by the choice of the precursor molecule, the structure of the SAM, the cross-linking and the transfer process. As the procedures that fabricate the CNMs are to a large extent compatible with nanofabrication protocols in industry, CNMs can be easily incorporated into electronic devices [68], into TEM grids as ultrathin carbon supports for high-contrast imaging [69], in separation technologies [64] or as protective coatings on chemically and mechanically sensitive technological components. This gives the material scientist a tool to produce membranes for technological applications. Here, a focus on material separation seems logical, as the manifold of diverse and specific separation problems is best addressed by a modular construction system of tailored CNMs. Beyond that the “similarity” of the CNM with biomembranes opens new opportunities for research. We can think of designing hybrid structures where biological membrane proteins are inserted into CNMs, creating an artificial 2D system with functional biological units. This could provide reference structures for the incorporation of single molecules into biotechnological sensors or devices. The future will show whether this will become a reality.

Acknowledgments

The work reviewed in this chapter was financially supported by the Volkswagenstiftung, the Deutsche Forschungsgemeinschaft, the German Bundesministerium für Bildung und Forschung. The research leading to these results has also received funding from the European Union Seventh Framework Programme under grant agreement n°604391 Graphene Flagship.

References

[1] Turchanin A, Gölzhäuser A. Carbon nanomembranes from self-assembled monolayers: functional surfaces without bulk. Prog Surf Sci 2012;87(5–8):108–62.

[2] Anselmetti D, Gölzhäuser A. Converting molecular monolayers into functional membranes. Angew Chem Int Ed 2014;53(46):12300–2.

[3] Rogers JA, Lagally MG, Nuzzo RG. Synthesis, assembly and applications of semiconductor nanomembranes. Nature 2011;477(7362):45–53.

[4] Wang EN, Karnik R. Water desalination: Graphene cleans up water. Nat Nano 2012;7(9):552–4.

[5] Geim AK, Novoselov KS. The rise of graphene. Nat Mater 2007;6(3):183–91.

[6] Angelova P, Vieker H, Weber N-E, Matei D, Reimer O, Meier I, et al. A universal scheme to convert aromatic molecular monolayers into functional carbon nanomembranes. ACS Nano 2013;7(8):6489–97.

[7] Cavallo F, Lagally MG. Semiconductors turn soft: inorganic nanomembranes. Soft Matter 2010;6(3):439–55.

[8] Vendamme R, Onoue SY, Nakao A, Kunitake T. Robust free-standing nanomembranes of organic/inorganic interpenetrating networks. Nat Mater 2006;5(6):494–501.

[9] (a) Yaghi OM, Li G, Li H. Selective binding and removal of guests in a microporous metal-organic framework. Nature 1995;378(6558):703–6; (b) Darbandi M, Arslan HK, Shekhah O, Bashir A, Birkner A, Wöll C. Fabrication of free-standing ultrathin films of porous metal-organic frameworks by liquid-phase epitaxy and subsequent delamination. Physica Status Solidi RRL 2010;4(8–9):197–199.

[10] Moore TT, Mahajan R, Vu DQ, Koros WJ. Hybrid membrane materials comprising organic polymers with rigid dispersed phases. AIChE J 2004;50(2):311–21.

[11] (a) Decher G. Fuzzy nanoassemblies: toward layered polymeric multicomposites. Science 1997;277(5330):1232–7; (b) Perez-Madrigal MM, Armelin E, Puiggali J, Aleman C. Insulating and semiconducting polymeric free-standing nanomembranes with biomedical applications. J Mater Chem B 2015;3(29):5904–32; (c) Sakamoto J, van Heijst J, Lukin O, Schlüter AD. Two-dimensional polymers: just a dream of synthetic chemists? Angew Chem Int Ed 2009;48(6):1030–69.

[12] Karan S, Jiang Z, Livingston AG. Sub-10 nm polyamide nanofilms with ultrafast solvent transport for molecular separation. Science 2015; 348(6241):1347–51.

[13] Lvov Y, Decher G, Möhwald H. Assembly, structural characterization, and thermal behavior of layer-by-layer deposited ultrathin films of poly(vinylsulfate) and poly(allylamine). Langmuir 1993;9(2):481–86.

[14] Raaijmakers MJ, Hempenius MA, Schön PM, Vancso GJ, Nijmeijer A, Wessling M, et al. Sieving of hot gases by hyper-cross-linked nanoscale-hybrid membranes. J Am Chem Soc 2014;136(1):330–5.

[15] Novoselov KS, Jiang D, Schedin F, Booth TJ, Khotkevich VV, Morozov SV, et al. Two-dimensional atomic crystals. Proc Natl Acad Sci USA 2005;102(30):10451–53.

[16] Novoselov KS, Fal'ko VI, Colombo L, Gellert PR, Schwab MG, Kim KG. A roadmap for graphene. Nature 2012;490:192–200.

[17] Colson JW, Woll AR, Mukherjee A, Levendorf MP, Spitler EL, Shields VB, et al. Oriented 2D covalent organic framework thin films on single-layer graphene. Science 2011;332(6026): 228–31.

[18] Schultz MJ, Zhang XY, Unarunotai S, Khang DY, Cao Q, Wang CJ, et al. Synthesis of linked carbon monolayers: films, balloons, tubes, and pleated sheets. P Natl Acad Sci USA 2008;105(21):7353–8.

[19] (a) Grill L, Dyer M, Lafferentz L, Persson M, Peters MV, Hecht S. Nano-architectures by covalent assembly of molecular building blocks. Nat Nanotechnol 2007;2(11):687–91; (b) Kissel P, Erni R, Schweizer WB, Rossell MD, King BT, Bauer T, et al. A two-dimensional polymer prepared by organic synthesis. Nat Chem 2012;4(4):287–91; (c) Bauer T, Zheng ZK, Renn A, Enning R, Stemmer A, Sakamoto J.et al. Synthesis of free-standing, monolayered organometallic sheets at the air/water interface. Angew Chem Int Ed 2011;50(34):7879–84; (d) Bally-Le Gall F, Friedmann C, Heinke L, Arslan H, Azucena C, Welle A. et al. free-standing nanomembranes based on selective CVD deposition of functional poly-p-xylylenes. ACS Nano 2015;9(2):1400–7; (e) Lackinger M. On-surface polymerization – a versatile synthetic route to two-dimensional polymers. Polym Int 2015;64(9):1073–8; (f) Schrettl S, Stefaniu C, Schwieger C, Pasche G, Oveisi E, Fontana Y. et al. Functional carbon nanosheets prepared from hexayne amphiphile monolayers at room temperature. Nat Chem 2014;6(6):468–76.

[20] Turchanin A, Beyer A, Nottbohm CT, Zhang X, Stosch R, Sologubenko AS, et al. One nanometer thin carbon nanosheets with tunable conductivity and stiffness. Adv Mater 2009;21:1233–7.

[21] Ward BW, Notte JA, Economou NP. Helium ion microscope: a new tool for nanoscale microscopy and metrology. J Vac Sci Technol B 2006;24(6):2871–4.

[22] (a) Bell DC, Contrast mechanisms and image formation in helium ion microscopy. Micros Microanal 2009;15(2):147–53; (b) Beyer A, Vieker H, Klett R, Meyer zu Theenhausen H, Angelova P, Gölzhäuser A, Imaging of carbon nanomembranes with helium ion microscopy. Beilstein J Nanotechnol 2015;6:1712–20.

[23] Geyer W, Stadler V, Eck W, Zharnikov M, Gölzhäuser A, Grunze M. Electron-induced crosslinking of aromatic self-assembled monolayers: negative resists for nanolithography. Appl Phys Lett 1999;75(16):2401–3.

[24] Turchanin A, El-Desawy M, Göelzhäuser A. High thermal stability of cross-linked aromatic self-assembled monolayers: Nanopatterning via selective thermal desorption. Appl Phys Lett 2007;90(5).

[25] Turchanin A, Weber D, Büenfeld M, Kisielowski C, Fistul MV, Efetov KB, et al. Conversion of self-assembled monolayers into nanocrystalline Graphene: structure and electric transport. ACS Nano 2011;5. DOI: 10.1021/nn200297n

[26] (a) Zheng Z, Nottbohm CT, Turchanin A, Muzik H, Beyer A, Heilemann, M, et al. Janus nanomembranes: a generic platform for chemistry in two dimensions. Angew Chem Int Ed 2010;49(45):8493–7; (b) Schnietz M, Turchanin A, Nottbohm CT, Beyer A, Solak HH, Hinze P, et al. Chemically functionalized carbon nanosieves with 1-nm thickness. Small 2009;5(23): 2651–5; (c) Nottbohm CT, Turchanin A, Gölzhäuser A, Metallization of organic monolayers:

electroless deposition of Cu onto cross-linked aromatic self-assembled monolayers. Z Phys Chem Int J Res Phys Chem Chem Phys 2008;222(5–6):917–26; (d) Völkel B, Kaltenpoth G, Handrea M, Sahre M, Nottbohm CT, Küller A, et al. Electrodeposition of copper and cobalt nanostructures using self-assembled monolayer templates. Surf Sci 2005;597(1–3):32–41; (e) Zheng Z, Zhang X, Neumann C, Emmrich D, Winter A, Vieker H, et al. Hybrid van der Waals heterostructures of zero-dimensional and two-dimensional materials. Nanoscale 2015;7(32):13393–7; (f) Beyer A, Turchanin A, Nottbohm CT, Mellech N, Schnietz M, Gölzhäuser A. Fabrication of metal patterns on freestanding graphenoid nanomembranes. J Vac Sci Technol B: Microelectron Nanometer Struct 2010;28(6):C6D5.

[27] Gölzhäuser A, Geyer W, Stadler V, Eck W, Grunze M, Edinger K, et al. Nanoscale patterning of self-assembled monolayers with electrons. J Vac Sci Technol B 2000;18(6):3414–18.

[28] (a) Küller A, Eck W, Stadler V, Geyer W, Gölzhäuser A. Nanostructuring of silicon by electron-beam lithography of self-assembled hydroxybiphenyl monolayers. Appl Phys Lett 2003;82(21):3776–8; (b) Kaltenpoth G, Völkel B, Nottbohm CT, Gölzhäuser A, Buck M. Electrode modification by electron-induced patterning of self-assembled monolayers. J Vac Sci Technol B 2002;20(6):2734–8.

[29] (a) Eck W, Stadler V, Geyer W, Zharnikov M, Gölzhäuser A, Grunze M. Generation of surface amino groups on aromatic self-assembled monolayers by low energy electron beams – A first step towards chemical lithography. Adv Mater 2000;12(11):805–8; (b) Chesneau F, Hamoudi H, Schuepbach B, Terfort A, Zharnikov M. Modification of self-assembled mono layers of perfluoroterphenyl-substituted alkanethiols by low-energy electrons. J Phys Chem C 2011;115(11):4773–82; (c) Frey S, Rong HT, Heister K, Yang YJ, Buck M, Zharnikov, M. Response of biphenyl-substituted alkanethiol self-assembled monolayers to electron irradiation: damage suppression and odd–even effects. Langmuir 2002;18(8):3142–50.

[30] Eck W, Küller A, Grunze M, Völkel B, Gölzhäuser A. Freestanding nanosheets from crosslinked biphenyl self-assembled monolayers. Adv Mater 2005;17(21):2583–7.

[31] Turchanin A, Tinazli A, El-Desawy M, Grossann H, Schnietz M, Solak HH, et al. Molecular self-assembly, chemical lithography, and biochemical tweezers: a path for the fabrication of functional nanometer-scale protein arrays. Adv Mater 2008;20(3):471–+.

[32] Gölzhäuser A, Eck W, Geyer W, Stadler V, Weimann T, Hinze P, et al. Chemical nanolithography with electron beams. Adv Mater 2001;13(11):806–809.

[33] Turchanin A, Käfer D, El-Desawy M, Wöll C, Witte G, Gölzhäuser A. Molecular mechanisms of electron-induced cross-linking in aromatic SAMs. Langmuir 2009;25(13):7342–52.

[34] (a) Cyganik P, Vandeweert E, Postawa Z, Bastiaansen J, Vervaecke F, Lievens P, et al. Modification and stability of aromatic self-assembled monolayers upon irradiation with energetic particles. J Phys Chem B 2005;109(11):5085–4; (b) Tai Y, Shaporenko A, Eck W, Grunze M, Zharnikov M. Depth distribution of irradiation-induced cross-linking in aromatic self-assembled monolayers. Langmuir 2004;20(17):7166–0; (c) Weidner T, Ballav N, Grunze M, Terfort A, Zharnikov M. Modification of biphenylselenolate monolayers by low-energy electrons. Physica Status Solidi B Basic Solid State Phys 2009;246(7):1519–28.

[35] Olsen C, Rowntree PA. Bond-selective dissociation of alkanethiol based self-assembled monolayers adsorbed on gold substrates, using low-energy electron beams. J Chem Phys 1998;108(9):3750–64.

[36] (a) Palmer RE, Rous PJ. Resonances in electron-scattering by molecules on surfaces. Rev Mod Phys 1992;64(2):383–440; (b) Illenberger E. Electron-attachment reactions in molecular clusters. Chem Rev 1992;92(7):1589–1609; (c) Illenberger E, Formation and evolution of negative ion resonances at surfaces. Surf Sci 2003;528(1–3):67–77.

[37] Silien C, Raeisaenen MT, Buck M, A Supramolecular Network as sacrificial mask for the generation of a nanopatterned binary self-assembled monolayer. Small 2010;6(3):391–4.

[38] Turchanin A, Schnietz M, El-Desawy M, Solak HH, David C, Gölzhäuser A. Fabrication of molecular nanotemplates in self-assembled monolayers by extreme-ultraviolet-induced chemical lithography. Small 2007;3(12):2114–9.
[39] Völkel B, Gölzhauser A, Müller HU, David C, Grunze M. Influence of secondary electrons in proximal probe lithography. J Vac Sci Technol B 1997;15(6):2877–81.
[40] Tobita S, Meinke M, Illenberger E, Christophorou LG, Baumgartel H, Leach S. Polycyclic aromatic-hydrocarbons – negative-ion formation following low-energy (0–15 eV) electron-impact. Chem Phys 1992;161(3):501–8.
[41] Hüfner S. Photoelectron spectroscopy principals and application, 3rd ed. Berlin: Springer, 2003.
[42] Laibinis PE, Whitesides GM, Allara DL, Tao YT, Parikh AN, Nuzzo RG. Comparison of the structures and wetting properties of self-assembled monolayers of n-alkanethiols on the coinage metal surfaces, copper, silver, and gold. J Am Chem Soc 1991;113(19):7152–67.
[43] Heister K, Zharnikov M, Grunze M, Johansson LSO, Ulman A. Characterization of X-ray induced damage in alkanethiolate monolayers by high-resolution photoelectron spectroscopy. Langmuir 2001;17(1):8–11.
[44] Woodruff DP. The interface structure of n-alkylthiolate self-assembled monolayers on coinage metal surfaces. Phys Chem Chem Phys 2008;10(48):7211–7221.
[45] (a) Edinger K, Grunze M, Wöll C. Corrosion of gold by alkane thiols. Berichte Der Bunsen-Gesellschaft-Phys Chem Chem Phys 1997;101(12):1811–15; (b) Yang GH, Liu GY, New insights for self-assembled monolayers of organothiols on Au(111) revealed by scanning tunneling microscopy. J Phys Chem B 2003;107(34):8746–59.
[46] (a) Maksymovych P, Voznyy O, Dougherty DB, Sorescu DC, Yates JT, Jr. Gold adatom as a key structural component in self-assembled monolayers of organosulfur molecules on Au(111). Prog Surf Sci 2010;85(5–8):206–40; (b) Maksymovych P, Sorescu DC, Yates JT, Jr. Gold-adatom-mediated bonding in self-assembled short-chain alkanethiolate species on the Au(111) surface. Phys Rev Lett 2006;97(14) 146103.
[47] Cossaro A, Mazzarello R, Rousseau R, Casalis L, Verdini A, Kohlmeyer A, et al. X-ray diffraction and computation yield the structure of alkanethiols on gold(111). Science 2008;321(5891):943–946.
[48] Matei DG, Muzik H, Gölzhäuser A, Turchanin A. Structural investigation of 1,1'-biphenyl-4-thiol self-assembled monolayers on Au(111) by scanning tunneling microscopy and low-energy electron diffraction. Langmuir 2012;28(39):13905–11.
[49] Azzam W, Cyganik P, Witte G, Buck M, Wöll C. Pronounced odd–even changes in the molecular arrangement and packing density of biphenyl-based thiol SAMs: A combined STM and LEED study. Langmuir 2003;19(20):8262–70.
[50] (a) Heimel G, Romaner L, Bredas J-L, Zojer E, Odd–even effects in self-assembled monolayers of omega-(biphenyl-4-yl)alkanethiols: A first-principles study. Langmuir 2008;24(2): 474–82; (b) Heimel G, Romaner L, Zojer E, Bredas J-L, Toward control of the metal-organic interfacial electronic structure in molecular electronics: a first-principles study on self-assembled monolayers of pi-conjugated molecules on noble metals. Nano Lett 2007;7(4):932–40.
[51] Charbonneau GP, Delugeard Y. Structural transition in polyphenyls. 5. Biphenyl – 3-Dimensional data and new refinement at 293 K. Acta Crystallogr Sect B-Struct Sci 1977;33(MAY13):1586–8.
[52] Matei DG, Weber, N-E, Kurasch S, Wundrack S, Woszczyna M, Grothe M, et al. Functional single-layer graphene sheets from aromatic monolayers. Adv Mater 2013;25(30):4146–51.
[53] (a) Feng X, Pisula W, Müllen K. Large polycyclic aromatic hydrocarbons: synthesis and discotic organization. Pure Appl Chem 2009;81(12):2203–24; (b) Wagner C, Kasemann D, Golnik C, Forker R, Esslinger M, Müllen K, et al. Repulsion between molecules on a metal: Monolayers and submonolayers of hexa-peri-hexabenzocoronene on Au(111). Phys Rev B 2010; 81(3):035423.
[54] Castner DG, Hinds K, Grainger DW. X-ray photoelectron spectroscopy sulfur 2p study of organic thiol and disulfide binding interactions with gold surfaces. Langmuir 1996;12(21):5083–86.

[55] Zharnikov M, Grunze M. Spectroscopic characterization of thiol-derived self-assembling monolayers. J Phys-Condens Matter 2001;13(49):11333–65.
[56] Kastler M, Pisula W, Wasserfallen D, Pakula T, Müllen K. Influence of alkyl substituents on the solution- and surface-organization of hexa-peri-hexabenzocoronenes. J Am Chem Soc 2005;127(12):4286–96.
[57] Vericat C, Vela ME, Benitez G, Carro P, Salvarezza RC. Self-assembled monolayers of thiols and dithiols on gold: new challenges for a well-known system. Chem Soc Rev 2010;39(5):1805–34.
[58] Zhang X, Beyer A, Gölzhäuser A Mechanical characterization of carbon nanomembranes from self-assembled monolayers. Beilstein J Nanotechnol 2011;2:826–33.
[59] Zhang X, Neumann C, Angelova P, Beyer A, Gölzhäuser A. Tailoring the mechanics of ultrathin carbon nanomembranes by molecular design. Langmuir 2014;30(27):8221–7.
[60] Lovatt AM, Publishers S, Shercliff HR, Withers PJ. Material selection and processing P. Southgate Publishers, 2000 http://www-materials.eng.cam.ac.uk/mpsite/physics/introduction.
[61] Zhang X, Waitz R, Yang F, Lutz C, Angelova P, Gölzhäuser A, et al. Vibrational modes of ultrathin carbon nanomembrane mechanical resonators. Appl Phys Lett 2015;106(6):063107.
[62] Gozde O-I, Anna Maria C, Karen KG. CVD of polymeric thin films: applications in sensors, biotechnology, microelectronics/organic electronics, microfluidics, MEMS, composites and membranes. Rep Prog Phys 2012;75(1):016501.
[63] Koenig SP, Wang L, Pellegrino J, Bunch JS. Selective molecular sieving through porous graphene. Nat Nano 2012;7(11):728–32.
[64] Ai M, Shishatskiy S, Wind J, Zhang X, Nottbohm CT, Mellech N, et al. Carbon nanomembranes (CNMs) supported by polymer: mechanics and gas permeation. Adv Mater 2014;26(21):3421–6.
[65] Wilfredo Y, Anja C, Jan W, Klaus-Viktor P. Nanometric thin film membranes manufactured on square meter scale: ultra-thin films for CO_2 capture. Nanotechnology 2010;21(39): 395301.
[66] D. Behling, K. Hattenbach, K. Ohlrogge, K.-V. Peinemann, J. Wind, U.S. Patent 4,994,094, 1991.
[67] S. P. Nunes, K.-V. Peinemann, in Membrane Technology in the Chemical Industry, (Eds: S. P. Nunes, K.-V. Peinemann), Wiley-VCH, Weinheim, Germany 2001, Ch. 1.
[68] Woszczyna M, Winter A, Grothe M, Willunat A, Wundrack S, Stosch R, et al. All-carbon vertical van der Waals heterostructures: non-destructive functionalization of Graphene for electronic applications. Adv Mater 2014;26(28):4831–7.
[69] (a) Nottbohm CT, Beyer A, Sologubenko AS, Ennen I, Hutten A, Rosner H, et al. Novel carbon nanosheets as support for ultrahigh-resolution structural analysis of nanoparticles. Ultramicroscopy 2008;108(9):885–92; (b) Rhinow D, Büenfeld M, Weber NE, Beyer A, Gölzhäuser A, Kühlbrandt W, et al. Energy-filtered transmission electron microscopy of biological samples on highly transparent carbon nanomembranes. Ultramicroscopy 2011;111(5):342–9.

Siegfried Eigler and Andreas Hirsch

2 Controlled Functionalization of Graphene by Oxo-addends

Abstract: The single carbon layer graphene and especially its oxidized derivatives, such as graphene oxide (GO), are in the focus of research that started already 150 years ago [1–6]. GO is a collective term for various single layers of graphene (with lattice defects) functionalized by oxo-addends. The type of oxo-groups is not defined, but epoxy and hydroxyl groups dominate the structure in addition to in-plane lattice defects on the percent scale. Those defects are rarely considered in chemical functionalization approaches and it is impossible to distinguish between functionalization of surface oxo-groups and in-plane oxo-groups.

This chapter focuses on functionalized derivatives of graphene with an almost intact carbon framework, termed "oxo-functionalized graphene" (oxo-G_1, index indicates the number of layers). Avoiding in-plane defects further allows the development of a controlled chemistry of graphene with oxo-addends. However, general approaches of conventional GO chemistry are summarized in a separate section.

Keywords: functionalization, oxo-functionalized graphene, graphene oxide

2.1 Introduction

Graphene is currently receiving tremendous attention as a new nanomaterial with high potential for future electronic, optical and sensor applications. This interest is based on a number of unprecedented properties, such as high electron mobility, electrical conductivity, transparency, flexibility and mechanical and chemical stability. Actually, graphene exists only on the surface. In order to take advantage of this exceptional profile of combinations it is of upmost importance to tune the solubility and processibilty of graphene and at the same time to explore methods for the mass production of graphene-based materials.

This is where wet-chemical functionalization and derivatization come into play. One of the oldest modifications of graphene is graphene oxide (GO). GO is produced by the oxidation of graphite. The resulting graphite oxide is very soluble in water and can be easily exfoliated to monolayer GO. One of the disadvantages of GO in the past was the ill-defined structure including impossible-to-heal σ-defects such as holes in the carbon-lattice plane. As a consequence, the challenge arose to create, isolate and characterize GO that does not contain σ-defects but only covalently bound oxo-functionalities. Such a material would combine the advantage of high solubility with the structural integrity of graphene, provided it can be defunctionalized to the parent two-dimensional carbon allotrope. The scope of this review is to introduce the field of oxo-functionalized graphene (oxo-G_1).

First a historical overview on graphite oxide and GO will be provided in Section 1. It will be demonstrated that they indeed constitute a rather old class of materials. A number of preparation methods will be discussed.

Section 2 is devoted to the structural characterization of GO. It will be outlined that normal GO is a polydisperse material that next to covalently bound oxo-functionalities also contains holes in the basal carbon lattice.

For a systematic study of the structure of GO suitable descriptors taking into account the number of layers, the substrate serving as support and the defect density are required. Such a rational nomenclature is pointed out in Section 4.

Section 5 is devoted to the oxo-functionalization of graphene. Particular emphasis is given to the recently accomplished preparation of graphite oxide and GO that is characterized by a very small amount of σ-defects. Moreover, the important role of intermediate organosulfates during the overall production process is explained.

In Section 6 analytical methods for the structural characterization are described. It is demonstrated that Raman spectroscopy, in particular statistical Raman spectroscopy, is the most important workhorse in this field.

The covalent and non-covalent chemistry of GO is summarized in Section 7. This includes binding of DNA, polymers, inorganic nanoparticles and small organic molecules. Also first approaches to C–C bond formations are discussed.

Section 8 is devoted to the controlled oxo-functionalization of graphene and also the most recently discovered wet-chemical back of GO reduction to high-quality graphene. In this regard the analysis of the stability of oxo-G_1 is outlined in detail. In order to screen the suitability of GO for practical applications such as floating gate memory devices, concepts for the synthetic modifications are also described.

This chapter shows that GO chemistry not only has enormous potential for the development of new high-performance materials but also provides insights to unprecedented chemical processes at the interface of solution and solid state chemistry.

2.2 History of GO and Graphite Oxide

Converting black graphite by oxidation was found to lead to yellow graphite oxide, also termed "graphitic acid," and research started already in 1840 [5, 7, 8]. The structure of the yellow compound was unexplored. Synthetic procedures developed with time and more effective procedures were described. Kohlschütter and Haenni summarized in 1919 many aspects of the chemistry of graphite oxide that had been repeated in recent years [9].

In 1855 Brodie synthesized graphite oxide in nitric acid with potassium chlorate as oxidant [10–14]. Several oxidation steps were, however, necessary and Gottschalk,

Berthold and Luzi determined many graphite oxide properties [15–18]. Staudenmaier found a two-step synthesis to graphite oxide; however, he also recognized that graphite oxide is similar to humic acids and not a defined compound [19–21]. Charpy published in 1909 a one-step synthesis for graphite oxide in sulfuric acid with potassium permanganate as oxidant [22]. Hummers and Offeman published a scalable method later in the 1950s and the protocol is most often termed "Hummers method" [23].

The properties of graphite oxide were explored in the 1960s by Hofmann, Fischer, Clauss, Eckel, Boehm and Scholz [6, 24–29], and Boehm published the laboratory curiosity of GO recently; especially the reduction product was described as a single layer of carbon [4]. Boehm also compared differently prepared graphite oxide and compared some properties [28]. The yield of graphite oxide and GO production can be improved by a pretreatment of graphite persulfate and phosphorus pentoxide [30]. In addition, phosphoric acid was claimed to improve the quality of GO [31].

2.3 Structure of GO

The structure of GO depends on the preparation conditions and can only be described in a generalized way. A molecular formula must be determined for each preparation batch. It should be clear that every sp^3-carbon of GO is a stereo center and molecular-like defined material can therefore not be expected. Graphite with a honeycomb network of C-atoms is used as starting material. During GO synthesis over-oxidation and thus formation of CO_2 are not avoided and that process leads to GO. As outlined later avoiding the over-oxidation keeps the carbon framework intact and thus oxo-G_1 is yielded. Most accepted is the structure model introduced by Lerf and Klinowski, termed "Lerf– Klinowski model." Nuclear magnetic resonance spectroscopy was used to clarify that structure [32–35]. Layers of graphite are oxidized by the oxidant and oxo-functional groups are introduced. Approximately 50% of C-atoms remain sp^2-carbon-forming islands that increase with time. Thus, the structure of GO is metastable and changes somewhat with time [36]. High-resolution transmission microscopy images at atomic resolution display preserved regions of ordered structures, but only with 1 nm diameter. About 80% of regions are disordered that were correlated to functionalized regions [37]. In addition, nanometer-sized holes were identified.

Epoxy and hydroxyl groups decorate both sides of the basal plane as major functional groups. Ketones and carboxyl groups are also introduced by over-oxidation. Those groups are formed by breaking C–C bonds and decorate edges of defects and flakes. In addition, in-plane hydroxyl groups must be considered as a structure motive [38–40]. Other species, such as unpaired spins, have been reported [41, 42], as well as endoperoxide groups. The latter were found as a source of toxicity toward cells [43]. Other structure models have been proposed by Dékány et al. [44], Nakajima–Matsuo

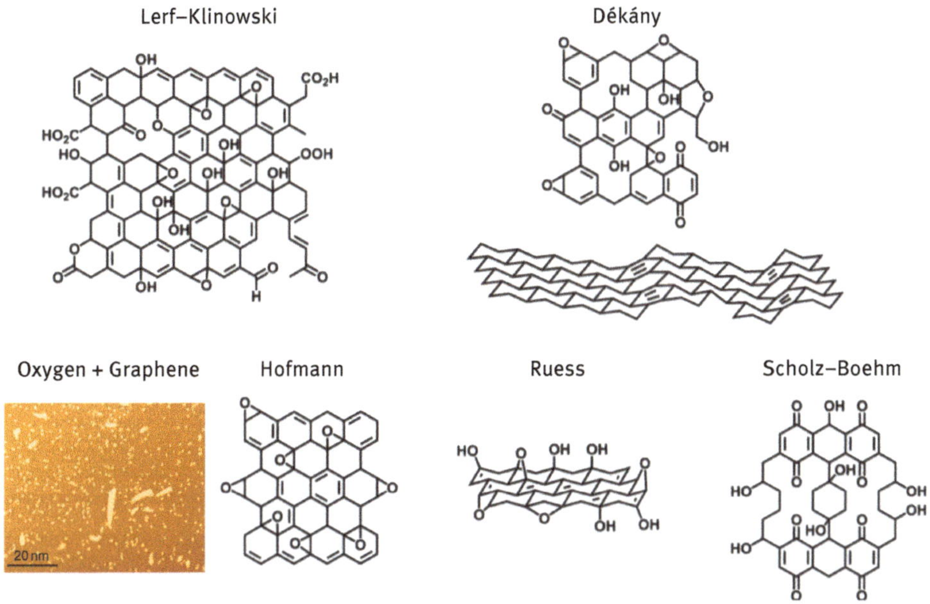

Figure 2.1: Proposed structure models of GO. Reproduced from Ref. [49] with permission from The Royal Society of Chemistry. The Hofmann model can be expected to be realized by exposure of graphene to oxygen atoms (scanning tunneling microscopy image with oxygen atoms covalently bound to graphene). Reprinted with permission from Macmillan Publishers Ltd: Nature Chemistry [50]. Copyright 2012.

[45], Hofmann [46], Ruess [47], Scholz–Boehm [48] and others (Figure 2.1), and all of those may be valid, depending on the preparation conditions [38].

2.3.1 Considerations about the Structure of GO

It can be concluded that GO implies a partially amorphous structure of in-plane defects and observation that were already made more than 100 years earlier [21]. Currently the formation mechanisms of GO is still under investigation [51]. It can be assumed that different applications may need a type of GO with structure defects and others work best with an intact carbon framework. However, developing a structure function relation requires controlling the structure of GO and thus, reliable analysis of the product is required to enable a controlled synthesis [52].

Metal impurities in GO were quantified by inductively coupled plasma mass spectrometry (MS). Iron, cobalt, nickel and manganese were found as residues from natural graphite and used reagents. Iron concentrations >1,000 ppm were found and other metals were detected in concentrations between roughly 5 and 20 ppm. Although these concentrations are low, catalytic effect may be due to metal impurities [53]. Contaminations may also stem from the aqueous work-up process, e.g., a functionalizing work-up procedure with dodecyl amine was recently introduced [54].

2.4 Nomenclature of Functionalized Graphene

Polydisperse materials, such as graphene and functionalized graphene prepared from graphite, are difficult to describe and therefore a systematic formula was introduced (eq. (2.1)) [55, 56].

$$S/^{s,d}G_n - (R)_f / A_f \tag{2.1}$$

Equation (2.1) is a generalized systematic formula to characterize graphene and functionalized graphene. Here, *S*: substrate; *s*: size of graphene; *d*: structural defect density of graphene within the carbon framework; *G*: graphene; *n*: number of layers of graphene; *R*: addend; *f*: degree of functionalization; *A*: non-covalently bound molecules; no *S*: reactions applied in dispersion.

The number of layers of graphene is indicated by the index *n* and G_1 relates to graphene, G_2 to bilayer graphene, G_3 to trilayer graphene and $G_{\text{few-layer}}$ or $G_{4\text{–}10}$ to few-layer graphene. Graphite can be described as (G_n). Graphene on substrates, such as SiO_2, is indicated as SiO_2/G_1 or on BN (boron nitride) as BN/G_1. The size of flakes can be determined by atomic force microscopy (AFM) or directly from solution by analytical ultracentrifugation [57]. Graphene with a size of flakes of about 5 μm on a SiO_2 surface can be described as $SiO_2/^{5\mu m}G_1$.

The density of defects can be determined by Raman spectroscopy [58]. This means, $^{0.1\%}G_1$ relates to one defect on 1,000 C-atoms. Structural defects can be missing or rearranged carbon atoms or sp^3-hybridized C-centers [59]. Other analytical tools are, however, necessary to distinguish these.

The regiochemistry of addends is currently not indicated and difficult to determine. Hydroxyl functionalization of graphene with every 20th C-atom functionalized would be indicated as $G_1\text{-}(OH)_{5\%}$. Adsorbed solvents, such as N-methyl pyrrolidone (NMP), are given in mass %, e.g., few-layer graphene with 30% of adsorbed NMP is indicated as $G_{\text{few-layer}}/NMP_{30\%}$.

2.5 Oxo-Functionalized Graphene

Graphene and oxide, as the name GO implies, would relate to the Hofmann model (Figure 2.1) [50]. Wet-chemistry, however, yields GO with a structure including manifold structure defects. Developing the chemistry of graphene with oxo-addends requires, however, an intact carbon framework. The analysis of defects in the carbon framework is therefore necessary and the formation of defects must be already avoided during oxidative functionalization. Two different approaches to oxo-G_1 derivatives are introduced in the following sections.

2.5.1 Oxo-G_1 from Graphite

Oxo-G_1 with an almost intact σ-framework of C-atoms was published recently [60]. The reaction sequence to oxo-G_1 is illustrated in Figure 2.2. Although potassium permanganate is used as oxidant, as described by Charpy in 1909 [22], the formation of permanent defects is largely avoided, maintaining the local temperature below 10°C, and also during aqueous work-up. Delamination of graphite oxide to oxo-G_1 proceeds in water and can be enhanced by sonication. Raman spectroscopy indicates a density of defects that is as low as 0.01% for the best quality of chemically reduced flakes. In oxo-G_1 every second carbon atom is sp^3-hybridized and functional groups are located on both sides of the basal plane. Functional groups at edges of flakes play a minor role. Thus, oxo-G_1 is a suitable precursor for wet-chemically synthesized graphene and for the development of a controlled chemistry.

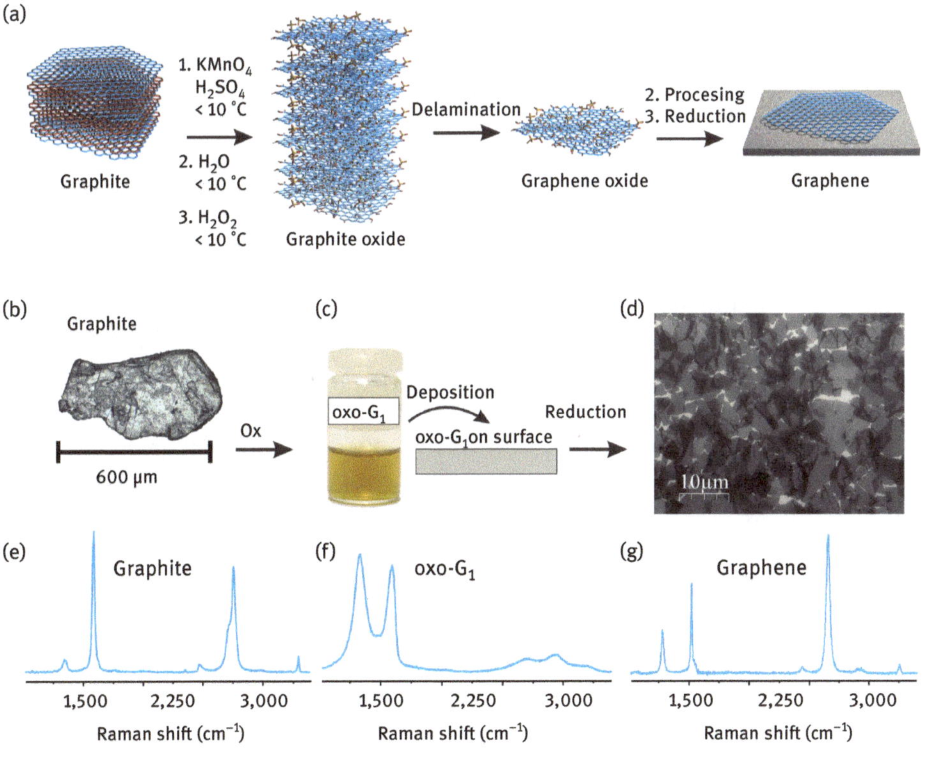

Figure 2.2: (a) Synthesis of GO with an almost intact carbon framework (oxo-G_1) and graphene, starting from graphite in sulfuric acid with potassium permanganate as the oxidant [56]. With permission from Wiley-VCH Verlag GmbH & Co. Copyright 2014, (b) reflected light microscope image of natural graphite (flake size: 600 µm), (c) aqueous dispersion of oxo-G_1 at 0.1 mg/ml, (d) scanning electron microscope image of graphene on Si/SiO_2 and Raman spectra of (e) graphite, (f) oxo-G_1 and (g) a flake of graphene obtained from oxo-G_1 by chemical reduction [60]. With permission from Wiley-VCH Verlag GmbH & Co. Copyright 2013.

After deposition of oxo-G_1 flakes onto Si/SiO_2 wafers, chemical reduction to graphene was conducted. The measurement of magnetoresistance and Hall effect in magnetic fields up to 14 T, at a temperature of 1.6 K, proved the graphene-like nature. Charge carriers are hole-like (charge carrier density $n = 1.6 \times 10^{12}$ cm^{-2}). The measured mobility values exceed 1,000 cm²/Vs. Furthermore, Shubnikov–de Haas (SdH) oscillations were observed and the Landau-level indices analysis confirmed the presence of SdH oscillations. They depend linearly on the inverse magnetic field as expected for SdH oscillations. The frequency of the oscillations plotted vs. the charge carrier density revealed the linear dependence due to the linear density of states. This observation is only expected for 2D graphene. An average quality of wet-chemically prepared graphene bears about 0.3% of defects but still provides mobility values of about 250 cm²/Vs. The residual defects are expected to be mainly point defects, since attempts to heal the defect by chemical vapor deposition process increased L_D values up to 18 nm [61].

2.5.2 Oxo-G_1 from Graphite Sulfate

Graphite sulfate, an acceptor-type intercalation compound, was described and investigated by Rüdorff and Hofmann in 1938. They found an idealized formula of $C_{24}(HSO_4)\cdot 2H_2SO_4$ (Figure 2.3(a)) [62]. Graphite sulfate is also an intermediate during the oxidation of graphite in sulfuric acid using permanganate as oxidant. Graphite sulfate can, however, be functionalized by hydroxyl groups and reverted back to graphene [63]. The positively charged C_{24} subunit can be reacted with water, leading to the formation of hydroxyl groups (Figure 2.3(a)). Single layers of oxo-G_1 can subsequently be isolated, and an AFM image is shown in Figure 2.3(b).

Raman spectra indicate by the broad D, G and 2D peaks a degree of chemical functionalization >3%, and thermogravimetric analysis (TGA) coupled with MS of oxo-G_1 reveals water as the major cleaved compound (Figure 2.3(c)). Elemental bulk analysis reveals finally an idealized formula of $C_{24}(OH)\cdot 2H_2O$. Thus, it seems likely that C_{24}^{+} reacts with water, forming a hydroxyl group on the graphene subunit. The degree of functionalization of this oxo-G_1 is about 4% and indicated as G_1-$(OH)_{4\%}$. However, since bulk analysis was conducted including few-layers the degree of pure single layers may differ to some extent. Flakes of G_1 on SiO_2 (SiO_2/G_1) can be obtained after reduction using vapor of hydriodic acid (HI) and trifluoroacetic acid (TFA). That reduction method quantitatively removes oxo-functional groups from the carbon lattice [64]. Raman analysis reveals that graphene with a density of defects of approximately 0.04% is obtained (Figure 2.3(d), $SiO_2/^{0.04\%}G_1$). Therefore, the defect density was assumably only 0.04% in oxo-G_1 ($^{0.04\%}G_1$-$(OH)_{4\%}$) and all individual Raman spectra of single layers of graphene show a sharp 2D peak ($\Gamma_{2D} = 33$ cm^{-1}, Figure 2.3(d) and (f)) [63].

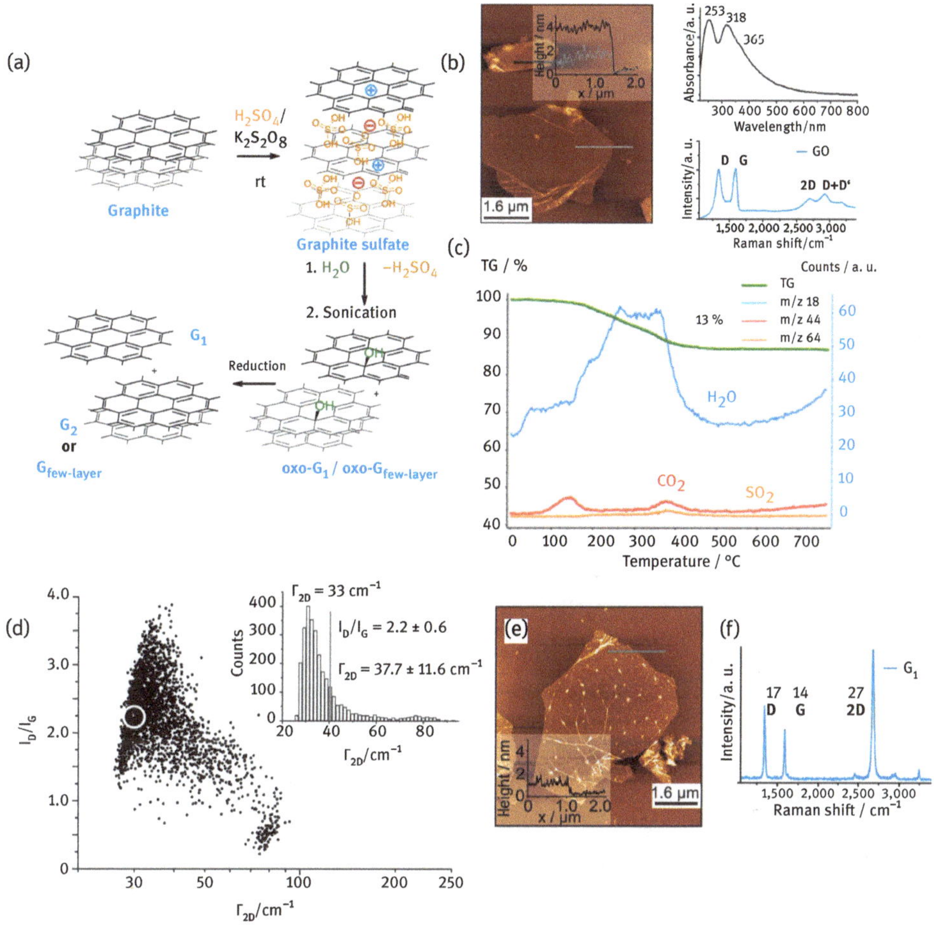

Figure 2.3: (a) Synthesis of graphene and few-layer graphene from graphite sulfate, followed by the reaction with water to yield oxo-G_1 and graphene after reduction, (b) AFM image, UV-vis spectrum and Raman spectrum of oxo-G_1, (c) TGA of oxo-G_1 coupled with MS, (d) statistical Raman analysis of a film of graphene flakes (inset: histogram of Γ_{2D}), (e) AFM image of graphene derived from G_1-$(OH)_{4\%}$, (f) Raman spectrum of graphene obtained from G_1-$(OH)_{4\%}$ after reduction; Γ_D, Γ_G and Γ_{2D} given in italic numbers [63].

2.5.3 Organosulfate in Oxo-G_1 and GO

Oxidation of graphite in sulfuric acid introduces oxo-groups, such as epoxy groups that can form sulfate esters and organosulfate. Boehm and Scholz suggested in 1966 that some of the sulfur content may be due to sulfonic acids or sulfate esters [28]. TGA found an answer to that question by coupling with MS. Organosulfate groups (sulfate

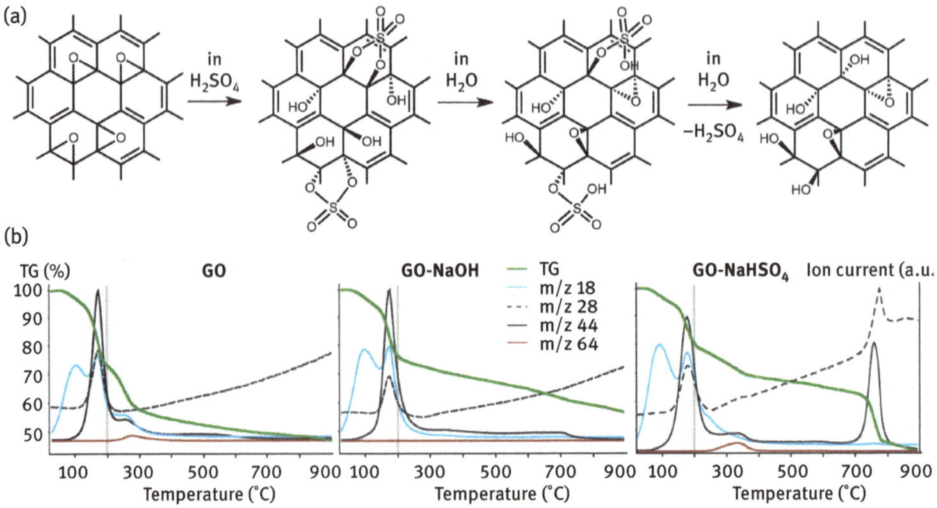

Figure 2.4: (a) Proposed formation of cyclic organosulfate groups in GO and hydrolytic cleavage to organosulfate groups and hydroxyl groups. Acid-catalyzed epoxide formation of cis-dihydroxy groups can be assumed [49, 66]. The structure of GO therefore bears epoxy, hydroxyl and organosulfate groups as major functional groups. Defects in GO are omitted and the structure model is also valid for oxo-G_1 [65], (b) TGA: left – GO; middle – GO treated with sodium hydroxide to cleave organosulfate groups; and right – GO treated with sodium hydroxide and mixed with sodium hydrogen sulfate [65]. With permission from Wiley-VCH Verlag GmbH & Co. Copyright 2013.

esters) were identified for GO and oxo-G_1 as structure motif, indicated by a weight-loss step at 200–300°C [65]. No sulfonic acids were found. The formation of organosulfate and partial and full hydrolysis is depicted in Figure 2.4(a) [66]. The organosulfate groups can be used to explain the acidity of oxo-G_1 and molecular architectures can be synthesized based on those groups, as shown in an example at the end of this chapter.

2.6 Raman Spectroscopy of Graphene, GO and Oxo-G_1

Raman spectroscopy became a characterization tool for probing the integrity of the graphene framework. Defects can be detected at concentrations between roughly 0.001% and 3%. Raman spectroscopy therefore helps in developing the controlled oxo-functionalization of graphene, because reaction conditions can be found that preserve the carbon framework. Raman spectroscopy of graphene and also the impact of defects are introduced and statistical Raman spectroscopy (SRS) and microscopy are explained, allowing a reliable characterization of prepared samples.

2.6.1 Raman Spectra of Graphene

Raman spectra of graphene have been published in 2006 [67]. The shape of the spectra differs from that of few $G_{few\text{-}layer}$, reduced GO (rGO) and graphite [68]. Typically, a laser-excitation wavelength at 532 nm is used [69]. Raman spectroscopy on graphene gives information about stress and strain, the doping level and others. The interpretation of Raman spectra has been developed in recent years [70–73]. Typical Raman spectra of graphene are depicted in Figure 2.5(a). The physical origin of the Raman bands can be explained by the electronic structure of graphene and can be found in

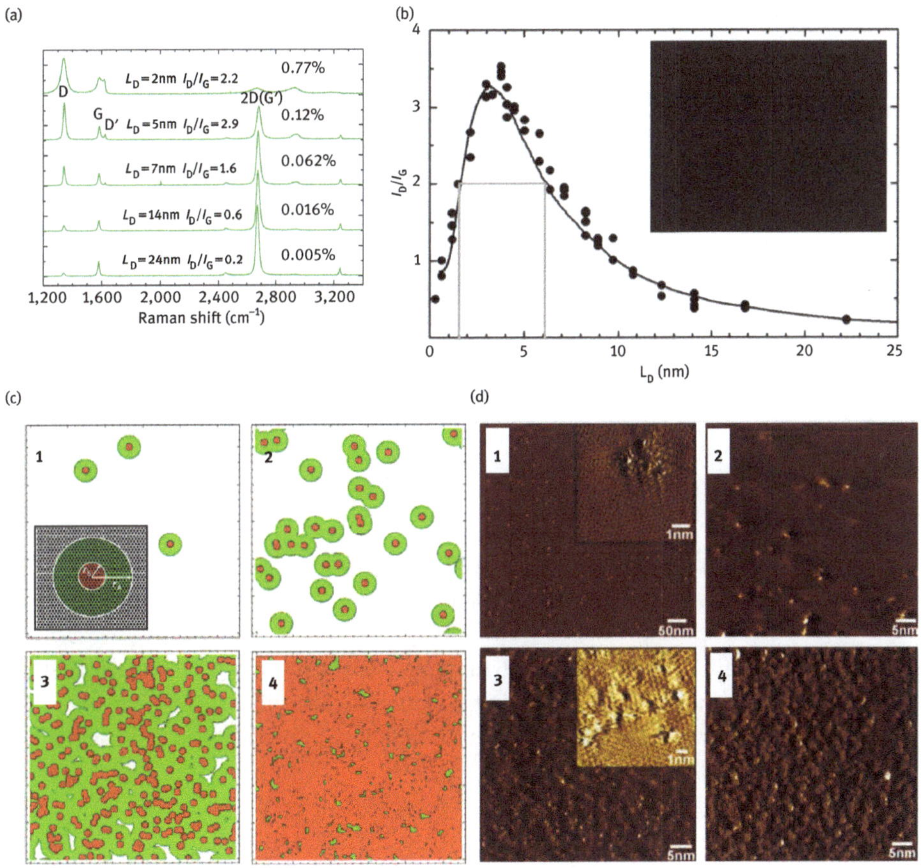

Figure 2.5: (a) Raman spectra of graphene with a variable amount of defects between 0.005% and 0.77%. Reprinted with permission from Ref. [69]. Copyright (2011) American Chemical Society. (b) The I_D/I_G data vs. the average distance L_D between defects. Defects generated by Ar^+ ion bombardment. (c) Activated region r_a (large radius) and disordered region r_s (small region). (d) Scanning tunneling microscope images (HOPG) highly ordered pyrolytic graphite sample subjected to Ar^+ ion bombardment (ion doses 10^{11} (1), 10^{12} (2), 10^{13} (3) and 10^{14} (4) Ar^+ ions per cm^2). Reprinted from Ref. [74]. Copyright 2010. With permission from Elsevier.

the literature, as well as information about Raman spectra of bilayer and few-layer graphene [58, 70–73]. Two major peaks at a Raman shift of about 1,580 cm^{-1} (G peak) and 2,700 cm^{-1} (2D peak) are found for graphene. The D peak at about 1,340 cm^{-1} needs activation by a defect, while the 2D peak at about 2,700 cm^{-1} is the corresponding overtone and needs no defect activation because of momentum conservation. The position of the D peak is dispersive due to the interaction of phonons and electrons. Its shift varies with excitation wavelength between roughly 1,310 cm^{-1} if excited by a red laser and 1,400 cm^{-1} if excited by a UV laser.

2.6.2 Raman Spectra of Graphene with Defects

Introducing defects into graphene leads to an evolving D peak and line broadening for all peaks (Figure 2.5(b)) [74]. The nature of defects can be any type of in-plane disorder, e.g., missing atoms or sp^3-carbon introduced by chemical functionalization [59, 75].

The evolution of Raman spectra was studied with Ar^+ ion bombardment on perfect graphene [74]. The intensity ratio of the D peak and the G peak (I_D/I_G) is plotted for increasing the distance of defects in Figure 2.5(b). The graph follows a relation. The maximum is at L_D = 3 nm. The full-width at half-maximum (Γ) of the peaks needs to be taken into account to discriminate, e.g., between L_D = 2 nm and L_D = 6 nm, respectively. In general, sharp peaks relate to higher L_D values and broad peaks to lower L_D values [69].

$$A_D = L_D^2 \frac{\sqrt{3}}{2} \tag{2.2}$$

$$\frac{A_D}{0.026195nm^2} = N = 33L_D^2 \tag{2.3}$$

One carbon atom fills an area of 0.026195 nm^2 and the area between defects (A_D) can be calculated according to eq. (2.2), assuming an ideal hexagonal packing of defects. The number of C-atoms (N) within this area can be calculated by eq. (2.3) (L_D in nm) [59].

2.6.3 Raman Spectra of GO, rGO, Oxo-G_1 and Related Graphene

Raman spectra of GO bear broad peaks, and the D peak and G peak are the major ones. The Γ value of the D peak is roughly 100 cm^{-1} (Figure 2.6). With reduction of GO the Γ values do almost not change, although the electrical conductivity increases by orders of magnitude [76, 77]. Moreover, the 2D peak, typical for graphene, does not evolve. It is well known that rGO bears extended amounts of lattice defects of approximately >5% and thus Raman spectroscopy is not sensitive at such high defect concentrations.

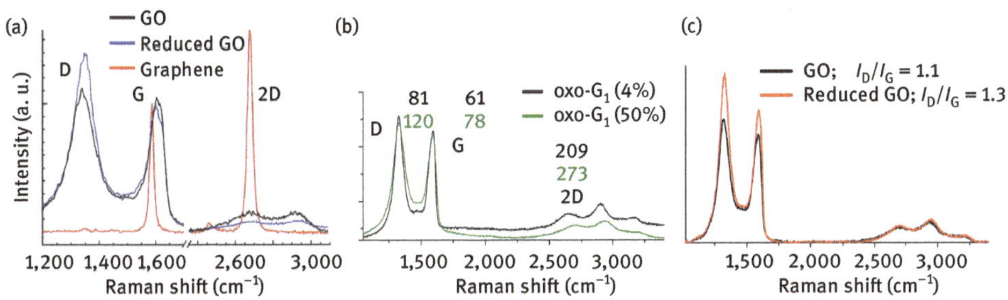

Figure 2.6: (a) Raman spectra of a monolayer of GO and rGO, compared to mechanically exfoliated graphene measured on a substrate of Si/SiO_2, normalized to the G-peak intensity. The I_D/I_G ratio of GO and rGO remains almost constant. Reproduced with permission from Ref. [76]. Copyright (2008) American Chemical Society. (b) Raman spectra of oxo-G_1 with a degree of functionalization of about 4% and 50%, respectively (Γ values are given in italic numbers) [63]. (c) Raman spectra of GO and rGO. The I_D/I_G ratio increases slightly, which indicates a minimal increase of quality. Reprinted from Ref. [78]. Copyright 2012. With permission from Elsevier.

Raman spectra are not expected to change at densities of defects between 4% and 50%. In Figure 2.6(b) Raman spectra of two differently oxidized layers of G_1 are displayed with peaks characterized by $\Gamma_D = 81\ cm^{-1}$, $\Gamma_G = 61\ cm^{-1}$ and $\Gamma_{2D} = 209\ cm^{-1}$ for G_1-$(OH)_{4\%}$ (black line) [63]. The Raman spectrum of G_1-$(OH)_{4\%}$ is superimposed to oxo-G_1 with about 50% of sp^3-carbon atoms [78]. The values of $\Gamma_D = 120\ cm^{-1}$, $\Gamma_G = 78\ cm^{-1}$ and $\Gamma_{2D} = 273\ cm^{-1}$ are about 30%–50% larger for GO compared to G_1-$(OH)_{4\%}$. However, up to now, there is no reliable correlation of such broad peaks to the degree of functionalization elaborated.

With reduction of oxo-G_1 permanent defect can be detected by Raman spectroscopy and the heterogeneity of samples becomes visible. The density of defects is between 0.03% and 1%, as depicted in Figure 2.7 [79]. The model of Lucchese and Cançado is used to determine the concentration of defects [79].

2.6.4 SRS and Microscopy

The reduction of flakes of oxo-G_1 leads to graphene with a heterogeneous quality that can be determined by scanning Raman microscopy (SRM) and analyzed by SRS (Figure 2.7). This allows reliably determining the quality of the carbon framework of graphene and oxo-G_1 of a whole sample. The D, G and 2D peaks can be fitted by Lorentz functions and the intensity of the peaks and the Γ values can be determined. These parameters can not only be determined for many spectra out of a scanned area [79].

A histogram of the I_G values (Figure 2.8(a)) lets discriminating between the substrate, edges of flakes, graphene and $G_{few\text{-}layer}$. A plot of I_G vs. Γ_G lets identifying the

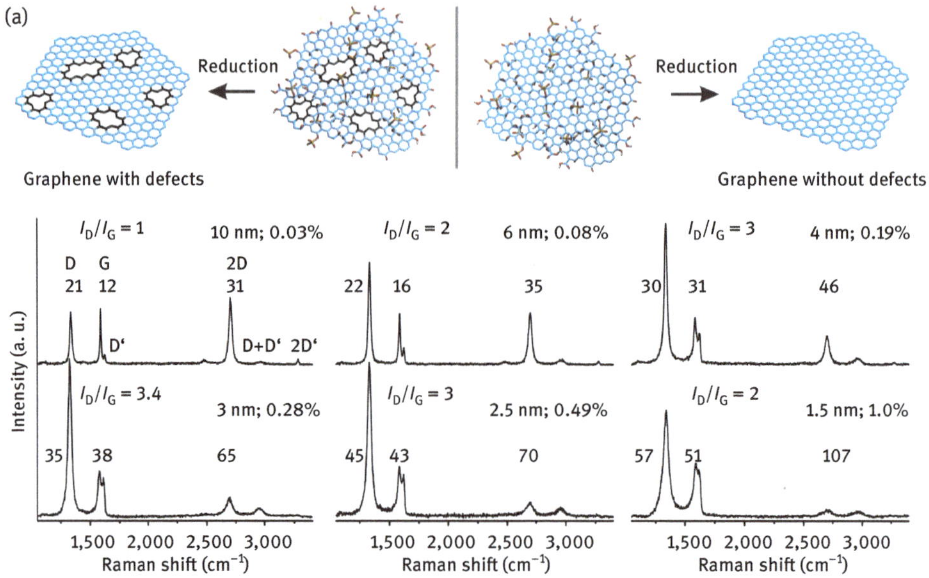

Figure 2.7: (a) Left: Illustration of the chemical structure of GO with structural defects on the %-scale. After chemical reduction, defects can be probed by Raman spectroscopy. Functional groups at defect sites are omitted for clarity. Right: oxo-G_1 with an intact carbon framework can be chemically reduced to intact graphene [56]. With permission from Wiley-VCH Verlag GmbH & Co. Copyright 2014. (b) Raman spectra of graphene from oxo-G_1 with a variable amount of defects between 0.03% and 1%. Γ_D, Γ_G and Γ_{2D} values are given in italic numbers. Reprinted with permission from Ref. [79]. Copyright (2014) American Chemical Society.

I_G values that relate to graphene. The Γ_G values <30 cm^{-1} relate to graphene with a density of defects about <0.3% (Figure 2.8(b)). Such values are not expected for G_2 and $G_{\text{few-layer}}$. The statistical information can be used for color-coding *x*–*y* maps as shown in Figure 2.8(c).

SRS and SRM are powerful tools to visualize the quality of graphene films (Figure 2.8(c), overlaid with AFM image). Moreover, the statistical approach can be used to determine reliable information about changes of the quality of the carbon framework of oxo-G_1 upon processing. Therefore, SRM and SRS are essential tools, developing the controlled chemistry of oxo-G_1.

2.7 Conventional Chemistry of GO

General approaches to functionalize GO are briefly summarized in this section. Further insights into the chemistry of GO and its functionalization protocols can be found in comprehensive review articles [7, 37, 56, 80–83]. The functionalization approaches of rGO are given in recent review articles [56, 84].

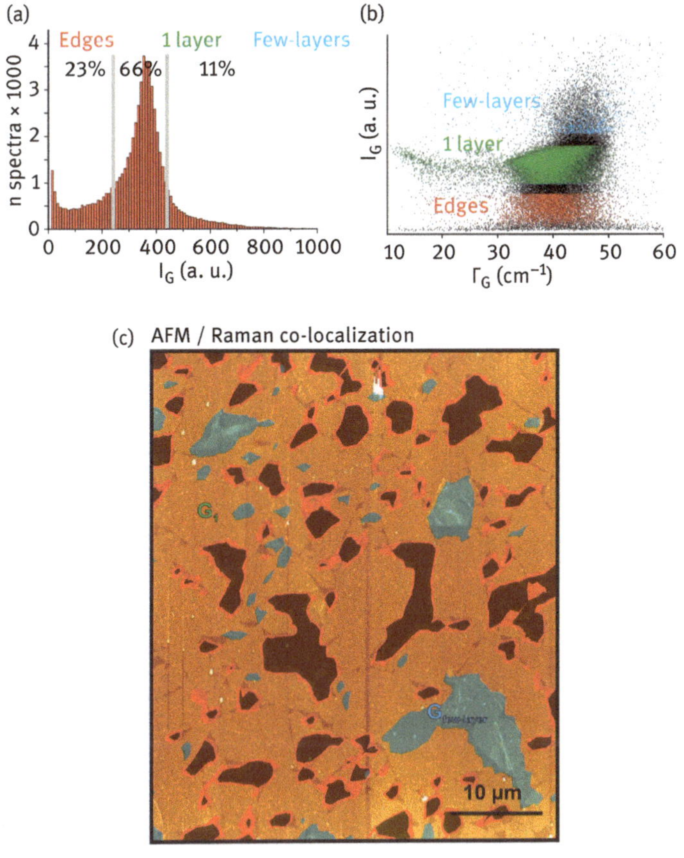

Figure 2.8: (a) Histogram of I_G illustrates the discrimination between G_1, $G_{few\text{-}layer}$ and edges. (b) plot of I_G vs. Γ_G; edges (bottom region), G_1 (middle region) and $G_{few\text{-}layer}$ (upper region). (c) SRM image of a film of graphene generated from oxo-G_1 superimposed with the corresponding AFM image [79]. Reprinted with permission from Ref. [79]. Copyright (2014) American Chemical Society.

2.7.1 Non-Covalent and Covalent Approaches

Non-covalent and covalent functionalization principles were exploited to functionalize GO. The residual π-system of GO interacts with π-conjugated molecules non-covalently (Figure 2.9). Also polar interactions, such as hydrogen bonding, can be exploited to adsorb molecules on GO, such as water in the simplest case. In accordance with these concepts, fluorescent dye-labeled single strand DNA (ssDNA) interacts with GO by polar and π-interactions which leads to the quenching of fluorescence. It could be observed that the fluorescence is restored upon addition of a complementary target, because of the stronger target–ssDNA interaction compared to ssDNA–GO interaction (Figure 2.9) [85].

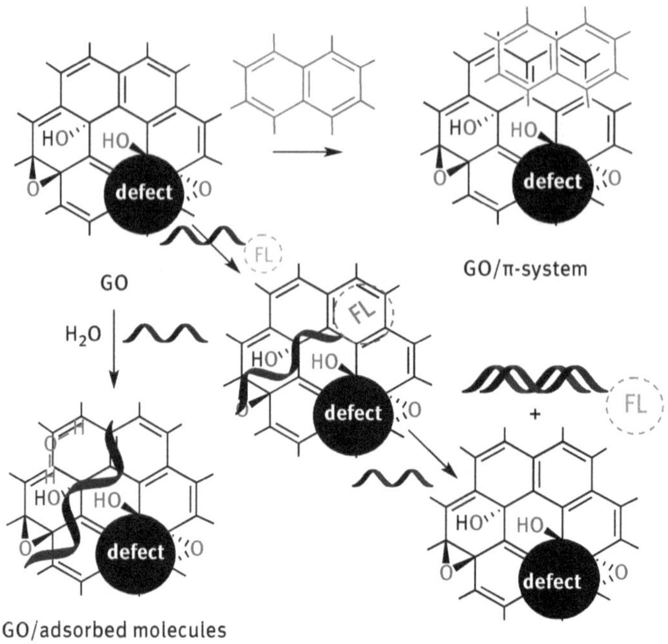

Figure 2.9: Non-covalent approaches to functionalized GO with π-interactions or polar interactions. Example of ssDNA conjugated with a fluorophor (FL, quenched) adsorbed on GO exploiting polar and π-interactions. Complementary target leads to desorption and restores fluorescence [85].

In another approach a dispersion of GO was first non-covalently stabilized by the polymer poly(diallyldimethylammonium chloride) (PDDA) and subsequent chemical reduction of the composite with sodium borohydride led to a stable dispersion of rGO/PDDA [86]. The interaction of the polymer with GO and rGO, respectively, yields the stable dispersion of single layers of rGO. RGO would have formed a precipitate without the stabilizing polymer. It can be assumed that the interactions of GO and PDDA are polar interactions of oxo-functional groups of GO with the cationic groups of PDDA. After reduction, the interaction of rGO and PDDA is due to van-der-Waals interactions and in part polar groups located at edges of defects that are responsible for polar interactions (Figure 2.10).

Moreover, it was reported that hydroxyl groups of GO sheets can be cross-linked by condensation with benzene-1,4-diboronic acid forming boronic esters. This approach yields a porous network that is attractive for gas adsorption [87, 88]. Furthermore, organic isocyanates were used for functionalization [89].

Porous materials are of significant importance for applications, such as for charge storage or adsorption materials. Aerogels can be synthesized by various methods and

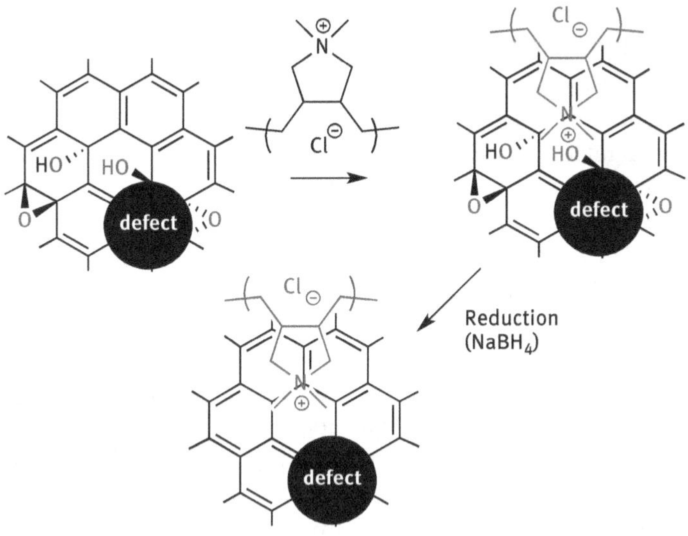

Figure 2.10: Non-covalent polar interaction of PDDA with GO as stabilizing agent, followed by deoxygenation with $NaBH_4$ to yield stabilized rGO [86].

those approaches are summarized in the recent literature [87, 88, 90, 91]. Moreover, ketones, which are functional groups of edges of flakes and edges of defect sites, were found to be reducible by zinc and an acid-forming C–H bonds [92]. Deuterium labeling leads to the conclusion that nascent hydrogen reduces functional groups on both sides of the carbon lattice. In contrast, the direct contact of ketones with zinc leads to the formation of C–H bonds. It can be assumed that selective engineering of functional groups at defect sites, e.g., by exchanging polar ketones to alkyl bonds, can modulate properties of rGO.

There are efforts to selectively react functional groups of GO. It was found that epoxy groups that are located at the basal plane are preferably reduced using a combination of In/InCl [93]. The π-system of GO can be increased by such a method and the residual functional groups make the product dispersible in solvents. However, the deoxygenation is also in part accompanied by the cleavage of hydroxyl groups and also metal impurities may be introduced. The direct reaction of amines with graphite oxide and GO, respectively, was also reported and this approach leads to partially reduced and functionalized materials [94, 95]. Although the products are dispersible in organic solvents, such as tetrahydrofuran, and although they can be mixed with polymers, the type of binding of amines to GO can only be assumed and functionalized polar edges of defect sites may play a crucial role. It can be expected that ketones at edges form hydrazones. Following this approach, stearyl amine-treated GO was incorporated in composite materials with styrene [96].

Ethylene diamine and GO can also react with each other and form partially reduced GO as a hydrogel. Freeze-drying yields a porous and solvent-free material that can be further processed by microwave irradiation to form a hydrogel with a density of only 3 mg/cm^3. The hydrogel has the ability to fully recover the volume after 90% of compression. Several review articles summarize approaches to functionalized and dispersible GO and rGO [37, 49, 97–99]. As described in the latter example, those materials provide distinct properties compared to as-synthesized GO.

Also nanoparticles (NPs) can be formed and deposited onto the sheets of GO and rGO, respectively, to form metal–graphene composites [100]. In a typical approach to composites, GO is dispersed in ethylene glycol together with, e.g., H_2PtCl_6 or $RuCl_3$ and after heating the mixture at 130°C the product is isolated and tested in applications. The electrocatalytic activity of those composites is beneficial for methanol or ethanol oxidation [100]. The NPs have typically a diameter less than 10 nm. Along these lines, the binding of Fe ions with GO was facilitated to prepare magnetic Fe_3O_4 particles that are bound to GO using NH_4OH to prepare the rGO hybrid (Figure 2.11) [101]. Those dispersed GO/NP materials can be attracted by a magnet, as illustrated in Figure 2.11(c). However, crucial questions about the binding and growth mechanism of particles arise. The interaction between rGO and the metal particle remains uncertain. The particles may be non-covalently bound or alternatively a metal–carbon bond may be formed. Residual oxygen moieties of rGO can also form metal–O–C bonds. It can be assumed that defects play a crucial role in those wet-chemical approaches to anchor NPs or to limit the NP growth.

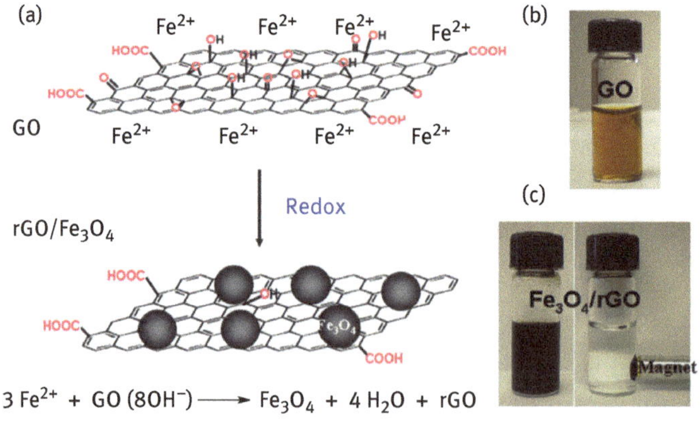

Figure 2.11: (a) GO functionalized with NPs. For instance, magnetic Fe_3O_4 NPs are prepared by redox reaction. Fe^{2+} ions are adsorbed on GO and the NPs are formed in the presence of NH_4OH at pH = 9. (b) dispersion of GO. (c) image of the dispersion after redox reaction. Reproduced from Ref. [101] with permission of The Royal Society of Chemistry.

Edge groups, such as ketones or carboxyl groups of GO, can be used for functionalization reactions [49, 102]. Most chemical functionalization approaches use protocols suitable for reactions with carboxyl groups that are activated and subsequently converted to esters or amides [103–105]. Activation agents for carboxylic acids, such as $SOCl_2$ or carbodiimides, were reported. Especially the reaction of $SOCl_2$ with alcohols, forming eventually instable C–Cl bonds on the surface according to the S_Ni mechanism, may be a concurrent reaction that cannot be ruled out in most approaches (Figure 2.12).

After activation of carboxylic acids, alcohols or amines are added to form esters and amides (Figure 2.12). Although this approach is manifold facilitated, it remains challenging to quantify the exact concentration of covalent bonds. Moreover, it is difficult to distinguish between carbonyl and carboxyl groups in X-ray photoelectron spectra (XPS), and therefore, quantification of ester groups and amides remains challenging, respectively. Although it is plausible that reactions can only occur at defect sites, proving the localization with precision remains challenging.

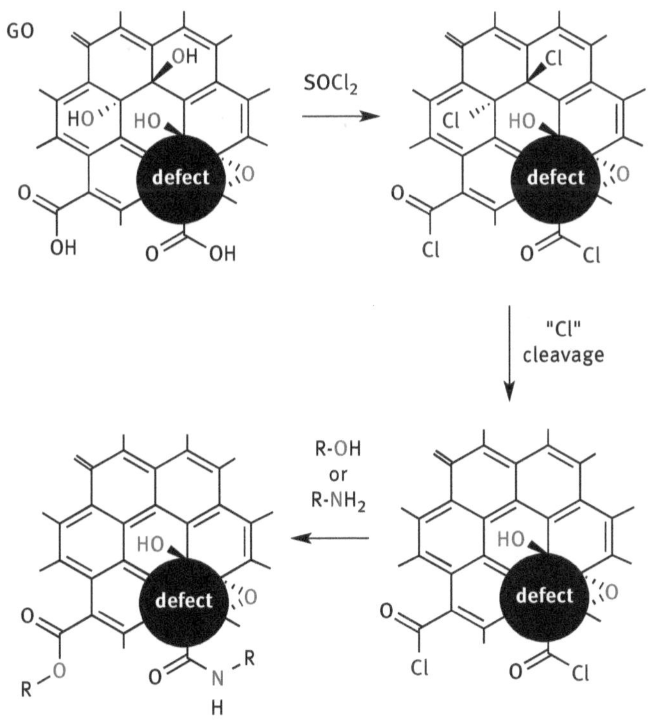

Figure 2.12: Functionalization of GO at carboxylic acids, located at edges of flakes or defect sites. Activation of carboxylic acids by thionyl chloride can also activate surface OH groups and cause elimination or other side reactions. Acid chlorides can subsequently react with alcohol or amines.

2.7.2 Considerations about Carbon–Carbon Bond Formation

It was also reported that C–C bonds from addends to the carbon framework of GO could be formed. This approach is facilitated because approximately 50% of C-atoms in GO are sp^2-hybridized. The formation of C–C bonds on the surface of graphene is highly desired for the synthesis of molecular architectures, as depicted in Figure 2.13.

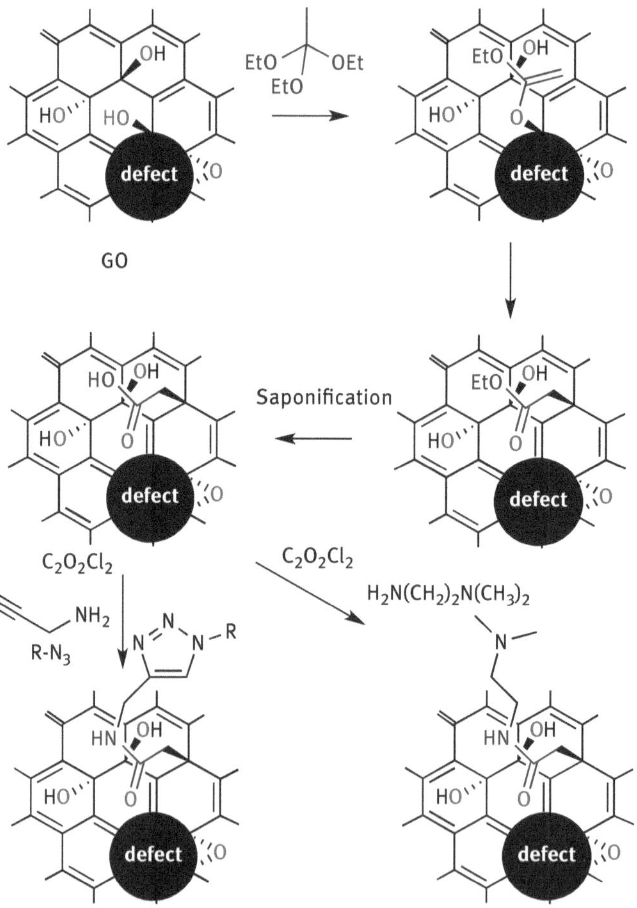

Figure 2.13: Illustration of the reaction of hydroxyl groups of GO to form C–C bonds by Johnson–Claisen rearrangement. The yielded ester can be saponified, activated and undergoes amidation with various amines, including propagyl amine that can be further exploited for alkine–azide reactions. The addition of various groups with ethylene glycol moieties or sulfonate groups was reported. Terminal amines can also be protonated to yield positively charged derivatives suitable for layer-by-layer assembly [113].

One approach toward C–C bond formation in GO was reported using partially reduced GO and the aim was to make soluble graphene. The concept involves the increase of the π-system by partial deoxygenation, followed by the addition of an aryldiazonium salt that also bears a sulfonic acid group to provide water solubility [106]. Although this approach is appealing and black dispersions were yielded, crucial questions could not be addressed at that time. The quantification of defects and the determination of the degrees of functionalization remained unaddressed. Moreover, it is plausible that aryl ether groups can be formed by the reaction of the aryldiazonium salt and residual hydroxyl groups at the surface or at edges [107]. Ether formation with OH groups at edges of defects may even be the major reaction and C–C bond formation could have been a side reaction. Edges of graphene flakes have been demonstrated to be more reactive than the graphene basal plane, an effect that is not well investigated for GO or rGO [108]. Moreover, the adsorption of aryldiazonium salts on carbon surfaces is known and such additional adsorption also alters the surface properties [109]. The reaction of carbon–nucleophiles or C–C bond formation by rearrangement reactions are reported as well [110–112]. Generally black dispersions are yielded, which give evidence for deoxygenation of the surface instead of functionalization. It can be expected that the addition of any addend to the π-system of GO would rather lead to a brighter material than a darker one. Anyhow, it was demonstrated that soluble defective graphene can be prepared, e.g., by a chemical approach using rearrangement reactions, based on the Claisen reaction, as illustrated in Figure 2.13 by the Johnson–Claisen approach [113, 114].

The triethyl orthoacetate is activated by an acid and then an alcohol group of GO forms first a C–O bond. The ether bond is then formed after ethanol cleavage and formation of the C=C double bond. In the second step a [3, 0] sigmatropic rearrangement occurs if a C=C double bond of the carbon framework was in the right conformation. Subsequently, carboxylic acids are formed under basic conditions, as illustrated in Figure 2.13. However, side reactions may also occur, because some carboxylic acids may also generate at plane defects under basic conditions and the influence of functional groups at defect sites on this reaction protocol is not clear. Nevertheless, it could be demonstrated that carboxylic acids can be formed and that the reduction product forms stable dispersions in water. Moreover, X-ray diffraction measurements indicate an increased sheet distance in films of flakes, in contrast to as-synthesized GO. The surface properties of GO can be widely altered by this approach (Figure 2.13). The formation of a series of amide bonds was reported as anchor for carbon-rich molecules or molecules with triple bonds. The triple bonds were even further used to make click reactions to functionalize the surface. However, from the perspective of organic chemistry, it seems to be illusive to prove the direct bond formation between one sp^2-C-atom of the carbon framework and the C-atom of the addend. Thus, analytical tools must be developed or improved to quantify the amount of formed C–C bonds and to evaluate the regiochemistry.

It is desirable to distinguish between edge functionalization, adsorption and functionalization on the intact carbon framework. The chemistry of GO is dominated by intrinsic properties of GO and functionalization is mostly based on C–O bond formation. In contrast to that the chemistry of graphene concentrates on C–C bond formation. Thus, these two worlds merge with C–C bond formation to GO. Overcoming structural defects in GO would possibly lead to graphene derivatives that cannot be distinguished from derivatives directly prepared from graphene.

2.8 Controlled Oxo-Functionalization of Graphene

Conventional approaches to functionalize GO do not account for the reactivity of in-plane defects. Recent reviews with focus on that type of chemistry can be found in the literature [7, 37, 56, 80–84]. Well-known reaction protocols are often applied on GO and the success of the reaction is evaluated by the analysis of, e.g., the performance in applications or by the stability of dispersions. GO and rGO have already performed well in many applications; however, it remains illusive to find structure property relations without knowing the used structure. This lack of knowledge hampers further development but overcoming this problem by controlled chemistry will enable further enhancement of applications.

The instability of the carbon framework of GO and oxo-G_1 during reactions is an important issue and the thermodynamically very stable CO_2 molecule might be formed, leading to an undefined structure. An instable starting material would exclude further functionalization in a controlled way. Thus, establishing the controlled chemistry of oxo-G_1 requires answers for those stability issues that are given in the next sections.

2.8.1 Stability of GO

Boehm and Scholz investigated the thermal decomposition of GO and found large variations [27]. Impurities and metal salts can dramatically decrease the thermal stability. The main weight loss due to decomposition of heated samples of GO (prepared by oxidation with permanganate) occurs between about 180°C and 200°C. The decomposition can proceed and is enhanced by potassium impurities and even explosive decomposition reactions were described [115–117].

TGA can detect the main decomposition temperature of GO detected by weight loss. Combing TGA with a mass spectrometer allows the identification of the evolving gases. Adsorbed water is released up to 120°C followed by CO_2 formation. The main weight loss occurs at a temperature of about 180°C. However, the decomposition of GO with lattice defects on the %-scale already starts at a temperature of 45°C.

This information was found by placing films of GO on ZnSe followed by temperature-dependent infrared spectroscopy. In contrast oxo-G_1 with 0.3% of defects in average is more stable as outlined below [118]. Spectra measured in transmission mode on ZnSe between 25°C and 150°C reveal a new absorption at 2,336 cm^{-1} that can be related to CO_2 trapped between layers with the maximum intensity at 120°C. The formed CO_2 is forming nanometer-sized blisters that were visualized in the films of GO during annealing using AFM. The CO_2 signal reveals the decomposition of GO with the carbon lattice as carbon source. The trapping of CO_2 prevents the detection by MS in the thermogravimetric experiment [118]. The CO_2 concentration could be determined and the results indicate that one CO_2 molecule is formed every 2 nm. Moreover, the role of water in the decarboxylation process of GO was investigated by ^{18}O labeling of water. Indeed, ^{18}O is incorporated into evolving CO_2, which could be detected by MS. That result can be explained by carbonyl chemistry by the formation of hydrates from ketones or carboxylic acids.

Dimiev et al. described the lack of stability of GO. Pristine graphite oxide, a type of graphite oxide that was treated with alcohols instead of water during work-up, was described to bear cyclic organosulfate groups that hydrolyze during water impact. Moreover, it was determined that there is approximately one carbonyl group on 10–12 carbon atoms. This high amount of carbonyl groups can only stem from functional groups at edges of defects and not only from functional groups of the edges of flakes that possess diameters of several micrometer. It was further determined that about one CO_2 molecule is formed on 35–55 carbon atoms during the synthesis. The formation of CO_2 and the formation of carbonyl groups within the carbon lattice require breaking of C–C bonds [66]. Dimiev et al. proposed a formation mechanism for CO_2 that originates from the carbon lattice of GO. The mechanism is illustrated in Figure 2.14, starting from vicinal diols **1**. The reaction of diols in GO with hydroxide causes C–C bond cleavage (**2, 3**) and further base treatment may form carboxylic acids and facilitate CO_2 formation (**8–10**). The carbon lattice can then be further degraded by subsequent reactions. The degradation of GO can lead to acidic functional groups and describe the acidity of conventional GO as illustrated by the reaction of **2** to **7**. A new dynamic structure model of GO was therefore proposed by Dimiev et al. that accounts also for the origin of acidity of GO, taking the degradation of the carbon lattice toward humic acid into account [119]. This instability of GO toward base treatment was further demonstrated to lead to oxidative debrides that are formed upon etching of flakes of GO. The disintegration of flakes upon base treatment is depicted in Figure 2.14.

Chemical reactions are generally driven by overcoming the activation barrier. Those reaction conditions must not be too harsh; otherwise the carbon framework becomes ruptured. Therefore, synthetic protocols that allow the transformation of functional groups must be developed, preserving the integrity of the carbon lattice.

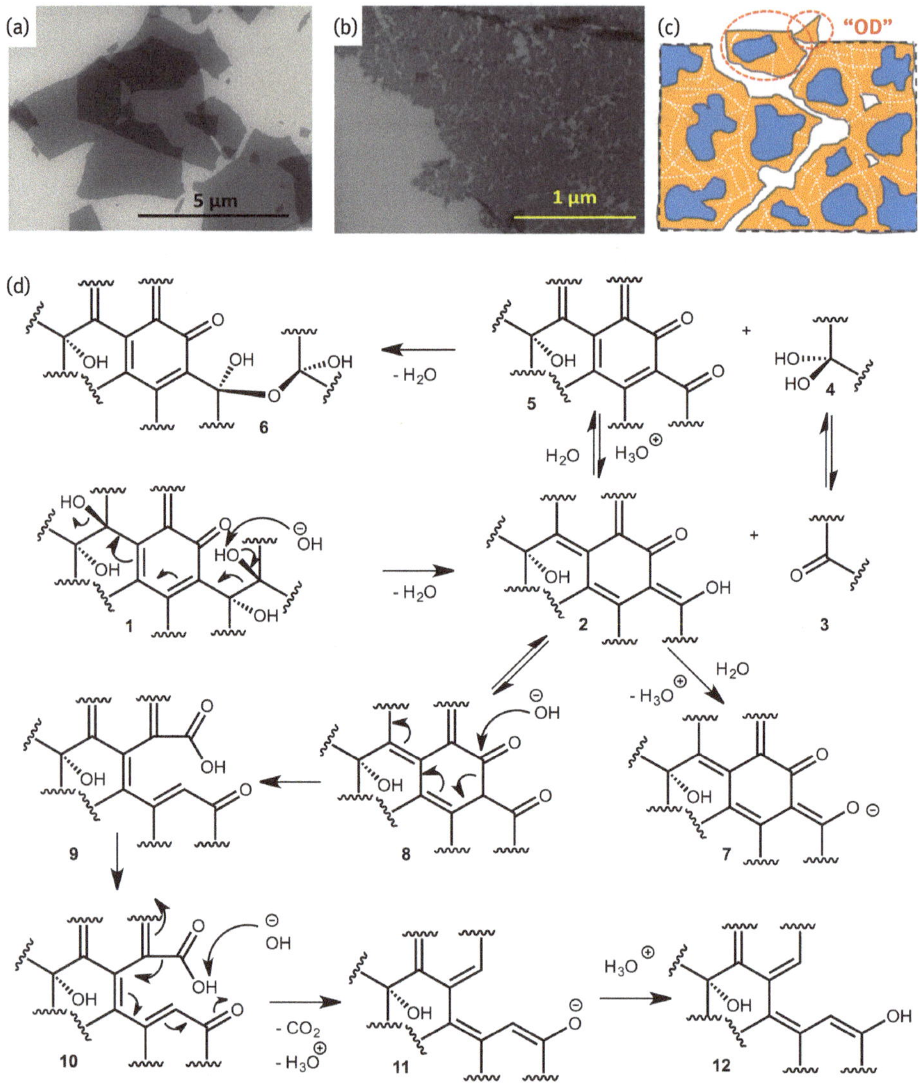

Figure 2.14: (a) Scanning electron microscope (SEM) image of flakes of GO with smooth edges. (b) SEM image of a base-treated GO flake that starts to disintegrate, forming oxidative debrides. (c) Schematic model of the disintegration of flakes of GO. Reprinted from Ref. [120]. Copyright 2015. With permission from Elsevier. (d) Proposed reaction schemes for the degradation of GO by water and base [66, 119, 120].

2.8.2 Stability of Oxo-G_1

Oxo-G_1 bears few defects, at least below 1%. The thermal stability was found to be higher than that of GO. However, since the carbon framework of oxo-G_1 was less damaged by the oxidation protocol than GO, SRS can be used to more precisely determine the thermal stability of the carbon framework.

Thus, SRS on thermally treated and subsequently chemically reduced films of flakes was conducted and the thermal stability of the σ-framework of C-atoms up to 100°C was identified (Figure 2.15(a)). The plot of I_D/I_G ratio vs. Γ_{2D} in Figure 2.15(b) illustrates the thermal stability. The Γ_{2D} values are between 60 and 70 cm^{-1} and remain almost constant up to 100°C. Heating oxo-G_1 to 150°C and subsequent reduction reveals the decomposition of the carbon framework. The Γ_{2D} values detected are between 90 and 300 cm^{-1}. The worst quality of thermally processed oxo-G_1 is measured for samples heated to 500°C. That type of graphene, better termed "thermally processed oxo-G_1," is amorphous and a 2D peak is hardly detectable with Γ_{2D} = 300 cm^{-1}. Thermal processing of oxo-G_1 to 1,000°C is found to partially increase the quality of the carbon lattice. However, the original quality is not recovered as quantified by SRS. The Γ_{2D} values are between 90 and 190 cm^{-1} and do not reach the low values of 60–70 cm^{-1} determined without thermal processing (Figure 2.15(b)).

It can be concluded that the carbon framework of oxo-G_1 is stable enough for chemical reactions. However, it must be kept in mind that chemicals can reduce the thermal stability of the carbon framework. This issue is discussed and illustrated in the synthetic modification section below.

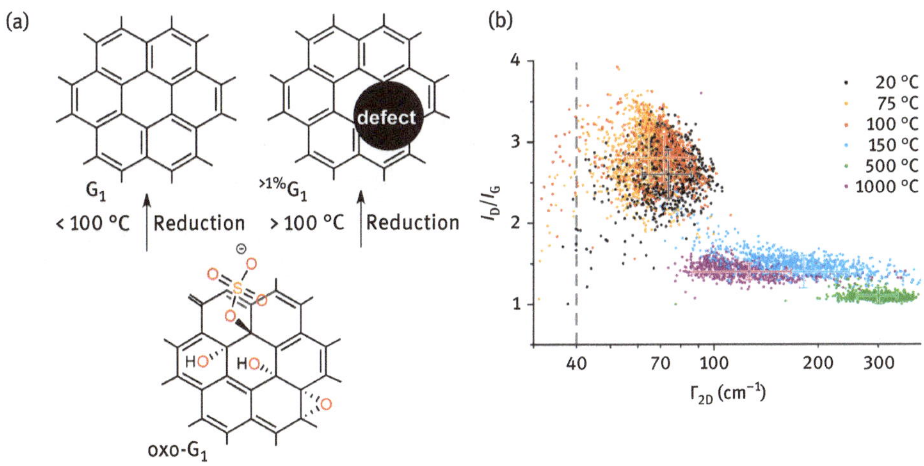

Figure 2.15: (a) The carbon framework of oxo-G_1 is thermally stable up to 100°C, but it is instable at higher temperatures [121]. (b) SRS of graphene derived from oxo-G_1 after thermal treatment [121].

2.8.3 Reduction of Oxo-G_1

SRS can also quantify the efficiency of reducing agents toward reduction of oxo-G_1 to graphene, because the density of defects is within the sensitivity of Raman spectroscopy [122]. In the case of GO, were defects exclude SRS, the success of the reduction is quantified by means of the C/O ratio, determined by XPS or by electrical conductivity measurements of films of flakes [123, 124]. The latter methods must be used for rGO with defects of several percent. However, with concentrations of defects far below 1%, as for oxo-G_1, contaminations may affect XPS results and SRS is beneficial.

The quality of generated graphene from oxo-G_1 can be determined by SRS and plots of I_D/I_G vs. Γ_{2D} values illustrate the results for four reducing methods (Figure 2.16(a)). The most efficient method identified is a mixture of HI and TFA. The slightly less efficient AS (vitamin C) can also produce graphene of high quality. However, as AFM images reveal, there remains a lot of AS adsorbed in the surface (Figure 2.16(b)). Vapor of hydrazine, which is often used in the literature for making rGO, is less efficient, and thermal processing of oxo-G_1 forms the worst quality of graphene. The latter method produced amorphous graphene, dominated by defect sites. The efficiency of reducing agents is thus ranked like HI/TFA > AS > hydrazine > 200°C [122].

Is the quality of the formed graphene limited by the efficiency of the reducing agent or by the quality of oxo-G_1? The study described above demonstrates the influence of different reducing agents on the quality of graphene; thus the reducing agent is the limit. However, the quality of graphene should also be limited by the in-plane

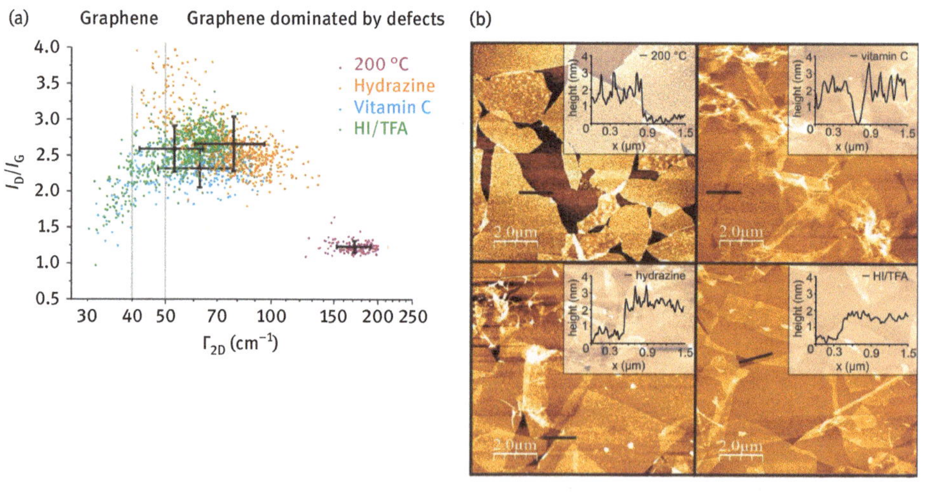

Figure 2.16: (a) Statistical Raman analysis of graphene from oxo-G_1 using different reducing agents. (b) AFM images of graphene from oxo-G_1 obtained after thermal reduction or reduction with vitamin C (ascorbic acid (AS)), hydrazine or HI/TFA [122].

defects of oxo-G_1. Oxo-G_1 is usually deposited on a Si/300 nm SiO_2 substrate by, e.g., Langmuir–Blodgett technique. After deposition, the film of oxo-G_1 flakes is reduced by the reducing agents, but the reducing agent can only access the upper surface of oxo-G_1. The lower side is blocked by the substrate [64]. Oxo-G_1 was therefore placed on the aqueous subphase with a reducing agent like AS placed in the subphase to ensure that the reducing agent has access to the lower side of the flakes of oxo-G_1 (Figure 2.17).

Now the reduction is also ensured to proceed from the lower side. The surface turns gray after several minutes and graphene starts to form and is floating on the aqueous subphase. After transfer of that graphene on a substrate the quantification of the reduction efficiency by SRS is allowed. The graphene, yielded after subphase reduction, can be further reduced by HI/TFA. The compared results reveal that the direct reduction of oxo-G_1 on the substrate without a reducing agent in the subphase

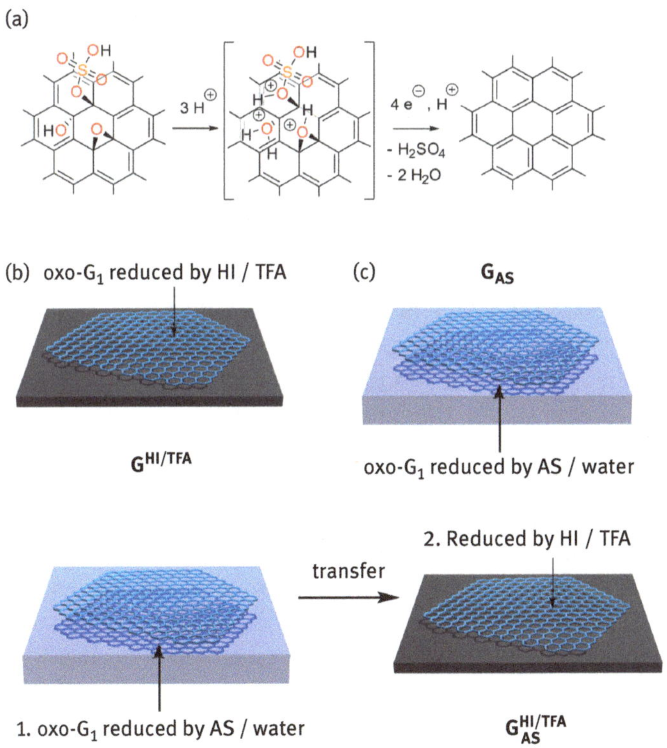

Figure 2.17: (a) Mechanism of the reduction of oxo-G_1 by an electron donor, such as iodide and a strong acid. Successive protonation and electron transfer steps remove oxygen functional groups to form graphene on a substrate. (b) Reduced oxo-G_1 on substrate by vapor of HI/TFA ($G^{HI/TFA}$), (c) reduced by AS from the reductive subphase (G_{AS}) and (d) reduced by the combination of both reduction methods ($G_{AS}^{HI/TFA}$) [64].

is the most efficient one. The two-step reduction process with the reducing agent in the subphase followed by HI/TFA reduction is less efficient. These results may indicate the hindrance of reaction pathways leading to graphene by adsorption of unreactive species.

A reduction mechanism was derived for the reduction of oxo-G_1, neglecting effects by defects. About every second carbon atom of oxo-G_1 is sp^3-hybridized (hydroxyl, epoxy and organosulfate groups) and there is a residual π-system. The functional groups can be protonated by the acid to form oxonium ions. It can be expected that the electron affinity of oxo-G_1 (π-system) increases with protonation to promote the electron transfer from iodide. Subsequently, water and sulfuric acid are cleaved to form graphene (Figure 2.17(a)). Since oxo-G_1 is a 2D material, electrons can be transferred to the π-system from the upper side. Cleaving functional groups from the lower side can be facilitated by protonation of oxo-groups between the substrate and the oxo-G_1 as a consequence of proton hopping from edges of flakes. It can be concluded that the reduction of oxo-G_1 is limited rather by in-plane using a potent reducing agent system such as HI/TFA than by the reducing agent [64].

2.8.4 Synthetic Modification of Oxo-G_1

The functionalization of graphene leads to novel graphene derivatives and also the modification of oxo-G_1. The current status of functionalization is summarized in the following. Full control over the type and density of functional groups, as well as their conversion, is not yet fully elaborate since the field of controlled chemical oxo-functionalization of graphene is a very young discipline. Hydroxyl, epoxy and organosulfate groups of oxo-G_1 are the major functional groups in addition to sp^2-carbon. The size of flakes is on the micrometer-scale and thus edges play a minor role. The density of in-plane defects is about 0.3% in average and thus edges of defects should also play a minor role.

The integrity of the carbon framework upon chemical reactions is necessary to elaborate novel preparation protocols that allow avoiding the rupture of the carbon framework [126]. It was demonstrated that functional groups of oxo-G_1, such as epoxy, hydroxyl and organosulfate, are stable in aqueous HCl at the pH of about 1 at 10°C (Figure 2.18). But at a temperature of 40°C, organosulfate groups are cleaved, while hydroxyl and epoxy groups remain chemically bound.

GO is decomposed by base treatment as outlined above. Thus, the chemical reactivity of oxo-G_1 against sodium hydroxide was of interest. The reaction conditions were found to be controllable to cleave organosulfate and possibly also epoxy groups to diols upon hydroxide treatment at 10°C. The SRS proves that the carbon framework remains stable under those reaction conditions. In contrast, treating oxo-G_1 with sodium hydroxide at 40°C not only cleaves functional groups, but also raptures the carbon framework and introduces therefore permanent defects into the carbon

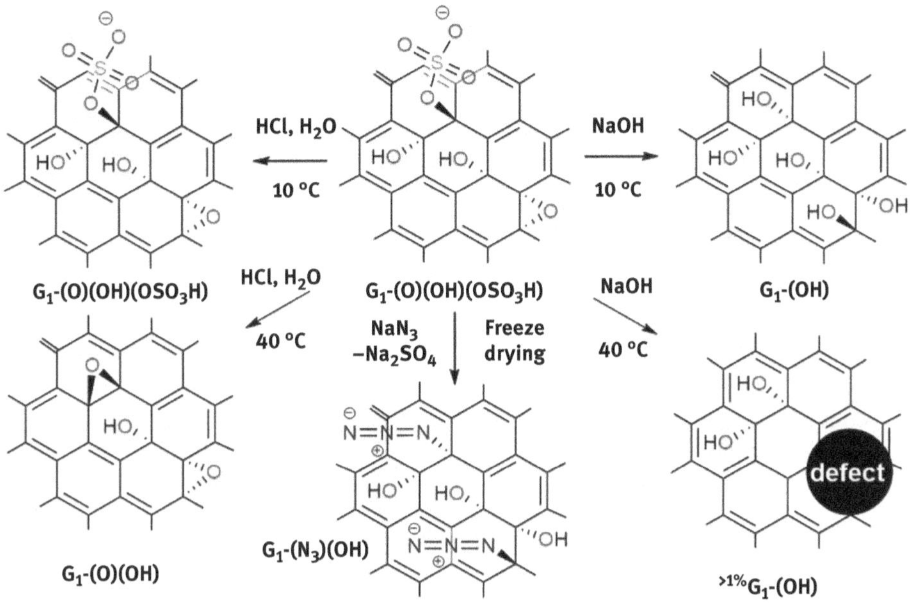

Figure 2.18: Illustration of the reactivity of oxo-G_1. Reactivity at 10°C and 40°C, respectively, upon treatment with HCl or NaOH is shown. Also the substitution of organosulfate by azide is possible under controlled reaction conditions [125].

framework (Figure 2.18). SRS indicates the rupture by the increase of Γ_{2D} from 72 to 120 cm^{-1}. It can be concluded that the reaction conditions must be well controlled to enable the chemistry on the surface of graphene, which leads to new derivatives of graphene. In that many derivatives, such as hydroxylated graphene (G_1-OH) can be prepared.

The reactivity of the organosulfate group of oxo-G_1 was investigated and it was found that azide groups can be introduced by substitution of organosulfate. This substitution reaction preserved thermally instable groups due to the mild reaction with sodium azide that proceeded in the solid state upon freeze-drying (Figure 2.18) [125]. The carbon–nitrogen bonds formation could be proved by Fourier transform infrared spectroscopy (FTIR), TGA coupled with MS, elemental analysis and ^{15}N nuclear magnetic resonance (NMR) spectroscopy measured in the solid state. Azide is predominantly located on the surface of oxo-G_1 and not on the edges. The C–N stretching vibration of azide was found at 2,123 cm^{-1} and shifted by 11 cm^{-1} upon binding of $^{15}N^{14}N_2$ to the carbon framework (Figure 2.19(a)). The substitution of organosulfate was also indicated by TGA (Figure 2.19(b)). The weight-loss step at 200–300°C is only detected for oxo-G_1 but not for the substitution product. The decomposition of azide was indicated by the signal of m/z 29, which is related to $^{15}N^{14}N$ formation (Figure 2.19(b)). Moreover, wet-chemically cleaved sulfate could be identified by precipitation of $BaSO_4$.

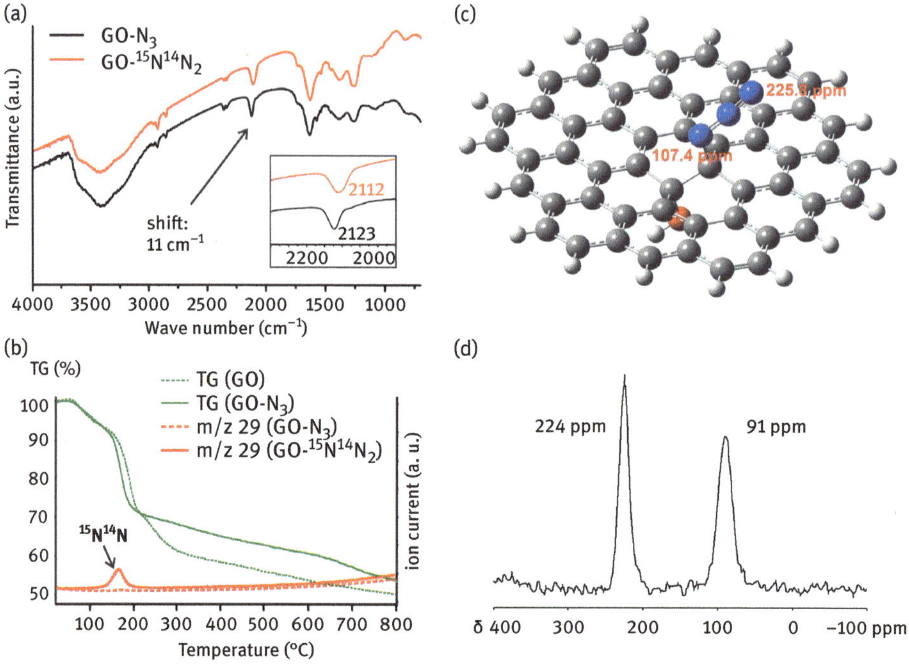

Figure 2.19: (a) Infrared spectra of GO-N_3 and GO-$^{15}N^{14}N_2$ (here GO is oxo-G_1). (b) TGA of GO and GO-$^{15}N^{14}N_2$, *m/z* 29 of GO-N_3 and GO-$^{15}N^{14}N_2$. (c) simplified model of GO-N_3 with an azide and a hydroxyl group connected to the carbon lattice in trans-configuration. ^{15}N NMR shifts are calculated by *ab initio* methods (107.4 ppm and 225.8 ppm), (d) ^{15}N ssNMR magic angle spinning spectrum of GO-$^{15}N^{14}N_2$ with two peaks (1:1 ratio) [125].

Also NMR spectroscopy in solids could be conducted on ^{15}N-labeled compounds (Figure 2.19(d)). Two signals at 107.4 and 225.8 ppm were found which relate to the two possible binding states of $^{15}N^{14}N^{14}N$ (R-$^{15}N^{14}N^{14}N$ or R^{14}-$N^{14}N^{15}N$). The two signals could also be confirmed by calculated predictions based on *ab initio* methods (Figure 2.19(c)). The quantification of N-content revealed one azide group on every 30th carbon atoms. The azide group in GO-N_3 is hydrolytically stable up to 60°C and thus, GO-N_3 can be expected as a suitable precursor for other reactions, opening the field of azide chemistry for graphene-based materials. In another approach oxo-groups are in part removed, while keeping the azide groups chemically bound, an approach that, if complete, would lead to graphene azide [127].

The following example demonstrates an oxo-G_1-based molecular architecture as floating gate memory device working at an exceptional low voltage of 3 V (Figure 2.20(e)) [128]. The synthesis and characterization are outlined in more detail in this example to illustrate the efforts necessary to prove the proposed structure on the complex graphene system.

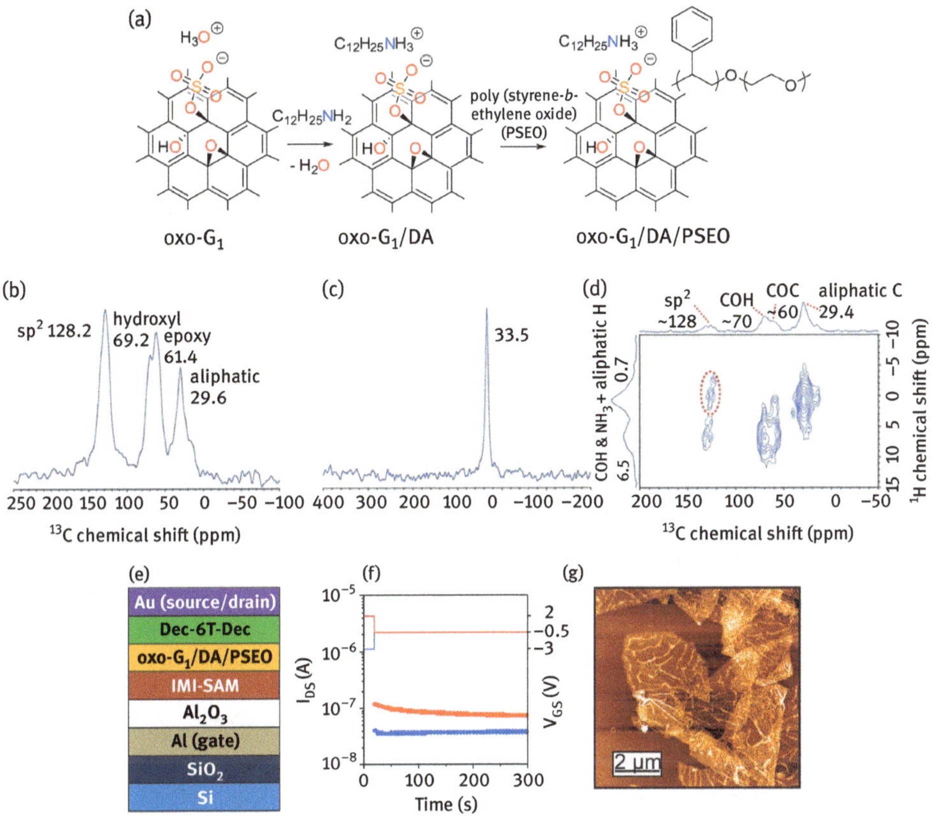

Figure 2.20: (a) Reaction scheme illustrating the reaction of oxo-G_1 with dodecylamine (oxo-G_1/DA) followed by non-covalent functionalization with a block copolymer of styrene and ethylene oxide (PSEO). Solid state NMR spectra of oxo-G_1/DA, (b) ^{13}C NMR, (c) ^{15}N NMR, (d) ^{1}H-^{13}C correlation NMR. (e) Device structure of a floating gate memory device with oxo-G_1/DA/PSEO as charge storage material. (f) Device characteristic; write signal at −3 V, erase signal at 2 V and read voltage at −0.5 V. (g) AFM image of flakes of oxo-G_1/DA/PSEO composite [128].

A very thin dielectric is necessary to separate the charge storage layer from the gate and the semiconductor to enable low-voltage operation. The used oxo-G_1 bears epoxy and hydroxyl groups in addition to organosulfate groups. The latter are complexed with hydronium ions and about 50% of C-atoms are sp^2-hybridized (Figure 2.20(a)). Titration of oxo-G_1 with dodecyl amine exchanges the hydronium ion by DA and the electrostatic complex of DA bound to organosulfate is formed.

This oxo-G_1/DA was thoroughly analyzed by combustion EA (elemental analysis), TGA (thermogravimetric analysis) coupled with FTIR, GC and MS. The data revealed that organosulfate and DA are structure motifs. TGA analysis reveals that SO_2 originates from organosulfate. Also signals of alkyl species could be detected. Moreover, the alkyl signals, identified by TGA–FTIR, stem from DA and could be detected by

GC–MS analysis due to the identification of dodecene, the expected elimination product of DA. In addition, ssNMR spectroscopy was performed (Figure 2.20(b–d)). The combination of ssNMR with *ab initio* calculations proves that alkyl-ammonium is the only one major nitrogen-containing species of the structure. The data are consistent with the proposed electrostatic binding of DA to organosulfate. The alkyl chain of DA was further found to interact with the π-system, as evidenced 2D ssNMR measuring ^{13}C-^{1}H correlation spectra (Figure 2.20(d)).

Single layers of the complex of oxo-G_1/DA are soluble in THF (tetrahydrofuran) and can therefore be mixed with soluble polymers. Here, the block PSEO was used. Thus, oxo-G_1/DA flakes become non-covalently functionalized and form the composite of oxo-G_1/DA/PSEO. The AFM image in Figure 2.20(g) proves the absence of free polymer particles. TGA gives further evidence for the decomposition of the adsorbed polymer at about 450°C and FTIR signals stemming from aryl groups could be traced. The block copolymer functions as dielectric layer with a thickness of only 1–2 nm on each side of the flakes. The oxo-G_1/DA/PSEO composite is also the charge storage material and enables the function of the floating gate memory device. Imidazolium-terminated self-assembled monolayer enables the deposition of the charge storage layer and efficient isolation of the substrate (Figure 2.20(e)). A sexithiophene derivative is used as semiconductor. A gate voltage of –3 V is sufficient to introduce charge carriers and the charge carriers can be erased at +2 V. Thus, information can be stored and the current can be “read” at −0.5 V (Figure 2.20(f)). Reference experiments reveal that the memory device working at 3 V can only be constructed with the defined composite of oxo-G_1/DA/PSEO. Other precursors or less-defined systems all fail, e.g., oxo-G_1, oxo-G_1/DA, defective GO, defective GO mixed with DA and PSEO, or reference systems that bear excess of PSEO. Other examples that relate to self-assembly processes and device fabrication with increased performance can be found in the literature [129–131].

2.9 Conclusions

The classical disciplines of organic chemistry and materials science mingle with conducting graphene chemistry. While functionalization of C_{60} is verified by molecular analysis, the functionalization of graphene requires novel methods. Especially the functionalization of graphene starting from graphite by oxo-addends requires well-controlled reaction conditions, not to over-oxidize and rupture the carbon framework. Avoiding in-plane lattice defects, the reactivity of functional groups on the basal plane can be explored and first experimental results are summarized in this chapter. Analytical tools were developed to probe reaction conditions and their impact on the integrity of the carbon framework. Introduced functional groups can be analyzed, e.g., by FTIR, EA, TGA coupled with FTIR, MS or GC–MS. In contrast to carbon nanotubes that are contaminated by paramagnetic impurities, functionalized graphene can be analyzed by ssNMR. Although that method is advancing, specialists are still required; however, with ssNMR more insights into the regiochemistry of addends will be gained. With controlling the chemistry of graphene, novel graphene derivatives will be accessible with specialized properties. It can be expected that those materials will enable new applications with increased performance.

References

[1] Dreyer DR, Ruoff RS, Bielawski CW. Ein Konzept und seine Umsetzung: Graphen gestern, heute und morgen. Angew Chem 2010;122(49):9524–32.

[2] Dreyer DR, Ruoff RS, Bielawski CW. From conception to realization: an historical account of graphene and some perspectives for its future. Angew Chem Int Ed 2010;49(49):9336–44.

[3] Boehm H-P. Graphen – wie eine Laborkuriosität plötzlich äußerst interessant wurde. Angew Chem 2010;122(49):9520–3.

[4] Boehm HP. Graphene – how a laboratory curiosity suddenly became extremely interesting. Angew Chem Int Ed 2010;49(49):9332–5.

[5] Eigler S. Transparent and electrically conductive films from chemically derived graphene. In: Mikhailov S, editor. Physics and applications of graphene – experiments. NewYork: InTech, 2011:109–30.

[6] Boehm HP, Clauss A, Fischer GO, Hofmann U. Dünnste Kohlenstoff-Folien. Z Naturforsch B 1962;17(3):150–3.

[7] Dreyer DR, Park S, Bielawski CW, Ruoff RS. The chemistry of graphene oxide. Chem Soc Rev 2010;39(1):228–40.

[8] Schafhaeutl C. Ueber die Verbindungen des Kohlenstoffes mit Silicium, Eisen und andern Metallen, welche die verschiedenen Gattungen von Gusseisen, Stahl und Schmiedeeisen bilden. J Prakt Chem 1840;21(1):129–57.

[9] Kohlschütter V, Zur HP. Kenntnis des graphitischen Kohlenstoffs und der Graphitsäure. Z Anorg Allg Chem 1919;105(1):121–44.

[10] Brodie BC. Note sur un nouveau procédé pour la purification et la désagrégation du graphite. Ann Chim Phys 1855;45:351–3.

[11] Brodie BC. On the atomic weight of graphite. Philos Trans R Soc London 1859;149:249–59.

[12] Brodie BC. On the atomic weight of graphite. Proc Roy Soc Lond 1859;10:11–12.

[13] Brodie BC. Sur le poids atomique du grafite. Ann Chim Phys 1860;59:466–72.

[14] Brodie BC. Ueber das Atomgewicht des Graphits. Ann Chem Pharm 1860;114(1):6–24.

[15] Gottschalk F. Beiträge zur Kenntniss der Graphitsäure. J Prakt Chem 1865;95(1):321–50.

[16] Berthelot M. Méthode universelle pour réduire et saturer d'hydrogène les composés organiques. Ann Chim Phys 1870;19(4er):392.

[17] Luzi W. Über Allotropie des amorphen Kohlenstoffes. Ber Dtsch Chem Ges 1892;25(1):1378–85.

[18] Ueber LW. Graphit und Graphitit. Ber Dtsch Chem Ges 1893;26(1):890–5.

[19] Staudenmaier L. Verfahren zur Darstellung der Graphitsäure. Ber Dtsch Chem Ges 1898;31(2):1481–7.

[20] Staudenmaier L. Verfahren zur Darstellung der Graphitsäure. Ber Dtsch Chem Ges 1899;32(2):1394–9.

[21] Staudenmaier L. Untersuchungen über den Graphit. Ber Dtsch Chem Ges 1899;32(3):2824–34.

[22] Charpy G. Sur la formation de l'oxyde graphitique et la définition du graphite. C R Hebd Séances Acad Sci 1909;148(5):920–3.

[23] William S, Hummers J, Offeman RE. Preparation of graphitic oxide. J Am Chem Soc 1958;80(6): 1339–39.
[24] Boehm HP, Clauss A, Fischer GO, Das HU. Adsorptionsverhalten sehr dünner Kohlen-stoff-Folien. Z Anorg Allg Chem 1962;316(3–4):119–27.
[25] Scholz W, Boehm HP. Die Ursache der Dunkelfärbung des hellen Graphitoxids. Z Anorg Allg Chem 1964;331(3–4):129–32.
[26] Scholz W, Boehm HP. Die thermische Zersetzung von Graphitoxyd. Naturwissenschaften 1964;51:160–60.
[27] Boehm HP, Der SW. Verpuffungspunkt" des Graphitoxids. Z Anorg Allg Chem 1965;335(1–2): 74–9.
[28] Boehm H-P SW. Vergleich der Darstellungsverfahren für Graphitoxyd. Liebigs Ann Chem 1966;691(1):1–8.
[29] Boehm HP, Eckel M, Scholz W. Über den Bildungsmechanismus des Graphitoxids. Z Anorg Allg Chem 1967;353(5–6):236–42.
[30] Kovtyukhova NI, Ollivier PJ, Martin BR, Mallouk TE, Chizhik SA, Buzaneva EV, et al. Layer-by-layer assembly of ultrathin composite films from micron-sized graphite oxide sheets and polycations. Chem Mater 1999;11(3):771–8.
[31] Marcano DC, Kosynkin DV, Berlin JM, Sinitskii A, Sun Z, Slesarev A, et al. Improved synthesis of graphene oxide. ACS Nano 2010;4(8):4806–14.
[32] He H, Riedl T, Lerf A, Klinowski J. Solid-state NMR studies of the structure of graphite oxide. J Phys Chem 1996;100:19954–8.
[33] Lerf A, Heb H, Riedl T, Forster M, Klinowskib J. ^{13}C and ^{1}H MAS NMR studies of graphite oxide and its chemically modified derivatives. Solid State Ionics 1997;101–103(2):857–62.
[34] He H, Klinowski J, Forster M, Lerf A. A new structural model for graphite oxide. Chem Phys Lett 1998;287:53–6.
[35] Lerf A, He H, Forster M, Klinowski J. Structure of graphite oxide revisited. J Phys Chem B 1998;102(23):4477–82.
[36] Kim S, Zhou S, Hu Y, Acik M, Chabal YJ, Berger C, et al. Room-temperature metastability of multilayer graphene oxide films. Nat Mater 2012;11(6):544–9.
[37] Chen D, Feng H, Li J. Graphene oxide: preparation, functionalization, and electrochemical applications. Chem Rev 2012;112(11):6027–53.
[38] Gao W, Alemany LB, Ci L, Ajayan PM. New insights into the structure and reduction of graphite oxide. Nat Chem 2009;1(5):403–8.
[39] Cai W, Piner RD, Stadermann FJ, Park S, Shaibat MA, Ishii Y, et al. Synthesis and solid-state NMR structural characterization of ^{13}C-labeled graphite oxide. Science 2008;321(5897): 1815–17.
[40] Casabianca LB, Shaibat MA, Cai WW, Park S, Piner R, Ruoff RS, et al. NMR-based structural modeling of graphite oxide using multidimensional ^{13}C solid-state NMR and Ab initio chemical shift calculations. J Am Chem Soc 2010;132(16):5672–6.
[41] Hou X-L, Li J-L, Drew SC, Tang B, Sun L, Wang X-G. Tuning radical species in graphene oxide in aqueous solution by photoirradiation. J Phys Chem C 2013;117(13):6788–93.
[42] Yang L, Zhang R, Liu B, Wang J, Wang S, Han MY, et al. Pi-conjugated carbon radicals at graphene oxide to initiate ultrastrong chemiluminescence. Angew Chem Int Ed 2014; 53(38):10109–13.
[43] Pieper H, Chercheja S, Eigler S, Halbig C, Filipovic MR, Mokhir A. Endoperoxides revealed as origin of the toxicity of graphene oxide. Angew Chem Int Ed 2016;55(1), 405-407; Angew. Chem. 2016;128(1), 413–416.
[44] Szabó T, Berkesi O, Forgó P, Josepovits K, Sanakis Y, Petridis D, et al. Evolution of surface functional groups in a series of progressively oxidized graphite oxides. Chem Mater 2006;18(11):2740–9.

[45] Nakajima T, Matsuo Y. Formation process and structure of graphite oxide. Carbon 1994;32(3):469–75.
[46] Clause A, Plass R, Boehm HP, Hofmann U. Untersuchungen zur Struktur des Graphitoxyds. Z Anorg Allg Chem 1957;291(5–6):205–20.
[47] Ruess G. Über das Graphitoxyhydroxyd (Graphitoxyd). Monatsh Chem 1947;76(3–5):381–417.
[48] Scholz W, Boehm HP. Betrachtungen zur Struktur des Graphitoxids. Z Anorg Allg Chem 1969;369(3–6):327–40.
[49] Dreyer DR, Todd AD, Bielawski CW. Harnessing the chemistry of graphene oxide. Chem Soc Rev 2014;43(15):5288–301.
[50] Hossain MZ, Johns JE, Bevan KH, Karmel HJ, Liang YT, Yoshimoto S, et al. Chemically homogeneous and thermally reversible oxidation of epitaxial graphene. Nature Chem 2012;4:305–9.
[51] Kang JH, Kim T, Choi J, Park J, Kim YS, Chang MS, et al. The hidden second oxidation step of Hummers' method. Chem Mater 2016;28(3), 756-764.
[52] Kozhemyakina NV, Eigler S, Dinnebier RE, Inayat A, Schwieger W, Hirsch A. Effect of the structure and morphology of natural, synthetic and post-processed graphites on their dispersibility and electronic properties. Fuller Nanotub Car N 2013;21(9):804–23.
[53] Wang L, Ambrosi A, Pumera M. Metal-free catalytic oxygen reduction reaction on heteroatom-doped graphene is caused by trace metal impurities. Angew Chem Int Ed 2013;52(51):13818–21.
[54] Feicht P, Kunz DA, Lerf A, Breu J. Facile and scalable one-step production of organically modified graphene oxide by a two-phase extraction. Carbon 2014;80:229–34.
[55] Koehler FM, Stark WJ. Organic synthesis on graphene. Acc Chem Res 2013;46(10):2297–306.
[56] Eigler S, Hirsch A. Chemistry with graphene and graphene oxide-challenges for synthetic chemists. Angew. Chem. 2014;126(30), 7852–7872.
[57] Walter J, Nacken TJ, Damm C, Thajudeen T, Eigler S, Peukert W. Determination of the lateral dimension of graphene oxide nanosheets using analytical ultracentrifugation. Small 2015;11(7):814–25.
[58] Ferrari AC, Basko DM. Raman spectroscopy as a versatile tool for studying the properties of graphene. Nat Nanotech 2013;8(4):235–46.
[59] Englert JM, Vecera P, Knirsch KC, Schafer RA, Hauke F, Hirsch A. Scanning-Raman-microscopy for the statistical analysis of covalently functionalized graphene. ACS Nano 2013;7(6):5472–82.
[60] Eigler S, Enzelberger-Heim M, Grimm S, Hofmann P, Kroener W, Geworski A, et al. Wet chemical synthesis of graphene. Adv Mater 2013;25(26):3583–7.
[61] Grimm SB, Schweiger M, Eigler S, Zaumseil J. High-quality reduced graphene oxide by CVD-assisted annealing. J Phys Chem C 2016;120(5), 3036–3041.
[62] Rüdorff W, Hofmann U. Über Graphitsalze. Z Anorg Allg Chem 1938;238(1):1–50.
[63] Eigler S. Graphite sulphate – a precursor to graphene. Chem Commun 2015;51(15):3162–5.
[64] Eigler S. Mechanistic insights into the reduction of graphene oxide addressing its surfaces. Phys Chem Chem Phys 2014;16(37):19832–5.
[65] Eigler S, Dotzer C, Hof F, Bauer W, Sulfur HA. Species in graphene oxide. Chem Eur J 2013;19(29):9490–6.
[66] Dimiev A, Kosynkin DV, Alemany LB, Chaguine P, Tour JM. Pristine graphite oxide. J Am Chem Soc 2012;134(5):2815–22.
[67] Ferrari AC, Meyer JC, Scardaci V, Casiraghi C, Lazzeri M, Mauri F, et al. Raman spectrum of graphene and graphene layers. Phys Rev Lett 2006;97(18):187401.
[68] Graf D, Molitor F, Ensslin K, Stampfer C, Jungen A, Hierold C, et al. Spatially resolved Raman spectroscopy of single- and few-layer graphene. Nano Lett 2007;7(2):238–42.

[69] Cançado LG, Jorio A, Ferreira EHM, Stavale F, Achete CA, Capaz RB, et al. Quantifying defects in graphene via Raman spectroscopy at different excitation energies. Nano Lett 2011;11(8):3190–6.
[70] Ni ZH, Yu T, Lu YH, Wang YY, Feng YP, Shen ZX. Uniaxial strain on graphene: Raman spectroscopy study and band-gap opening. ACS Nano 2008;2(11):2301–5.
[71] Wang YY, Ni ZH, Yu T, Shen ZX, Wang HM, Wu YH, et al. Raman studies of monolayer graphene: the substrate effect. J Phys Chem C 2008;112(29):10637–40.
[72] Das A, Pisana S, Chakraborty B, Piscanec S, Saha SK, Waghmare UV, et al. Monitoring dopants by Raman scattering in an electrochemically top-gated graphene transistor. Nat Nanotech 2008;3(4):210–15.
[73] Casiraghi C, Hartschuh A, Qian H, Piscanec S, Georgi C, Fasoli A, et al. Raman spectroscopy of graphene edges. Nano Lett 2009;9(4):1433–41.
[74] Lucchese MM, Stavale F, Ferreira EH, Vilani C, Moutinho MV, Capaz RB, et al. Quantifying ion-induced defects and Raman relaxation length in graphene. Carbon 2010;48(5):1592–7.
[75] Koehler FM, Jacobsen A, Ensslin K, Stampfer C, Stark WJ. Selective chemical modification of graphene surfaces: distinction between single- and bilayer graphene. Small 2010;6(10):1125–30.
[76] Jung I, Dikin DA, Piner RD, Ruoff RS. Tunable electrical conductivity of individual graphene oxide sheets reduced at "low" temperatures. Nano Lett 2008;8(12):4283–7.
[77] Eda G, Chhowalla M. Chemically derived graphene oxide: towards large-area thin-film electronics and optoelectronics. Adv Mater 2010;22(22):2392–415.
[78] Eigler S, Dotzer C, Hirsch A. Visualization of defect densities in reduced graphene oxide. Carbon 2012;50(10):3666–73.
[79] Eigler S, Hof F, Enzelberger-Heim M, Grimm S, Müller P, Hirsch A. Statistical-Raman-microscopy atomic-force-microscopy on heterogeneous graphene obtained after reduction of graphene oxide. J Phys Chem C 2014;118(14):7698–704.
[80] Allen MJ, Tung VC, Kaner RB. Honeycomb carbon: a review of graphene. Chem Rev 2010;110(1):132–45.
[81] Wu D, Zhang F, Liu P, Feng X. Two-dimensional nanocomposites based on chemically modified graphene. Chem Eur J 2011;17(39):10804–12.
[82] Huang X, Qi X, Boey F, Zhang H. Graphene-based composites. Chem Soc Rev 2012;41(2):666–86.
[83] Zhu Y, Murali S, Cai W, Li X, Suk JW, Potts JR, et al. Graphene and graphene oxide: synthesis, properties, and applications. Adv Mater 2010;22(35):3906–24.
[84] Park J, Yan M. Covalent functionalization of graphene with reactive intermediates. Acc Chem Res 2013;46(1):181–9.
[85] Lu CH, Yang HH, Zhu CL, Chen X, Chen GN. A graphene platform for sensing biomolecules. Angew Chem Int Ed 2009;48(26):4785–7.
[86] Wang S, Yu D, Dai L, Chang DW, Baek JB. Polyelectrolyte-functionalized graphene as metal-free electrocatalysts for oxygen reduction. ACS Nano 2011;5(8):6202–9.
[87] Burress JW, Gadipelli S, Ford J, Simmons JM, Zhou W, Yildirim T. Graphene oxide framework materials: theoretical predictions and experimental results. Angew Chem Int Ed 2010;49(47):8902–4.
[88] Burress JW, Gadipelli S, Ford J, Simmons JM, Zhou W, Yildirim T. Graphene oxide framework materials: theoretical predictions and experimental results. Angew Chem 2010;122(47):9086–8.
[89] Stankovich S, Piner RD, Nguyen ST, Ruoff RS. Synthesis and exfoliation of isocyanate-treated graphene oxide nanoplatelets. Carbon 2006;44(15):3342–7.
[90] Sun H, Xu Z, Gao C. Multifunctional ultra-flyweight, synergistically assembled carbon aerogels. Adv Mater 2013;25(18):2554–60.
[91] Chabot V, Higgins D, Yu A, Xiao X, Chen Z, Zhang J. A review of graphene and graphene oxide sponge: material synthesis and applications to energy and the environment. Energ Environ Sci 2014;7(5):1564–96.

[92] Sofer Z, Jankovsky O, Libanska A, Simek P, Novacek M, Sedmidubsky D, et al. Definitive proof of graphene hydrogenation by Clemmensen reduction: use of deuterium labeling. Nanoscale 2015;7(23):10535–43.

[93] Chua CK, Pumera M. Regeneration of a conjugated sp(2) graphene system through selective defunctionalization of epoxides by using a proven synthetic chemistry mechanism. Chem Eur J 2014;20(7):1871–7.

[94] Bourlinos AB, Gournis D, Petridis D, Szabó T, Szeri A, Dékány I. Graphite oxide chemical reduction to graphite and surface modification with primary aliphatic amines and amino acids. Langmuir 2003;19(15):6050–5.

[95] Yang H, Shan C, Li F, Han D, Zhang Q, Niu L. Covalent functionalization of polydisperse chemically-converted graphene sheets with amine-terminated ionic liquid. Chem Commun 2009;19(26):3880–2.

[96] Beckert F, Rostas AM, Thomann R, Weber S, Schleicher E, Friedrich C, et al. Self-initiated free radical grafting of styrene homo- and copolymers onto functionalized graphene. Macromolecules 2013;46(14):5488–96.

[97] Kim J, Cote LJ, Huang J. Two dimensional soft material: new faces of graphene oxide. Acc Chem Res 2012;45(8):1356–64.

[98] Loh KP, Bao Q, Ang PK, Yang J. The chemistry of graphene. J Mater Chem 2010;20(12): 2277–89.

[99] Kuila T, Bose S, Mishra AK, Khanra P, Kim NH, Lee JH. Chemical functionalization of graphene and its applications. Prog Mater Sci 2012;57(7):1061–105.

[100] Dong L, Gari RRS, Li Z, Craig MM, Hou S. Graphene-supported platinum and platinum–ruthenium nanoparticles with high electrocatalytic activity for methanol and ethanol oxidation. Carbon 2010;48(3):781–7.

[101] Xue Y, Chen H, Yu D, Wang S, Yardeni M, Dai Q, et al. Oxidizing metal ions with graphene oxide: the in situ formation of magnetic nanoparticles on self-reduced graphene sheets for multifunctional applications. Chem Commun 2011;47(42):11689–91.

[102] Liu Z, Robinson JT, Sun X, Dai H. PEGylated nanographene oxide for delivery of water-insoluble cancer drugs. J Am Chem Soc 2008;130(33):10876–7.

[103] Zhang X, Huang Y, Wang Y, Ma Y, Liu Z, Chen Y. Synthesis and characterization of a graphene–C_{60} hybrid material. Carbon 2009;47(1):334–7.

[104] Xu Y, Liu Z, Zhang X, Wang Y, Tian J, Huang Y, et al. A Graphene Hybrid Material Covalently Functionalized with Porphyrin: Synthesis and Optical Limiting Property. Adv Mater 2009;21(12):1275–9.

[105] Yu D, Yang Y, Durstock M, Baek JB, Dai L. Soluble P3HT-grafted graphene for efficient bilayer-heterojunction photovoltaic devices. ACS Nano 2010;4(10):5633–40.

[106] Si Y, Samulski ET. Synthesis of water soluble graphene. Nano Lett 2008;8(6):1679–82.

[107] DeTar DF, Kosuge T. Mechanisms of diazonium salt reactions. VI. The reactions of diazonium salts with alcohols under acidic conditions; evidence for hydride transfer1. J Am Chem Soc 1958;80(22):6072–7.

[108] Sharma R, Baik JH, Perera CJ, Strano MS. Anomalously large reactivity of single graphene layers and edges toward electron transfer chemistries. Nano Lett 2010;10(2):398–405.

[109] Pinson J, Podvorica F. Attachment of organic layers to conductive or semiconductive surfaces by reduction of diazonium salts. Chem Soc Rev 2005;34(5):429–39.

[110] Xiao L, Liao L, Liu L. Chemical modification of graphene oxide with carbethoxycarbene under microwave irradiation. Chem Phys Lett 2013;556:376–9.

[111] Collins WR, Lewandowski W, Schmois E, Walish J, Swager TM. Claisen rearrangement of graphite oxide: a route to covalently functionalized graphenes. Angew Chem Int Ed 2011;50(38): 8848–52.

[112] Collins WR, Schmois E, Swager TM. Graphene oxide as an electrophile for carbon nucleophiles. Chem Commun 2011;47(31):8790–2.
[113] Sydlik SA, Swager TM. Functional graphenic materials via a Johnson–Claisen rearrangement. Adv Funct Mater 2013;23(15):1873–82.
[114] Johnson WS, Werthemann L, Bartlett WR, Brocksom TJ, Li T-T, Faulkner DJ, et al. Simple stereoselective version of the Claisen rearrangement leading to trans-trisubstituted olefinic bonds. Synthesis of squalene. J Am Chem Soc 1970;92(3):741–3.
[115] Kim F, Luo JY, Cruz-Silva R, Cote LJ, Sohn K, Huang JX. Self-propagating domino-like reactions in oxidized graphite. Adv Funct Mater 2010;20(17):2867–73.
[116] Krishnan D, Kim F, Luo J, Cruz-Silva R, Cote LJ, Jang HD, et al. Energetic graphene oxide: challenges and opportunities. Nano Today 2012;7(2):137–52.
[117] Qiu Y, Guo F, Hurt R, Kulaots I. Explosive thermal reduction of graphene oxide-based materials: mechanism and safety implications. Carbon 2014;72:215–23.
[118] Eigler S, Dotzer C, Hirsch A, Enzelberger M, Müller P. Formation and decomposition of CO_2 intercalated graphene oxide. Chem Mater 2012;24(7):1276–82.
[119] Dimiev AM, Alemany LB, Tour JM. Graphene oxide. Origin of acidity, its instability in water, and a new dynamic structural model. ACS Nano 2013;7(1):576–88.
[120] Dimiev AM, Polson TA. Contesting the two-component structural model of graphene oxide and reexamining the chemistry of graphene oxide in basic media. Carbon 2015;93:544–54.
[121] Eigler S, Grimm S, Hirsch A. Investigation of the thermal stability of the carbon framework of graphene oxide. Chem Eur J 2014;20(4):984–9.
[122] Eigler S, Grimm S, Enzelberger-Heim M, Müller P, Hirsch A. Graphene oxide efficiency of reducing agents. Chem Commun 2013;49(67):7391–3.
[123] Chua CK, Pumera M. Chemical reduction of graphene oxide: a synthetic chemistry viewpoint. Chem Soc Rev 2014;43(1):291–312.
[124] Pei S, Cheng H-M. The reduction of graphene oxide. Carbon 2012;50(9):3210–28.
[125] Eigler S, Hu Y, Ishii Y, Hirsch A. Controlled functionalization of graphene oxide with sodium azide. Nanoscale 2013;5(24):12136–9.
[126] Eigler S, Grimm S, Hof F, Hirsch A. Graphene oxide: a stable carbon framework for functionalization. J Mater Chem A 2013;1(38):11559–62.
[127] Halbig C, Rietsch P, Eigler S. Towards the synthesis of graphene azide from graphene oxide. Molecules 2015;20(12):21050–7.
[128] Wang Z, Eigler S, Ishii Y, Hu Y, Papp C, Lytken O, et al. A facile approach to synthesize an oxo-functionalized graphene/polymer composite for low-voltage operating memory devices. J Mater Chem C 2015;3(33):8595–604.
[129] Wang Z, Eigler S, Halik M. Scalable self-assembled reduced graphene oxide transistors on flexible substrate. Appl Phys Lett 2014;104(24):243502.
[130] Kirschner J, Wang Z, Eigler S, Steinrück HP, Jager CM, Clark T, et al. Driving forces for the self-assembly of graphene oxide on organic monolayers. Nanoscale 2014;6(19):11344–50.
[131] Wang Z, Mohammadzadeh S, Schmaltz T, Kirschner J, Khassanov A, Eigler S, et al. Region-selective self-assembly of functionalized carbon allotropes from solution. ACS Nano 2013;7(12):11427–34.

Yasutomo Segawa, Akiko Yagi and Kenichiro Itami

3 Chemical Synthesis of Cycloparaphenylenes

3.1 Introduction

Since carbon nanotubes (CNTs) were first discovered by Iijima in 1991 [1], the synthesis of CNTs has attracted the interest of many scientists because of their outstanding physical properties as well as their potential applications in technology [2–7]. Single-walled CNTs can be considered as rolled-up structures of graphene sheets. The manner in which the graphene sheet is wrapped is given by the chiral index, which is represented by a pair of numbers (n, m) as shown in Figure 3.1. The structures corresponding to $m = 0$, $n = m$ and $n > m > 0$ are called "zigzag," "armchair" and "chiral" CNTs, respectively. The chiral index (n, m) can be used to deduce a variety of electrical, optical, magnetic and mechanical properties of the CNTs. For example, (n, m) CNTs in which $n - m$ is zero or a multiple of three are metallic, whereas other CNTs are semiconducting [8, 9]. Therefore, the controlled, chirality-specific synthesis of CNTs is important for imparting defined macroscopic material properties.

Ring-shaped structures can be obtained by "slicing" CNTs perpendicular to the main axis. These ring-shaped aromatic molecules, called "carbon nanorings," represent the sidewall segment of CNTs with a specific chirality. Representative structures of carbon nanorings, cycloparaphenylene (CPP), cycloparaphenylene-2,6-naphthylene and cyclacene are shown in Figure 3.2. Thus, these carbon nanorings can be considered as ideal templates or building blocks for the construction of uniform CNTs. In this chapter, syntheses of carbon nanorings, especially CPP and its derivatives, are reviewed.

3.2 Synthetic Efforts toward CPPs

Carbon nanorings consisting solely of n benzene rings connected via the *para* position are called [n]cycloparaphenylenes ([n]CPPs, Figure 3.3). Although synthetic studies on CPP had already begun in the 1930s, the synthesis of CPP was not accomplished for almost 80 years. Figure 3.3 shows representative early synthetic attempts toward CPPs.

In 1934, Parekh and Guha reported an attempt toward the synthesis of CPPs [10]. Their strategy was to synthesize CPP by the copper-mediated desulfurization of dithia[1_2]paracyclophane (**1a**). Attempts involving similar large macrocycles **1b** and **2** were reported by the group of Vögtle in 1984 [11]. However, CPP could not be synthesized by these methods. In 1993, Vögtle and coworkers proposed three synthetic approaches toward CPP (Figure 3.3). Their first approach involved the synthesis of cyclohexane-inserted CPP **4** by magnesiation followed by the copper-mediated

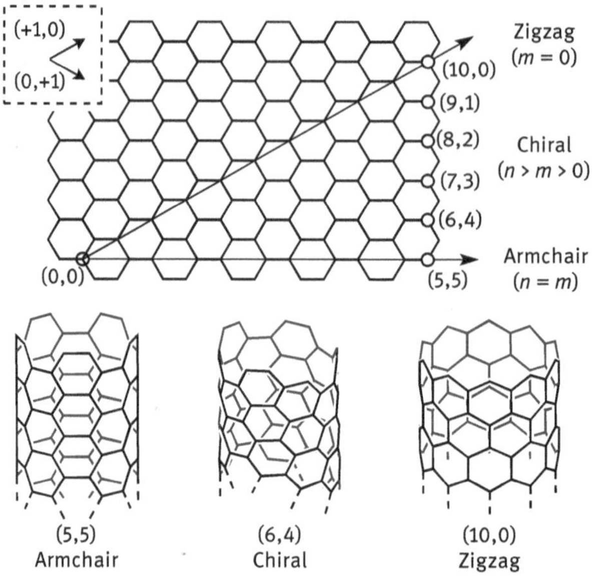

Figure 3.1: Chiral index of CNTs.

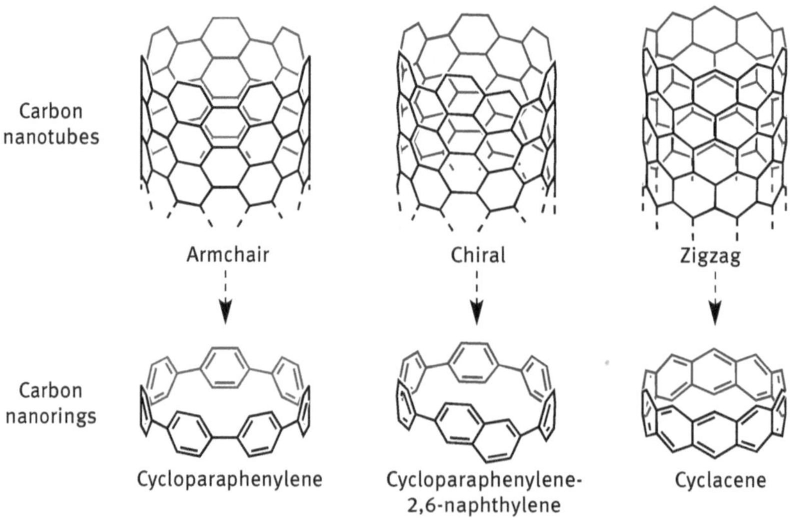

Figure 3.2: Structures of CNTs and carbon nanorings.

homocoupling of *cis*-1,4-bis(4-halophenyl)cyclohexanes **3a** and **b**. However, this approach only resulted in the formation of acyclic oligomers and polymers. The second approach involved the Diels–Alder reaction of macrocyclic enyne compounds **7a** and **b**. Although the preparation of **7a,b** by a Wittig reaction between **5** and **6a,b** was successful, the [4+2] cycloaddition did not occur. They also attempted the synthesis of

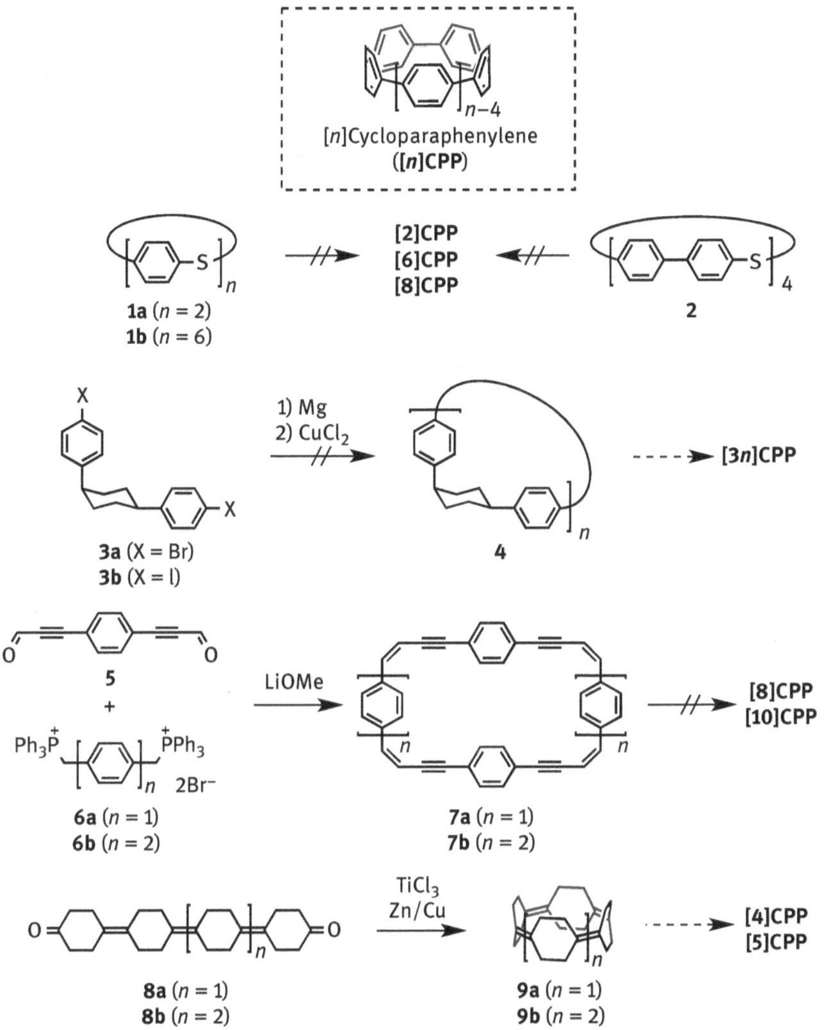

Figure 3.3: Synthetic attempts toward CPPs.

macrocycle **9b** as a possible precursor to [5]CPP based on the synthesis of **9a** from **8a** reported by McMurry et al. [13]. The McMurry coupling of diketone **8b** under the influence of $TiCl_3$/Zn produced **9b**. However, **9b** was only detected in mass spectrometry, and further transformation to [5]CPP was not investigated.

In 2008, Bertozzi and Jasti reported the first synthesis of CPP [14]. A few months later, the Itami group reported the first size-selective synthesis of CPP [15], and Yamago reported an alternative approach in 2010 [16]. Three groups of researchers under Jasti, Itami and Yamago have developed a synthetic methodology for the selective or random formation of CPPs, and [n]CPPs with $n = 5–16, 18$ have now been synthesized.

3.3 Synthetic Strategies toward CPPs

The biggest difficulty in the synthesis of [*n*]CPPs is the formation of a macrocycle, which usually has high levels of ring strain because of the connection of benzene rings in the *para* position. The Jasti/Bertozzi, Itami and Yamago groups addressed this problem with the initial formation of unstrained macrocyclic precursors. Three of their synthetic strategies and the mechanisms of the key steps are shown in Figure 3.4. The method employed by Jasti and Bertozzi uses L-shaped building blocks consisting of phenyl-substituted cyclohexadiene units. Because 1,4-dimethoxycyclohexadiene moieties can be converted into benzene rings via reductive aromatization, the L-shaped units are regarded as triphenylene precursors. Sequential coupling reactions of the L-shaped units afforded macrocycles with cyclohexadiene moieties occupying the corners, thus generating the CPP precursors. In the final step, these cyclohexadiene moieties were subjected to reductive aromatization to furnish the corresponding CPPs. Itami, on the other hand, used *cis*-1,4-diphenyl-1,4-alkoxycyclohexanes as triphenylene precursors. The macrocycles obtained by coupling the L-shaped units were subsequently transformed into CPPs by one-pot deprotection and oxidative aromatization. Yamago applied Bäuerle's *cis*-platinum complex method [17] to unstrained CPP precursors. Initially, macrocyclic platinum-bridged biphenyls were synthesized, and a subsequent reductive elimination between the biphenyl ligands generated the CPPs. Reductive elimination of the two aryl units from platinum was promoted by oxidizing the platinum center with an oxidant. These methods have been developed

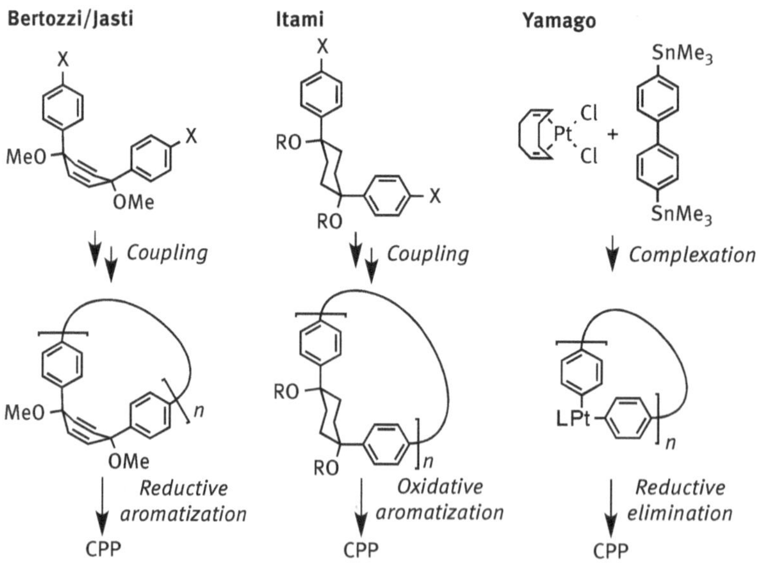

Figure 3.4: Synthetic strategies toward CPPs of three research groups.

by the three above-mentioned research groups and applied to the synthesis of a series of CPPs and related carbon nanorings.

3.4 Synthesis of [5]–[12] and [18]CPP by Bertozzi and Jasti

The first synthesis of CPP was achieved by Bertozzi, Jasti and coworkers in 2008 (Figure 3.5) [14]. For the synthesis of the macrocyclic CPP precursor, L-shaped unit **10a** was prepared from monolithiated 1,4-diiodobenzene and *p*-benzoquinone, and subsequent borylation of **10a** led to **10b** (B(pin) = 4,4,5,5-tetramethyl-1,3,2-dioxab-orolan-2-yl). The palladium-catalyzed cross-coupling of **10a** and **10b** afforded three different macrocycles **11a, 12a** and **13.** Finally, reductive aromatization of the isolated macrocycles **11a, 12a** and **13** with lithium naphthalenide (LiNaph) afforded [9]CPP [12] CPP and [18]CPP.

Later, Jasti succeeded in the selective synthesis of [6]–[12]CPP. The strategy toward [7]–[12]CPP relied on sequential Suzuki–Miyaura cross-coupling involving three L-shaped units **10b–d** (Figure 3.6) [18, 19]. For this purpose, silyl-protected 4-(4′-bromophenyl)phenol **14** underwent oxidative dearomatization with iodobenzene diacetate (PIDA), resulting in the formation of ketone **15a.** Subsequently, **15a** was treated with 4-chlorophenyllithium to yield the unsymmetrical L-shaped unit **10c,** which was borylated to afford **10d.** Using the two L-shaped units **10c** and **d**, the corresponding macrocyclic precursors for CPPs were synthesized by orthogonal Suzuki–Miyaura cross-coupling reactions. The coupling of L-shaped units **10b–d** with linear units (**18** or **19**) afforded macrocycles **11a, 12a** and **20–23,** which were converted into the corresponding CPPs by reductive aromatization using LiNaph or sodium naphthalenide (NaNaph). Synthesis of [6]CPP was first accomplished in 2012 (Figure 3.7) [20]. Ketone **15a** was combined with silyl-protected biphenyl unit **24** to afford **25a.** The *tert*-butyldimethylsilyl group on **25a** was removed by tetrabutylammonium fluoride (TBAF) to give **25b,** and the phenol moiety was the oxidized by PIDA. Nucleophilic

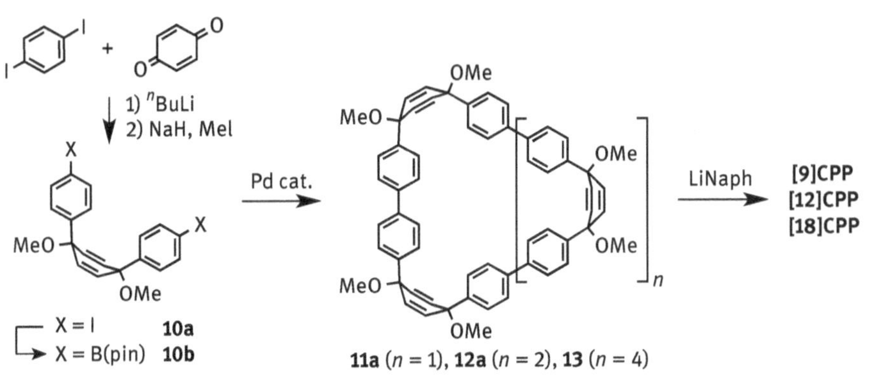

Figure 3.5: Synthesis of [9]CPP, [12]CPP and [18]CPP.

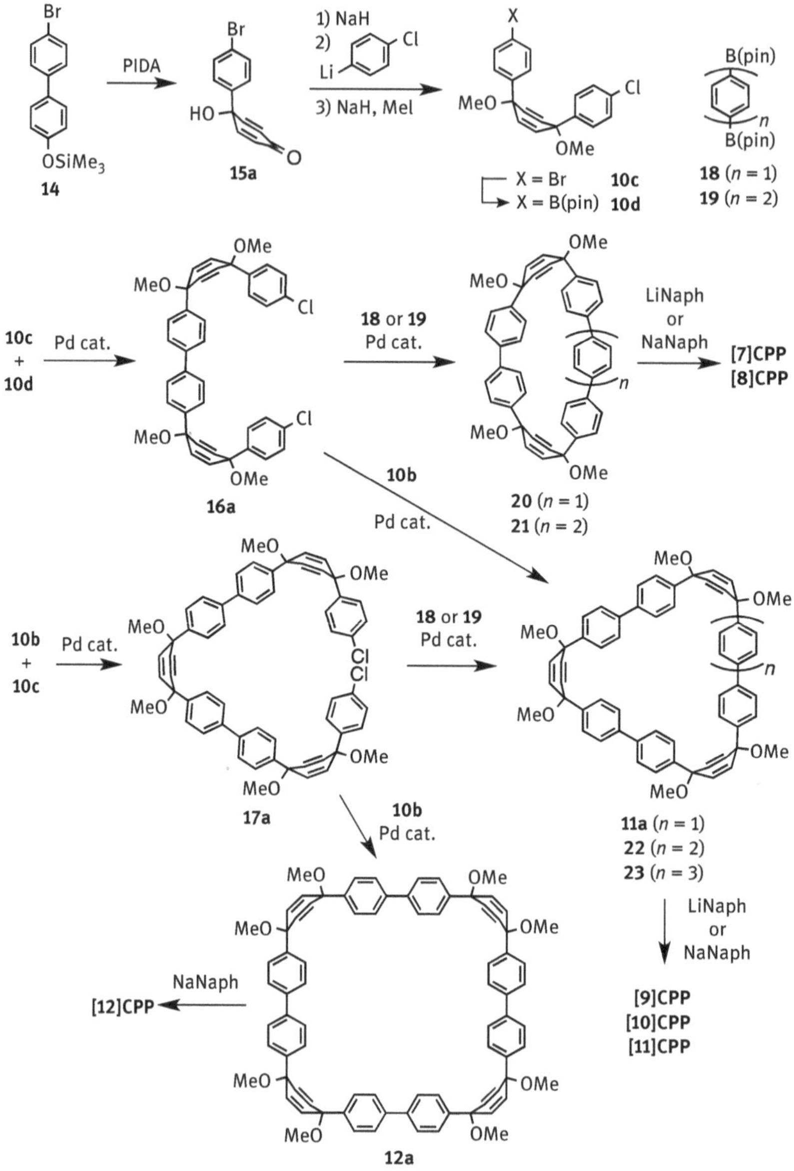

Figure 3.6: Selective synthesis of [7]–[12]CPP.

addition of 4-bromophenyllithium yielded U-shaped unit **26a**, and subsequent coupling with 1,4-diborylbenzene **18** afforded **27**, which furnished [6]CPP after reductive aromatization.

For [8]CPP and [10]CPP, Jasti was able to establish a gram-scale synthesis (Figure 3.8) [21]. The synthesis of macrocycle **28a** was achieved by cross-coupling

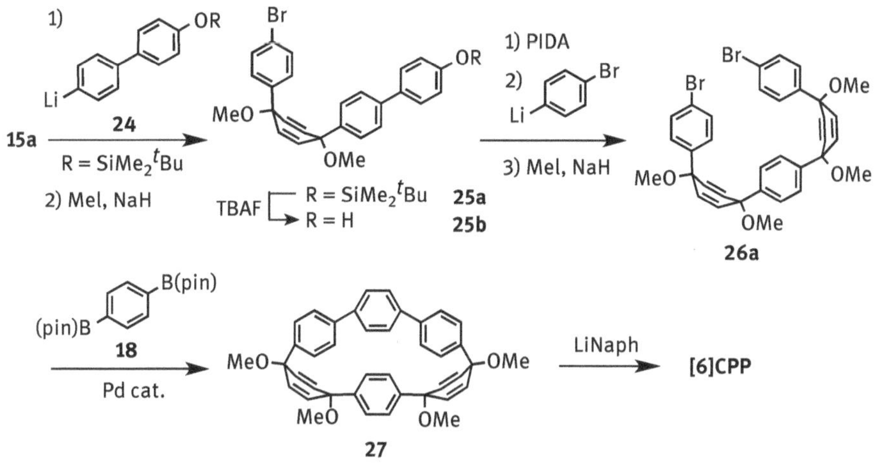

Figure 3.7: Selective synthesis of [6]CPP.

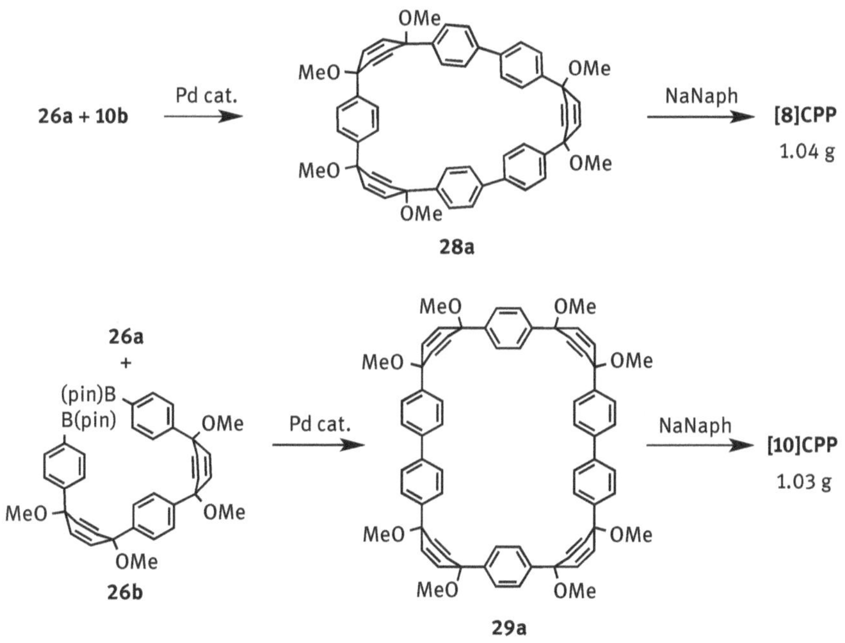

Figure 3.8: Gram-scale synthesis of [8]CPP and [10]CPP.

of L-shaped unit **10b** and U-shaped unit **26a**, followed by reductive aromatization, which afforded 1.04 g of [8]CPP. Correspondingly, cross-coupling of **26a** and **26b** afforded 1.03 g of [10]CPP through the macrocycle **29a**.

Interestingly, Jasti and Yamago independently published the synthesis of [5]CPP, the smallest CPP isolated to date, almost simultaneously. In Jasti's [5]CPP synthesis

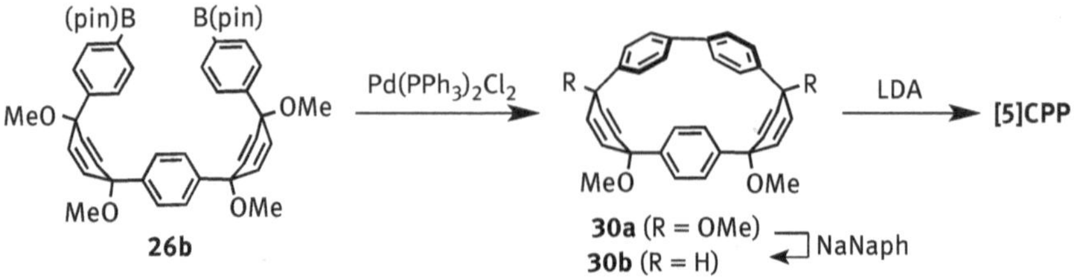

Figure 3.9: Synthesis of [5]CPP.

(Figure 3.9) [22], homocoupling reaction between the C–B bonds of **26b** provided macrocycle **30a**. Treatment of **30a** with NaNaph promoted partial demethoxylation to afford **30b**, and further reductive aromatization of **30b** with lithium diisopropylamide (LDA) afforded [5]CPP.

3.5 Synthesis of [7]–[16]CPP by Itami

In 2009, Itami reported the first size-selective synthesis of [12]CPP (Figure 3.10) [15]. For the synthesis of [12]CPP, L-shaped unit **31a** was firstly synthesized from

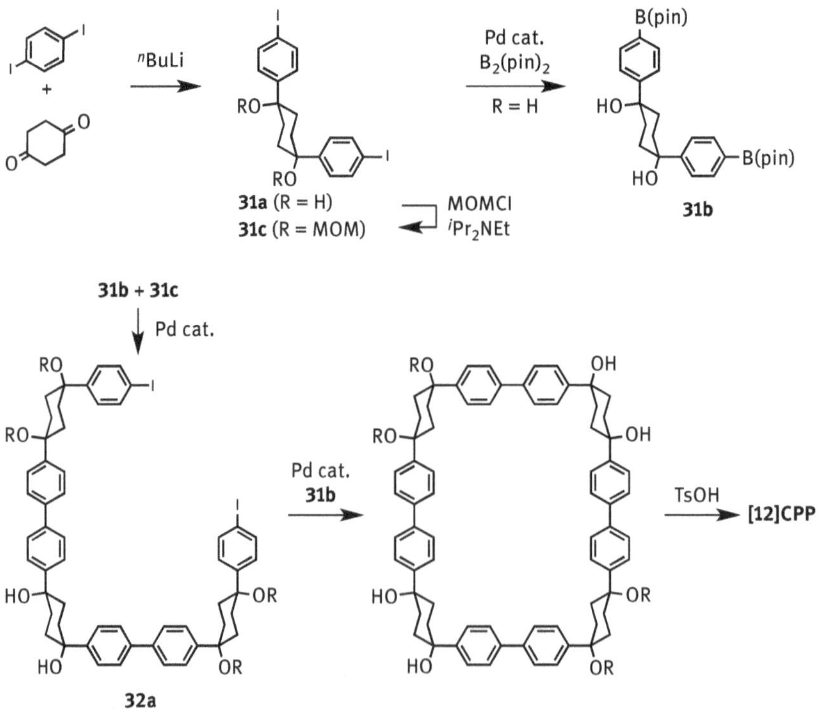

Figure 3.10: Selective synthesis of [12]CPP (R = MOM unless otherwise noted).

1,4-diiodobenzene and cyclohexane-1,4-dione. Then, for the subsequent Suzuki–Miyaura coupling, **31a** was converted into borylated L-shaped unit **31b** and MOM-protected L-shaped unit (MOM = methoxymethyl) **31c**. Palladium-catalyzed cross-coupling of **31b** and an excess of **31c** selectively afforded acyclic C-shaped unit **32a**, which was subsequently coupled with **31b** to furnish the macrocycle **33a**. After **33a** was synthesized, the cyclohexane units were deprotected, dehydrated and dehydrogenated in a *p*-toluenesulfonic acid (TsOH)-mediated one-pot reaction with microwave irradiation under air, thus producing [12]CPP.

While the synthetic route to [12]CPP has the advantage of being selective, several reaction steps are required. Therefore, Itami developed a concise method for the synthesis of [9]CPP and [12]CPP in 2011 (Figure 3.11) [23, 24]. The homocoupling reaction of L-shaped units **31c** or brominated derivative **31d**, promoted by Ni(cod)$_2$/bpy (cod = 1,5-cyclooctadiene, bpy = 2,2′-bipyridyl), afforded macrocycles **34a** and **33b**, which are easily separated by silica gel column chromatography. The final aromatization step was also improved using sodium hydrogen sulfate ($NaHSO_4$) as an acid and *m*-xylene/DMSO (DMSO = dimethyl sulfoxide) as solvent, thus forgoing the necessity of microwave heating. As a result [9]CPP and [12]CPP were synthesized in four steps from commercially available materials. This concise method can be scaled up easily, and consequently large amounts of [9]CPP and [12]CPP are now commercially available.

The Itami group also established a size-selective synthesis for [7]–[16]CPP by a combination of L-shaped building blocks with linear units. First, the size-selective

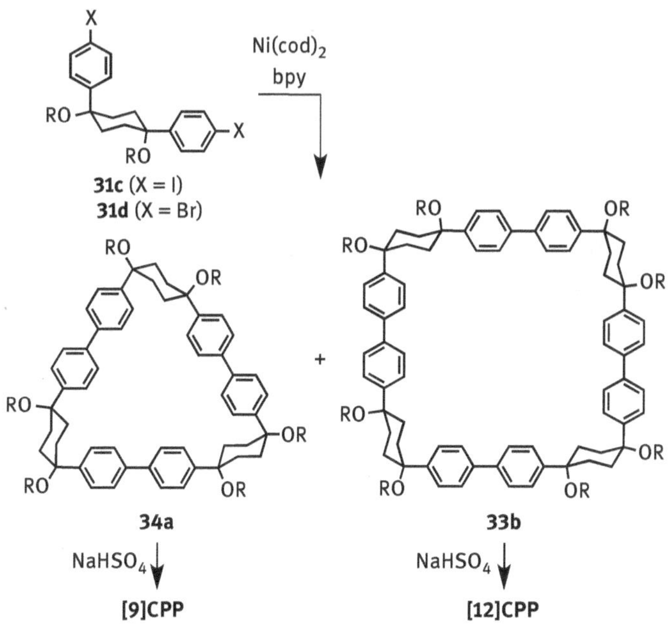

Figure 3.11: Synthesis of [9]CPP and [12]CPP (R = MOM).

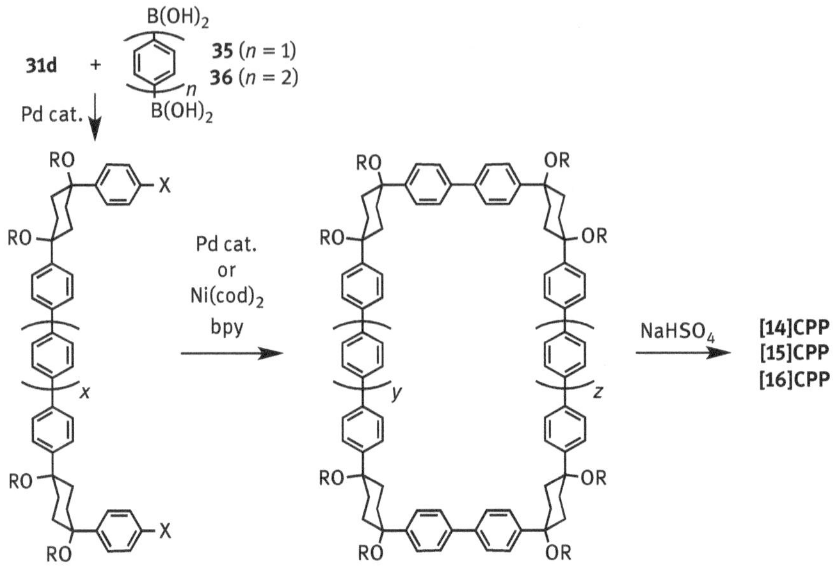

37a ($x = 1$, X = Br)
37b ($x = 1$, X = B(pin))
38a ($x = 2$, X = Br)
38b ($x = 2$, X = B(pin))

39 ($y = z = 1$) : **37a** + **37b** (Pd), **37a** + **37a** (Ni)
40 ($y = 1, z = 2$) : **37b** + **38a** (Pd)
41 ($y = z = 2$) : **38a** + **38a** (Pd), **38a** + **38a** (Ni)

Figure 3.12: Synthesis of [14]–[16]CPP (R = MOM).

synthesis of large CPPs, [14]– [16]CPP, was accomplished by assembling L-shaped unit **31d** with 1,4-diborylbenzene **35** and 4,4′-diborylbiphenyl **36** to yield U-shaped units **37a** and **38a**, respectively (Figure 3.12) [25]. The bromo substituents on the U-shaped units **37a** and **38a** were converted into boryl groups by palladium-catalyzed Miyaura borylation, resulting in the formation of **37b** and **38b**. Cross-coupling reactions between suitable pairs among **37a,b** and **38a,b** afforded the corresponding rectangular macrocycles **39**–**41** (**39**: **37a** + **37b**, **40**: **37b** + **38a**, **41**: **38a** + **38b**). Alternatively, macrocycles **39** and **41** can also be obtained by the nickel(0)-mediated homocoupling of **37a** and **38a**, respectively [26]. Final oxidative aromatization of **39**–**41** with $NaHSO_4$ yielded [14]–[16]CPP. In addition, the size-selective synthesis of [9]–[11] and [13]CPP was also accomplished in 2012 (Figure 3.13) [26]. C-shaped units **32b** and **42**, obtained from **31b,d** and 1,4-dibromobenzene, were used as precursors for [9]–[11]CPP. Connecting the C–Br terminals of **32b** furnished **34a** and subsequent aromatization led to [9]CPP, whereas cross-coupling of **32b** with 1,4-diborylbenzene yielded **43**, which was converted to [10]CPP. Assembling **31b,d** and 1,4-dibromobenzene afforded **44** through the macrocyclization of **42**, which was then converted into [11]CPP. The coupling reaction of U-shaped unit **37b** with L-shaped unit **31d** yielded **45**; the terminal C–Br moieties in **45** were connected using $Ni(cod)_2$/bpy to give **46** as a precursor for [13]CPP. The aromatization of **46** was accelerated by the addition of *o*-chloranil as an oxidant.

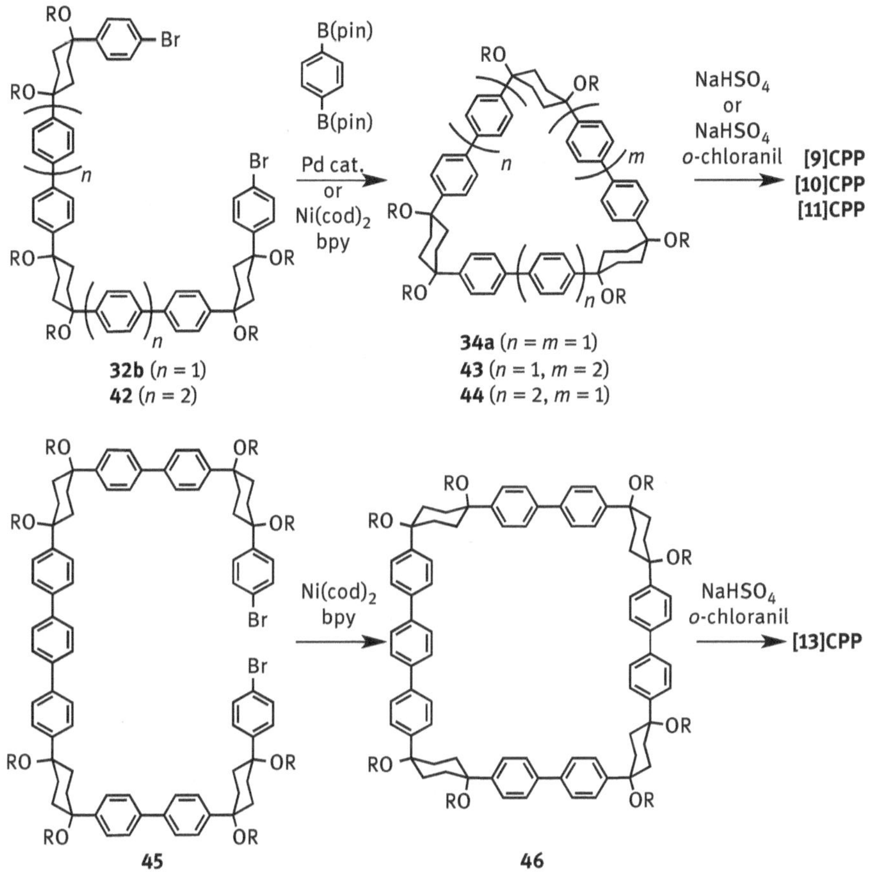

Figure 3.13: Selective synthesis of [9]–[11] and [13]CPP (R = MOM).

In 2014, the size-selective synthesis of smaller CPPs, [7]CPP and [8]CPP, was achieved (Figure 3.14) [27]. Because [*n*]CPP with $n < 9$ cannot be synthesized by assembling the previously described L-shaped units, Itami synthesized a new smaller L-shaped unit **47a**, which can be considered a biphenyl-convertible unit. The two-fold addition of dilithiated L-shaped unit **31d** to **47a** afforded C-shaped unit **48**, and subsequent homocoupling of the terminal groups in **48** afforded macrocycle **49**, which was used as the precursor for [7]CPP. Cross-coupling of **48** and 1,4-diborylbenzene yielded **50**, which was subsequently converted to [8]CPP. Selective synthesis of [10]CPP was also improved using the small L-shaped unit **47a** (Figure 3.14) [28]. Nucleophilic addition of dilithiated 1,4-dibromobenzene to **47a** and subsequent MOM-protection generated U-shaped unit **51**. The Ni(cod)$_2$/bpy-mediated homocoupling of **51** afforded rectangular macrocycle **52**, which was converted into [10]CPP by acid-mediated

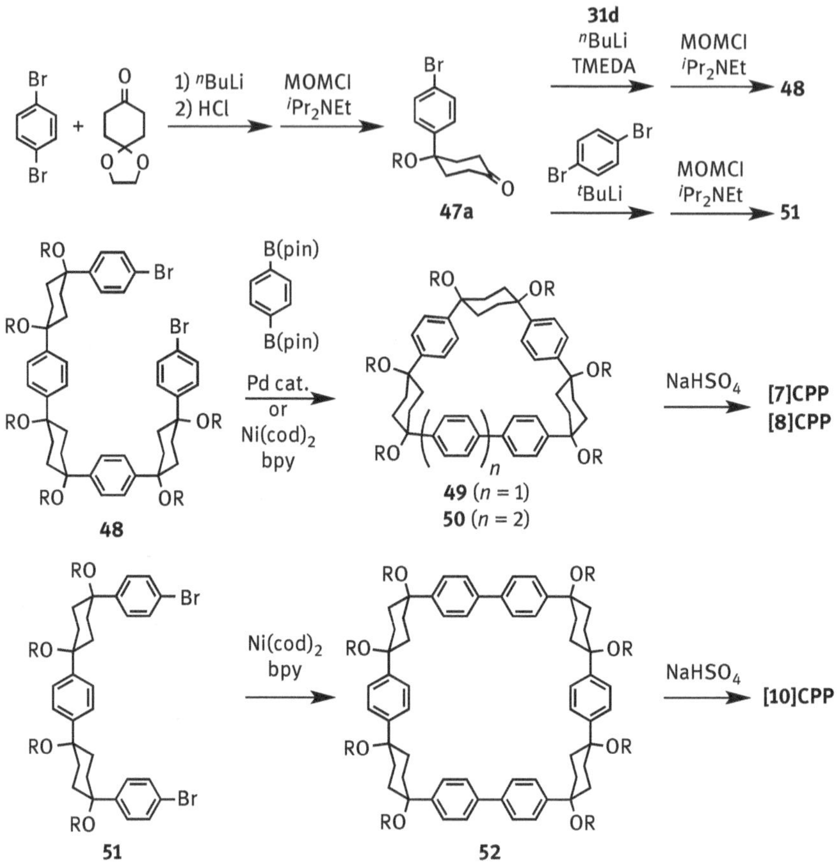

Figure 3.14: Selective synthesis of [7], [8] and [10]CPP (R = MOM).

aromatization. With this study, Itami completed the comprehensive size-selective synthesis of [7]–[16]CPP.

3.5.1 Synthesis of [5]–[13] and [16]CPP by Yamago

The Yamago group reported an alternative approach toward [8]CPP by applying Bäuerle's cycloarene synthesis using platinum-containing macrocycles (Figure 3.15) [16]. Complexation of 4,4′-bis(trimethylstannyl)biphenyl **53** and Pt(cod)Cl_2 afforded macrocycle **54a**. Because the ligand exchange on platinum is reversible, tetraplatinum complex **54a** was selectively formed as it represents the least strained isomer. Reductive elimination of the aryl groups from the platinum center promoted by ligand exchange from cod to dppf (dppf = 1,1′-bis(diphenylphosphino)ferrocene), followed by oxidation with bromine resulted in the formation of [8]CPP.

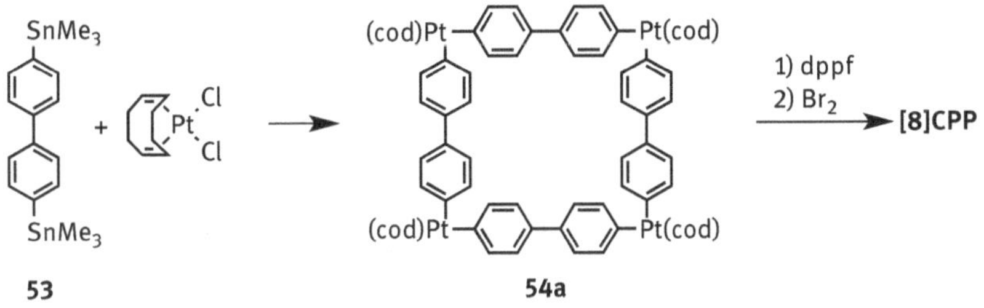

Figure 3.15: Selective synthesis of [8]CPP.

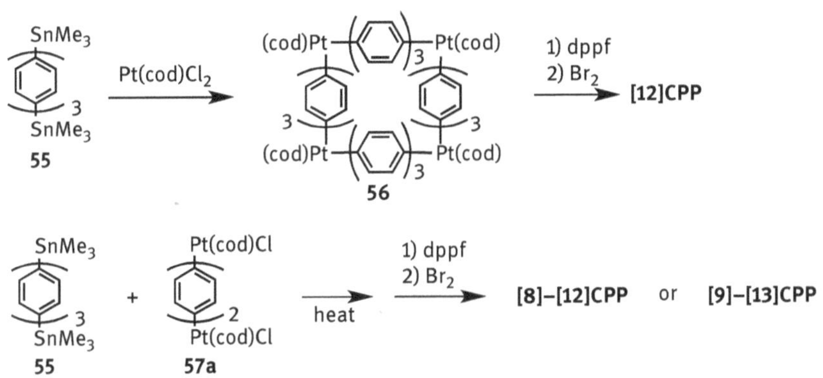

Figure 3.16: Selective and random synthesis of CPPs.

In 2011, Yamago reported both the random and selective synthesis of CPPs via the platinum complexation method (Figure 3.16) [29]. For example [12]CPP was synthesized selectively using bis(trimethylstannyl)terphenyl **55** via macrocyclic platinum complex **56.** In addition, the attempted reaction between **55** and diplatinum unit **57a** was expected to selectively furnish [10]CPP. However, [8]–[13]CPP were obtained randomly, depending on the reaction time at elevated temperatures. Yamago achieved an unexpected size-selective synthesis of [10]CPP (Figure 3.17) [30]. The homocoupling reaction of L-shaped platinum complex **59** was attempted to obtain cyclic dimers, trimers and tetramers as precursors for [8]CPP, [12]CPP and [16]CPP, respectively. When $Ni(cod)_2$/bpy and silver tetrafluoroborate were used for the homocoupling and reductive elimination, respectively, [8]CPP, [10]CPP, [12]CPP and [16]CPP were generated randomly. When $Pd(dba)_2$ (dba = dibenzylideneacetone) was used instead of $Ni(cod)_2$ [10]CPP was obtained selectively. In 2013, the selective synthesis of [6]CPP and [8]CPP was achieved and the method for the preparation of [10]CPP was improved [31]. The reaction of diplatinum complexes **60, 57b** and **61** with 4-bromophenyllithium afforded U-shaped diplatinum complexes **62–64**, and the subsequent homocoupling of **62–64** afforded macrocycles **65, 54b** and **66.** [6]CPP, [8]CPP and [10]CPP were obtained in the reductive elimination promoted by Br_2 (for [8]CPP and [10]CPP) or by XeF_2 (for [6]CPP).

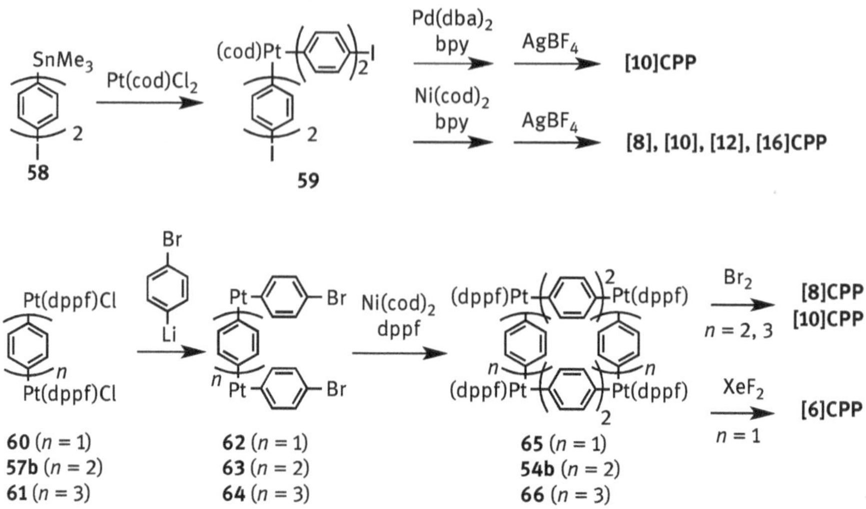

Figure 3.17: Synthesis of [6], [8], [10], [12] and [16]CPP.

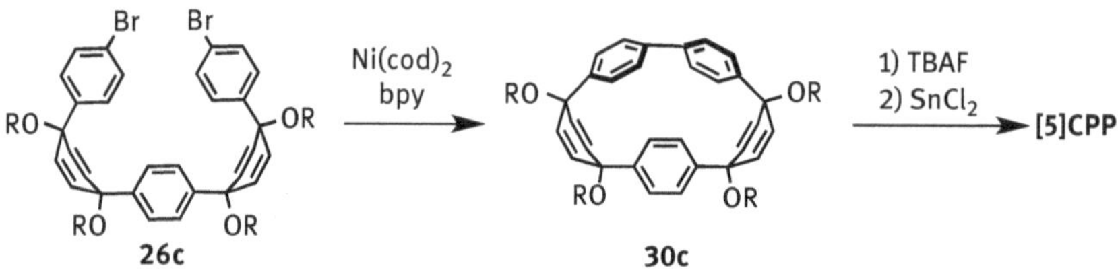

Figure 3.18: Selective synthesis of [5]CPP (R = $SiEt_3$).

As mentioned in the synthesis of [5]CPP by Jasati, Yamago also achieved the synthesis of [5]CPP (Figure 3.18) [32]. They used the triethylsilyl group instead of the methyl group for hydroxy group protection. The nickel(0)-mediated homocoupling of **26c** afforded macrocycle **30c**. Removal of the triethylsilyl group and reduction of **30c** with tin(II) chloride furnished [5]CPP.

After the synthesis of [5]CPP, Yamago applied the tin(II)-promoted reductive aromatization method to the selective synthesis of [7]–[12]CPP (Figure 3.19) [33]. First the building blocks were prepared: arylcyclohexadienone (**15a,15b**), symmetric and asymmetric L-shaped units with triethylsilyl protection (**10e–h**) and U-shaped units (**26c,d**). C-shaped unit **17b** was synthesized from **10e** via dilithiation and nucleophilic addition to **15a**, followed by protection. U-shaped unit **16b** was synthesized by the Suzuki–Miyaura cross-coupling reaction of **10f** and **10h**. The macrocycles used as the precursors of [7]–[12]CPP were synthesized as follows: cross-coupling of **26c** and **10g** (**67** as [7]CPP precursor), homocoupling of **17b** (**28b** as [8]CPP precursor), cross-coupling of **16b** and **10g** (**11b** as [9]CPP precursor), homocoupling of **16b** (**29b**

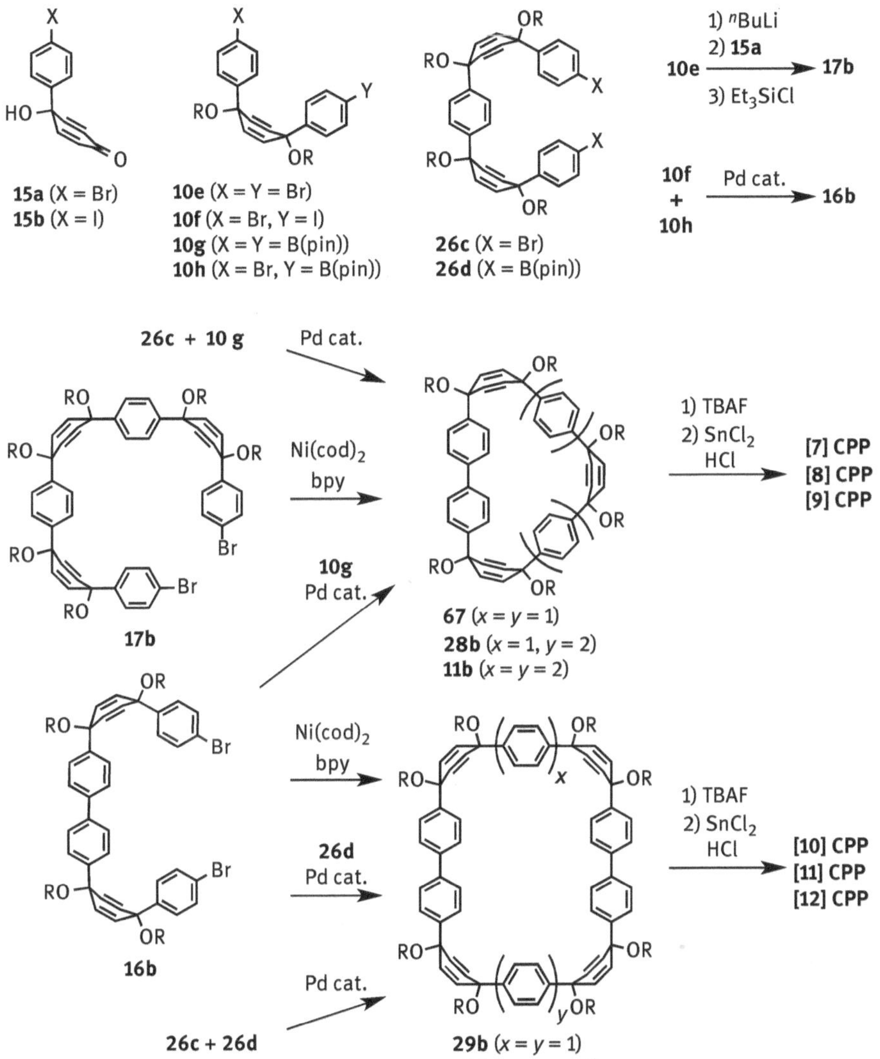

Figure 3.19: Selective synthesis of [7]–[12]CPP (R = $SiEt_3$).

as [10]CPP precursor), cross-coupling of **16b** and **26d** (**68** as [11]CPP precursor) and cross-coupling of **26c** and **26d** (**12b** as [12]CPP precursor). Each precursor was subjected to TBAF-mediated deprotection and subsequent reductive aromatization to selectively afford [7]–[12]CPP.

Yamago's [6]CPP synthesis was a hybrid of the platinum method and the reductive aromatization method (Figure 3.20) [34]. The transmetallation reaction of dimetallated

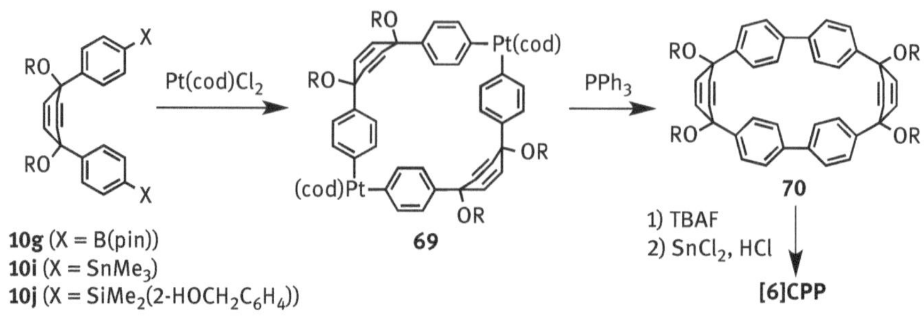

Figure 3.20: Selective synthesis of [6]CPP (R = $SiEt_3$).

L-shaped units **10g**, **10i** and **10j** to the platinum(II) complex produced macrocyclic diplatinum complex **69**. Reductive elimination promoted by triphenylphosphine proceeded to yield strained macrocycle **70** as a precursor to [6]CPP. Removal of the silyl groups of **70** followed by reductive aromatization afforded [6]CPP.

3.6 Synthesis of Armchair Carbon Nanorings

The previously reported synthetic strategies for [*n*]CPPs can also be applied to the synthesis of a variety of CPP-related, strained ring-shaped compounds. When substituted benzene rings, polycyclic aromatic hydrocarbons (PAHs) or heteroarenes are used in addition to benzene rings, a range of CPP derivatives can be obtained. Soon after Jasti, Itami and Yamago developed CPP synthesis methods, a variety of carbon nanorings were synthesized by them and other groups. In this section, carbon nanorings, which represent segments of armchair CNTs, are described. Armchair carbon nanorings can be subdivided into multiarylCPPs, CPP dimers and π-extended carbon nanorings.

Jasti synthesized tetraphenylated [12]CPP in 2012 (Figure 3.21) [35], whereby an L-shaped unit with four 4-*n*-butylphenyl groups (**71**) was prepared from tetrabromoquinone, 4-*n*-butylphenylboronic acid and 1,4-diiodobenzene. Macrocycle **72** was obtained by stepwise cross-coupling of **71** with the L-shaped unit bearing boryl and chloro groups (**10d**) and diboryl L-shaped unit **10b**. Finally, the reductive aromatization of **72** mediated by NaNaph and followed by quenching with iodine furnished tetraaryl [12]CPP **73**.

Several multiarylCPPs were synthesized by Nishiuchi and Müllen and were used as potential precursors for the synthesis of belt-shaped molecules. In 2012, dodecaaryl [9]CPP was synthesized using Jasti's method (Figure 3.22) [36]. Initially, L-shaped units with four phenyl or 4-*tert*-butylphenyl groups **74a,b** were prepared. Nickel(0)- or copper(I)-mediated homocoupling reactions of **74a,b** furnished macrocycles, which were converted into dodecaaryl [9]CPP **75a,b** via reductive aromatization using $TiCl_4$/$LiAlH_4$. In 2014, a series of multiarylCPPs **75c–f** were prepared (Figure 3.22) [37].

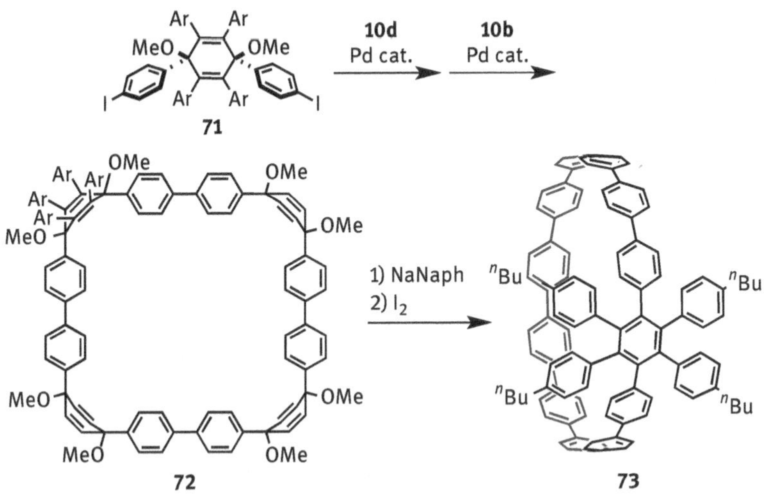

Figure 3.21: Synthesis of tetraaryl [12]CPP **73** (Ar = 4-*n*-butylphenyl).

Nishiuchi and Müllen synthesized multiphenyl-substituted L-shaped units **74c–f**, which were subjected to a coupling reaction and aromatization to form carbon nanorings **75c–f**. Nishiuchi and Müllen also conducted the synthesis of methylated or ethylene-bridged nanorings **75g,h** in 2015 (Figure 3.22) [38]. Although they attempted oxidative intramolecular cyclization, or the Scholl reaction, on nanorings **75a–f** to

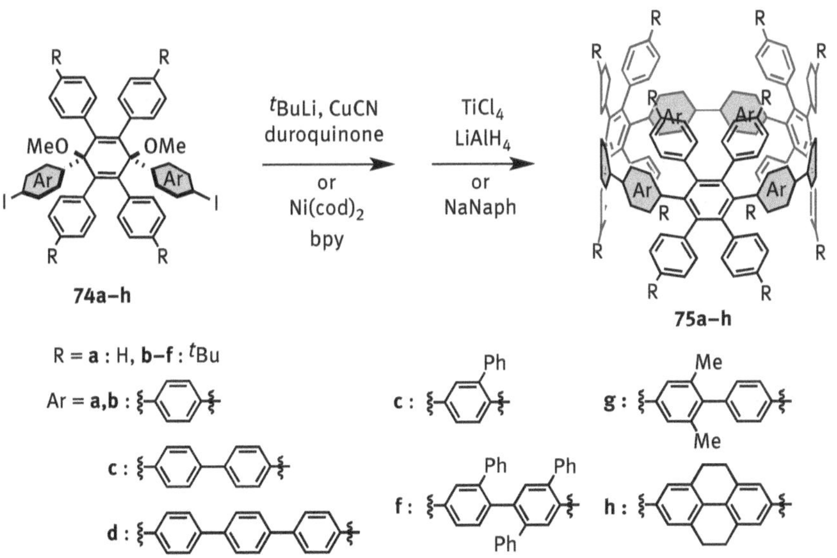

Figure 3.22: Synthesis of multiarylCPPs **75a–h**.

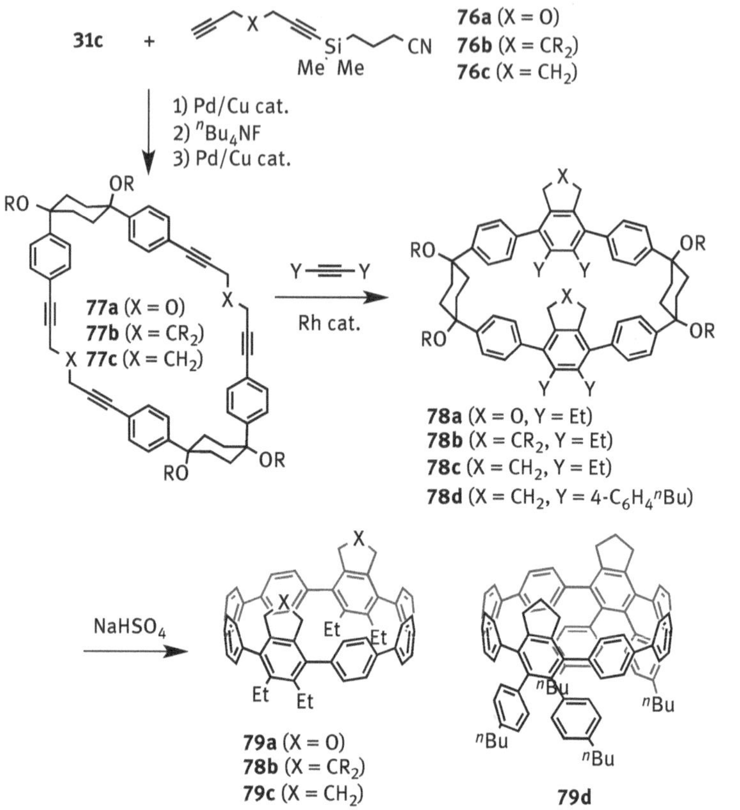

Figure 3.23: Synthesis of substituted [8]CPP derivatives **79a–d** (R = MOM).

obtain belt-shaped molecule, the desired product was not obtained because of undesired 1,2-phenyl shift or other side reactions.

In 2014, Wegner reported the synthesis of [8]CPP derivatives **79a–d** containing tetraaryl [8]CPP derivative **79d** (Figure 3.23) [39]. Itami's L-shaped unit **31c** and monoprotected diynes **76a–c** were connected by sequential Sonogashira coupling and deprotection to form unstrained macrocycles **77a–c**. The diyne moieties of **77a–c** were converted into benzene rings by rhodium-catalyzed [2+2+2] cycloaddition reaction with 3-hexyne or di(4-*n*-butylphenyl)acetylene. The resulting macrocycles **78a–d** were finally subjected to acid-mediated aromatization to obtain [8]CPP derivatives **79a–d**.

When two CPPs are connected, an ideal belt-shaped molecule or the so-called carbon nanobelt can be obtained. As a first step toward the synthesis of such carbon nanobelts, the synthesis of CPP dimers was examined. In 2012, Jasti succeeded in the synthesis of *p*-phenylene- and 1,5-naphthylene-bridged [8]CPP dimers **80** and **81**, respectively (Figure 3.24) [40]. By using a bromine-containing

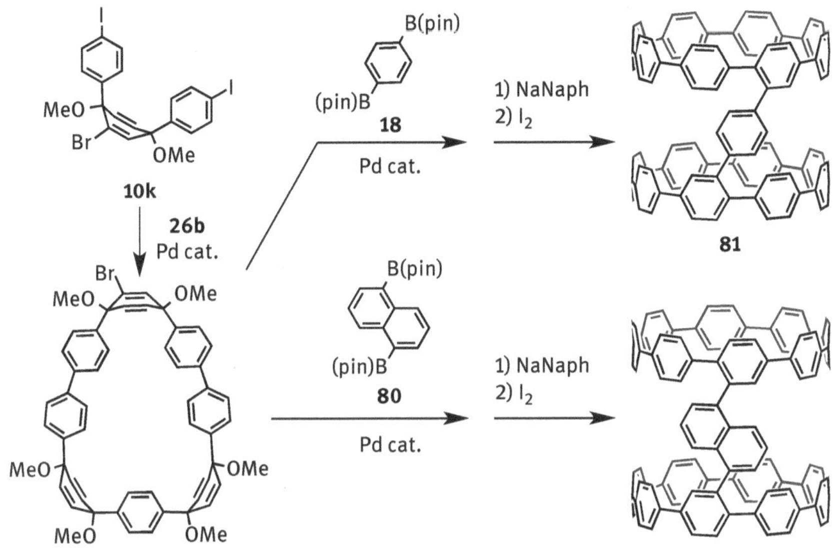

Figure 3.24: Synthesis of *p*-phenylene- or 1,5-naphthylene-bridged [8]CPP dimers **81**, **82**.

L-shaped unit (**10k**) along with the previously reported U-shaped unit **26b**, bromine-containing macrocycle **28c** was prepared. Subsequently, two **28c** macrocycles were connected with 1,4-diborylbenzene (**18**) or 1,5-diborylnaphthalene (**80**) by the cross-coupling reaction, followed by reductive aromatization to form [8] CPP dimers **81** and **82**. Jasti reported that directly connected [8]CPP dimer could not be obtained from **28c**.

In 2014, Itami accomplished the synthesis of a directly connected CPP dimer through the bottom-up synthesis of chloroCPP. Considering the tolerance required for the reaction conditions of CPP synthesis such as palladium-catalyzed cross-coupling and acid-mediated aromatization, the chloro group is expected to be an efficient handle for further synthetic manipulation. The preparation of chloro [10]CPP was achieved by a simple modification of a previously reported route for [10]CPP (Figure 3.25) [41]. Instead of 1,4-diborylbenzene **18**, which was used for the synthesis of [10]CPP, 1,4-diboryl-2-chlorobenzene **83** was used as the chloro-containing linear unit. A cross-coupling reaction of 1,4-diboryl-2-chlorobenzene and C-shaped unit **32b** formed chloro-containing macrocycle **34b**. The subsequent aromatization of **34b** furnished chloro [10]CPP **84**, which was the first example of a monosubstituted CPP. Finally, a homocoupling reaction of **84** with $Ni(cod)_2$/bpy afforded the directly connected [10]CPP dimer **85**.

Next, π-extended carbon nanorings including *p*-phenylene-PAH hybrid rings and all-PAH rings are presented. In 2013, PAH-containing ring was reported by Swager's

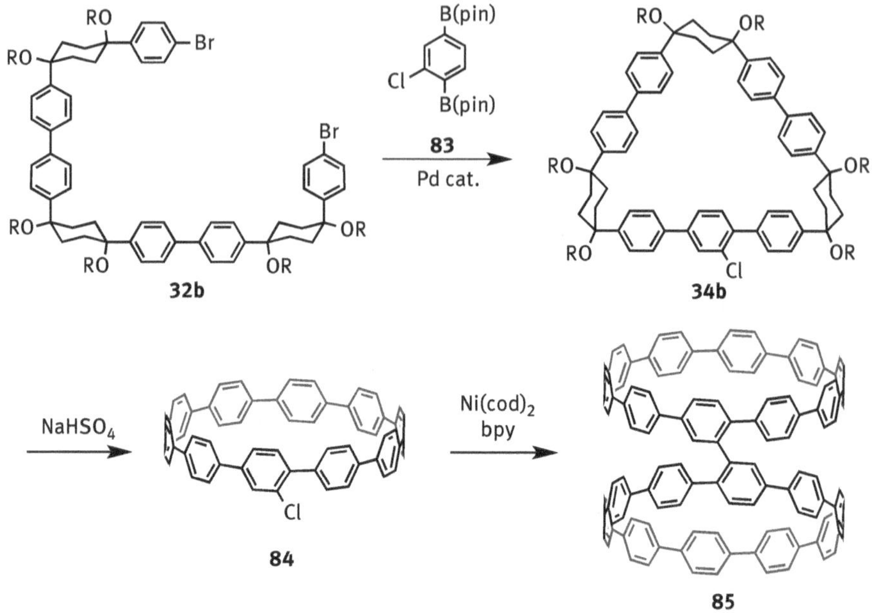

Figure 3.25: Synthesis of chloro [10]CPP **84** and [10]CPP dimer **85** (R = MOM).

group, who applied Itami's "cyclohexane" method to the synthesis of carbon nanoring **88**, bearing naphthalene and benzene rings (Figure 3.26) [42]. Naphthalene-containing L-shaped unit **86** was prepared from 1,4-dibromonaphthalene and 1,4-cyclohexanedione. Subsequently, macrocycle **85** was obtained by the nickel(0)-mediated homocoupling of **86**. Oxidative aromatization of cyclic tetramer **87** with TsOH yielded cyclo [4]paraphenylene [8]1,4-naphthylene **88**.

In 2014, Wang synthesized a carbon nanoring consisting of naphthalene and benzene via an alternative route (Figure 3.27) [43]. *cis*-Configurated L-shaped unit **89a** was selectively obtained by the Diels–Alder reaction of 1,4-bis(4-bromophenyl) butadiene and *p*-benzoquinone. Homocoupling reaction of **89a** afforded macrocycle **90**, and cyclo [6]paraphenylene [3]1,4-naphthylene **91** was obtained by the oxidative aromatization of the cyclohexadiene moiety using 2,3-dichloro-5,6-dicyano-*p*-benzoquinone (DDQ).

Synthesis of the pyrene-containing carbon nanoring, cyclo [12]paraphenylene [2]2,7-pyrenylene (**94**) was reported by Itami in 2014 (Figure 3.28) [44]. The synthetic route was based on the synthesis of [16]CPP. Macrocycle **93** was obtained by the homocoupling reaction of pyrene-containing U-shaped units, which was generated by the cross-coupling reaction of 2,7-diborylpyrene **92** and L-shaped unit **31d**. Oxidative aromatization of **93** under high temperatures produced cyclo [12]paraphenylene [2]2,7-pyrenylene **94**.

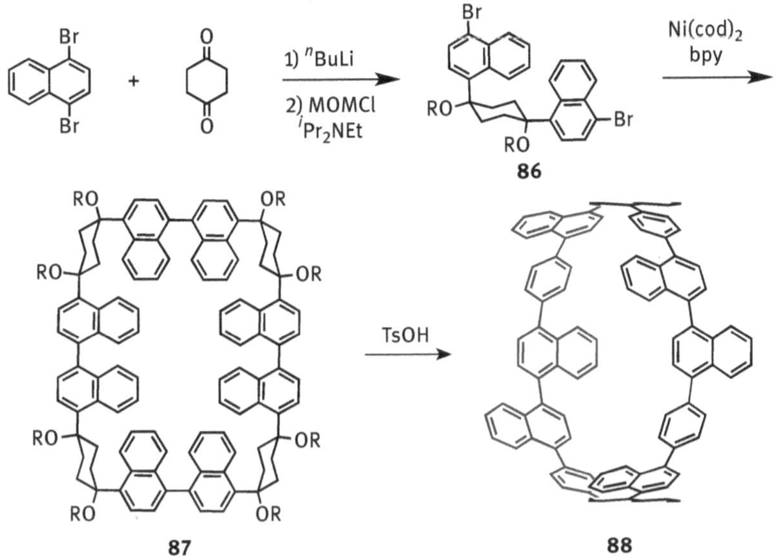

Figure 3.26: Synthesis of cyclo [4]paraphenylene [8]1,4-naphthylene **88** (R = MOM).

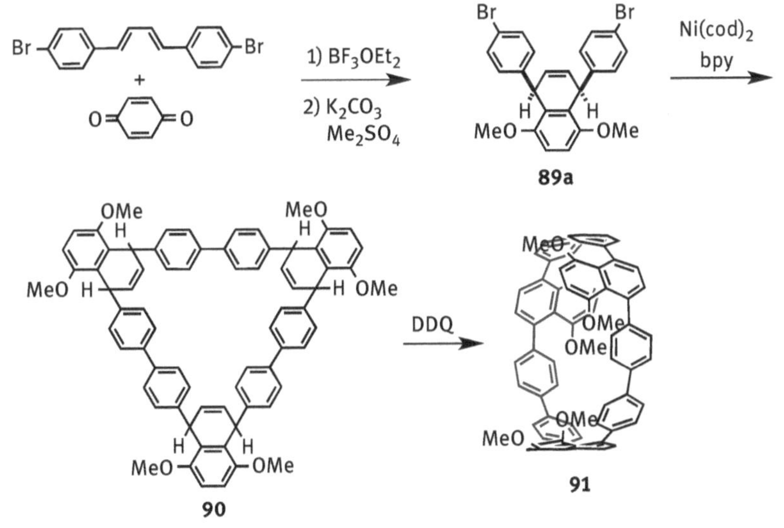

Figure 3.27: Synthesis of cyclo [6]paraphenylene [3]1,4-naphthylene **91**.

In 2012, Itami reported the synthesis of [9]cyclo-1,4-naphthylene **97**, which consists exclusively of naphthalene rings (Figure 3.29) [45]. For the generation of **97**, a synthesis based on a reductive aromatization pathway was selected. Firstly, L-shaped unit **95** was prepared from 1,4-dibromonaphthalene and 1,4-naphthoquinone. Nickel(0)-mediated macrocyclization of **95** yielded cyclic trimer **96**, which was subsequently

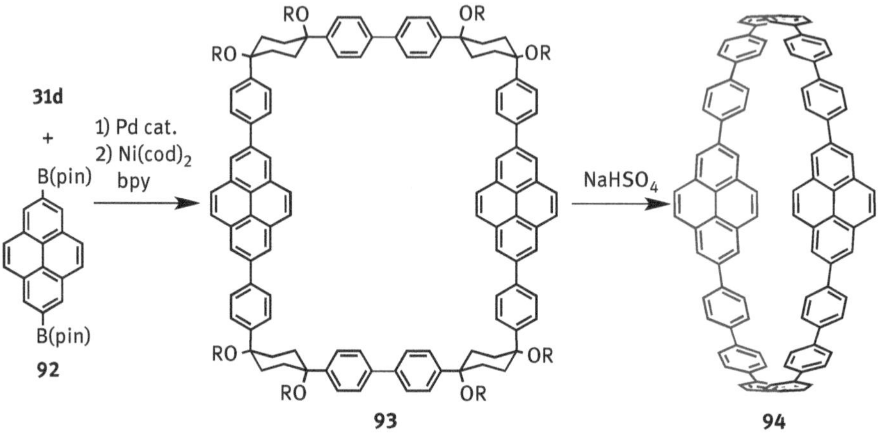

Figure 3.28: Synthesis of cyclo [12]paraphenylene [2]2,7-pyrenylene **94**.

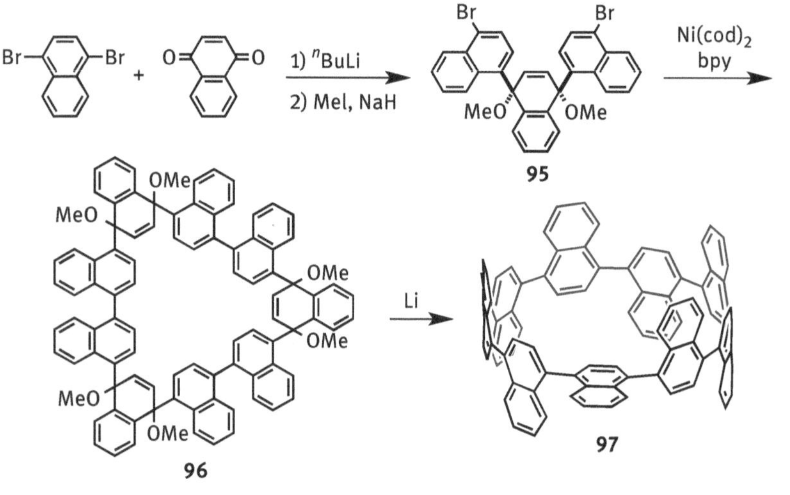

Figure 3.29: Synthesis of [9]cyclo-1,4-naphthylene **97**.

converted into **97** by reductive aromatization with lithium metal. Interestingly, previously reported aromatization conditions, e.g., the use of LiNaph, were not suitable for this reaction.

In 2014, Yamago reported the synthesis of [4]cyclo-2,7-pyrenylene **100** (Figure 3.30) [46]. In order to increase the solubility of the synthetic intermediates, 4,5,9,10-tetrahydropyrene was used instead of pyrene. Macrocyclic tetraplatinum complex **99**, consisting of four tetrahydropyrene rings, was synthesized from 2,7-distannyl-4,5,9,10-tetrahydropyrene **98** by using Yamago's macrocyclization method. The reductive elimination of **99** promoted by triphenylphosphine provided a

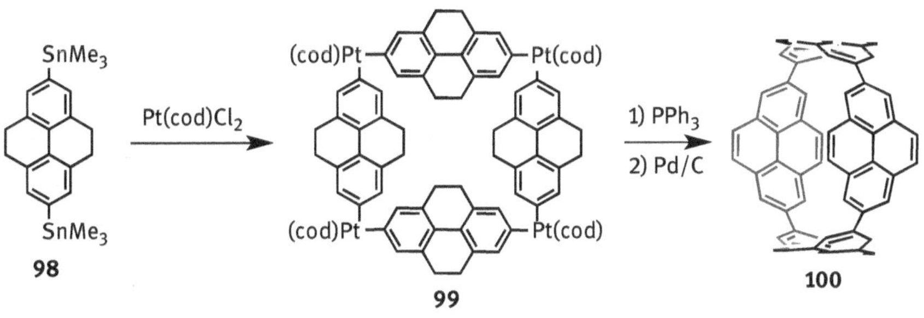

Figure 3.30: Synthesis of [4]cyclo-2,7-pyrenylene **100**.

ring-shaped structure, which was subsequently dehydrogenated using palladium on carbon to afford [4]cyclo-2,7-pyrenylene **100**.

3.6.1 Synthesis of Chiral and Zigzag Carbon Nanorings

In this section, the syntheses of segments of chiral and zigzag CNTs (chiral/zigzag carbon nanorings) are summarized. A possible explanation for the lower number of reported examples of chiral/zigzag carbon nanorings relative to that of armchair carbon nanorings could be the inherently lower symmetry of the former with respect to the latter.

In 2011, Itami reported the first synthesis of a chiral carbon nanoring, representing a (15,14)CNT segment (Figure 3.31) [47]. A modular method for the synthesis of [14]–[16]CPP was used to synthesize naphthalene-containing macrocycle **102**, employing L-shaped unit **31d**, U-shaped unit **37b** and 2,6-diborylnaphthalene **101** as

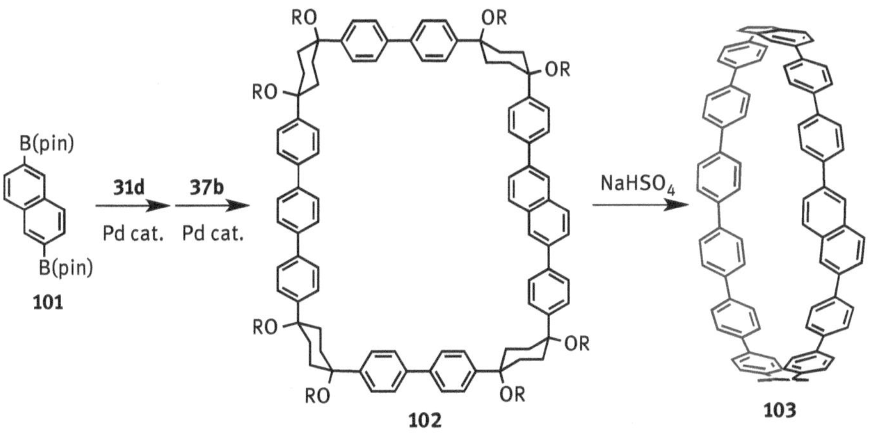

Figure 3.31: Synthesis of [13]cycloparaphenylene-2,6-naphthylene **103**.

the chirality-inducing units. Cycloparaphenylene-2,6-naphthylene **103** was obtained by the subsequent oxidative aromatization of **102.** Itami mentioned that many possible segments of chiral CNTs can be obtained by varying the acene unit and/or the number of paraphenylene moieties. For example, segments of (*n*+2,*n*+1), (*n*+3,*n*+1) and (*n*+4,*n*+1)CNTs can be obtained by inserting a 2,6-naphtylene, 2,6-anthrylene or 2,8-tetracenylene unit, respectively, into the [*n*]CPP ring. As the 2,6-naphthylene moiety in **103** can easily rotate at ambient temperature, a separation of the enantiomers of **103** was not possible.

In 2011–2013, Isobe synthesized carbon nanorings solely consisting of chrysenes or anthanthrylenes as shown in Figure 3.32 [48–50]. Using Yamago's method for the synthesis of CPPs, macrocyclic tetraplatinum complexes **105a–c** were prepared by the transmetalation of Pt(cod)Cl_2 with diborylated PAHs **104a–c**, which can be synthesized from chrysene and pigment red, respectively. In contrast to Yamago's original macrocyclization that includes the use of Pt(cod)Cl_2 and distannylarenes, Isobe's modification employs CsF or K_3PO_4 as promoters for the transmetalation of the diborylarenes. Treating macrocycles **105a–c** with triphenylphosphine at elevated temperatures provided a mixture of rotational isomers of carbon nanorings **106a–c.** One of the rotational isomers of [4]cyclo-2,8-chrysenylene (**106a**), as well as that of [4]cyclo-2,8-anthanthrylene (**106b**), represents armchair (10,10)CNT segments, and the other types of rotational isomers represent (11,9)CNT and (12,8)CNT segments.

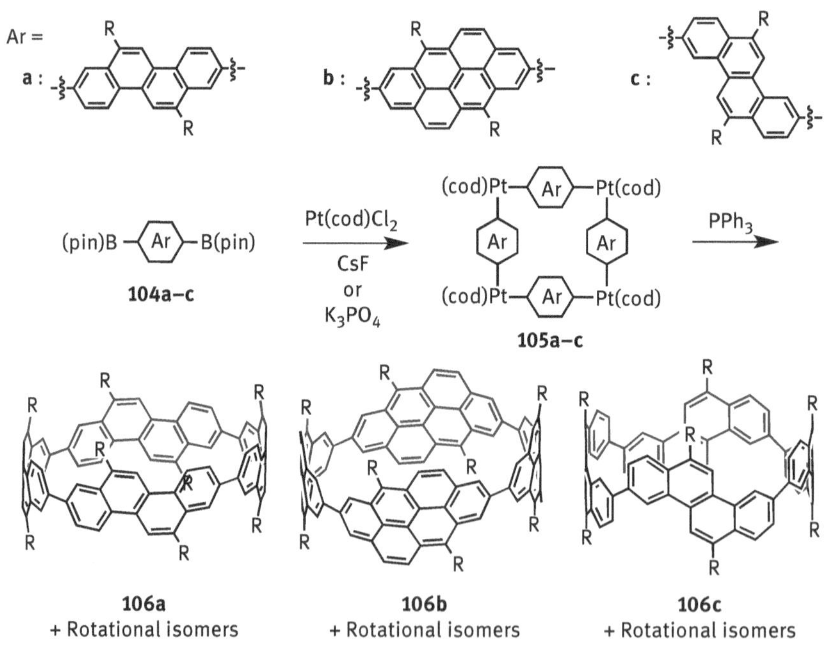

Figure 3.32: Synthesis of [4]cyclo-2,8-chrysenylene **106a** [4],cyclo-2,8-anthanthrylene **106b** and [4]cyclo-3,9-chrysenylene **106c** (R = *n*-hexyl (**a–c**) or triisopropylsilylethynyl (**b**)).

In addition, each isomer consists of two enantiomers, whose separation was also successfully accomplished. One of the rotational isomers of [4]cyclo-3,9-chrysenylene (**106c**) can be regarded as a sidewall segment of zigzag (16,0)CNTs (Figure 3.32) [50].

3.7 Synthesis of Heteroatom-Containing Carbon Nanorings

By including heteroarenes in the components of carbon nanorings, heteroatom-containing carbon nanorings can be obtained. The first synthesis of a CPP derivative containing heteroatoms was reported by Itami and coworkers in 2012. They synthesized a pyridine-containing carbon nanoring by using 5,5′-dibromo-2,2′-bipyridyl (**107**) as a building block (Figure 3.33) [51]. Stepwise Suzuki–Miyaura cross-coupling reaction of **107** and diborylated U-shaped unit **37b** generated macrocycle **108**, which bears four pyridine rings. Aromatization reaction mediated by $NaHSO_4$ yielded cyclo [14] paraphenylene [4]2,5-pyridylene (**109a**). The pyridine-containing ring **109a** showed

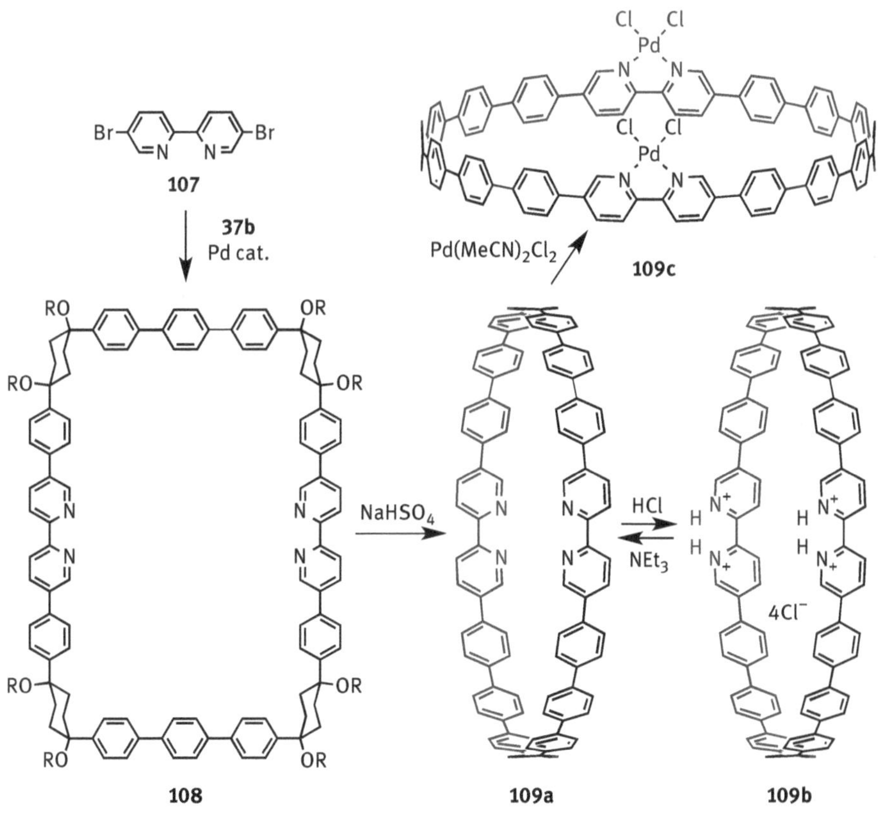

Figure 3.33: Synthesis of cyclo [14]paraphenylene [4]2,5-pyridylene **109a** and its reactions with acid and palladium(II) (R = MOM).

halochromic property because of reversible protonation (**109b**) and deprotonation. Complexation of **109a** to palladium(II) also occurred, generating dipalladium complex **109c**.

Itami also synthesized anthraquinone-containing carbon nanoring **112a** as the first donor–acceptor carbon nanoring (Figure 3.34) [52]. Diborylated 9,10-anthraquinone **110** was connected to U-shaped unit **51**, which was used for high-yielding selective synthesis of [10]CPP, to obtain an acyclic intermediate. Then, nickel-mediated homocoupling reaction produced anthraquinone-containing macrocycle **111**, which was successfully converted into carbon nanoring **112a** by an oxidative aromatization reaction. By using malononitrile, **112a** can be easily derivatized to quinodimethane-containing ring **112b**. Because the fluorescent color of **112a,b** changed, depending on the polarity of solvent (solvatofluorochromism), it can be concluded that anthraquinone and tetracyanoanthraquinodimethane behaved as π-acceptors and parahenylene as a π-donor in **112a,b**.

The oxidative aromatization route was applied to the synthesis of thiophene-containing rings (Figure 3.35) [53]. Itami and coworkers prepared a new L-shaped unit containing two ethynyl groups **113** as the starting material. Macrocycles were generated by the copper-catalyzed homocoupling reaction of terminal alkynes. Cyclic tetramer (**114**), pentamer (**115**) and hexamer (**116**) were obtained together with a cyclic dimer, trimer and acyclic oligomers. The butadiyne moieties of **114**–**116** were then converted into thiophene rings by treating them with sodium sulfide to yield thiophene-containing precursors **117**–**119**. Finally, acid-mediated aromatization reaction of **117**–**119** produced cycloparahenylene-2,5-thienylenes **120**–**122**.

Wang and coworkers synthesized ring-shaped molecules consisting of benzene, naphthalene and thiophene rings by using their aromatization method (Figure 3.36) [54]. Their L-shaped unit **89a** was borylated by Miyaura borylation to form **89b**, which

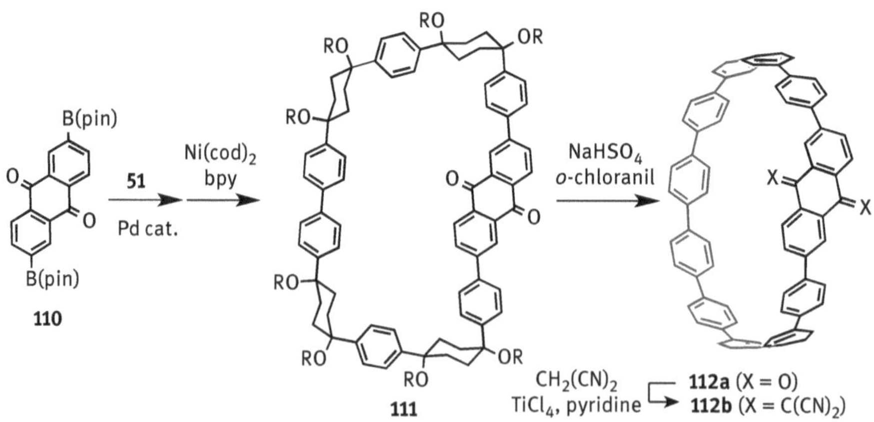

Figure 3.34: Synthesis of cyclo [10]paraphenylene-9,10-anthraquinon-2,6-ylene **112a** and its tetracyanoquinodimethane derivative **112b** (R = MOM).

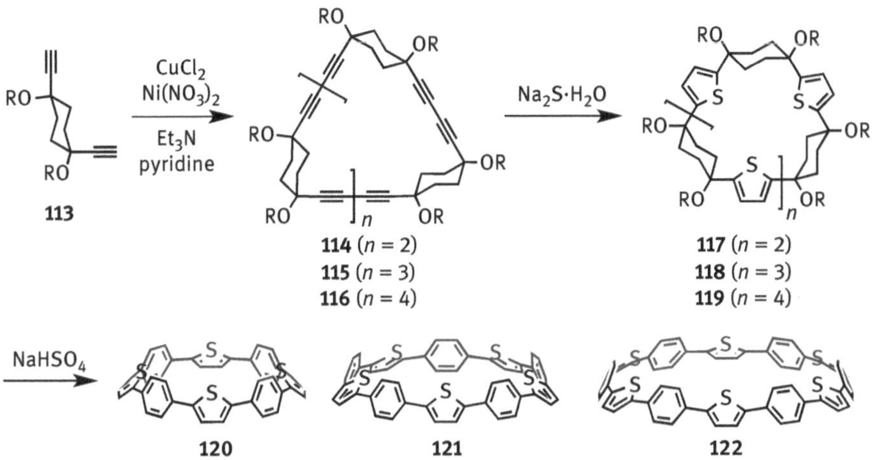

Figure 3.35: Synthesis of [4]–[6]cycloparaphenylene-2,5-thienylene **120–122** (R = MOM).

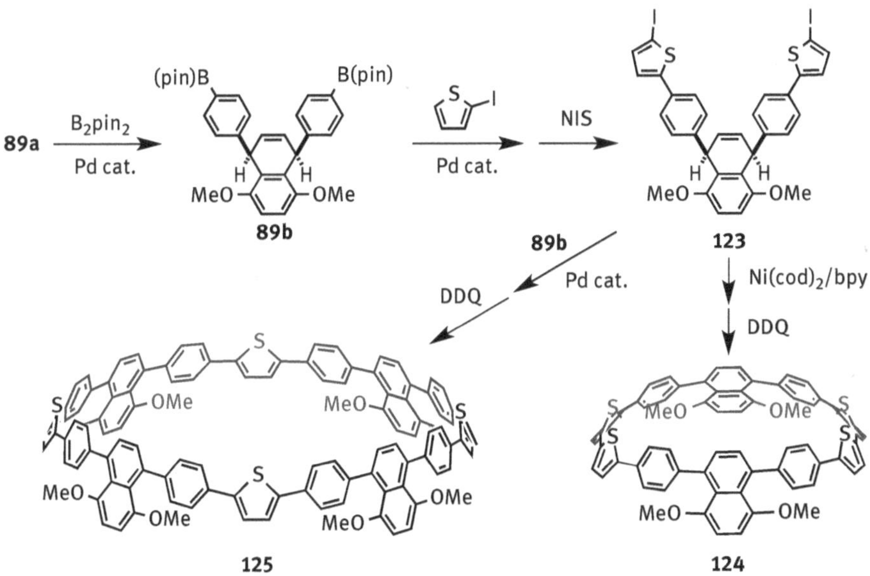

Figure 3.36: Synthesis of cycloparaphenylene-1,4-naphthylene-2,5-thienylene **124** and **125**.

was then coupled with 2-iodothiophene. The thiophene moieties were then subjected to iodination, which generated thiophene-attached L-shaped unit **123**. The homocoupling reaction of **123** and oxidative aromatization of the resulting cyclic dimer produced cycloparaphenylene-1,4-naphthylene-2,5-thienylene **124**. Larger ring **125** was synthesized by the sequential Suzuki–Miyaura cross-coupling reaction of **123** with **89b** and DDQ-mediated oxidative aromatization reactions.

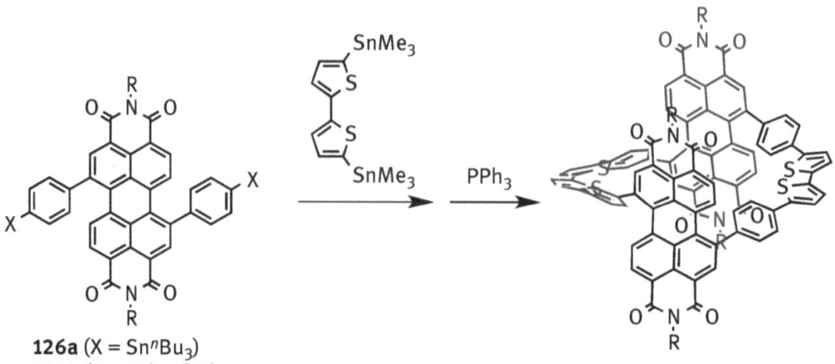

Figure 3.37: Synthesis of PDI-containing nanoring **127** (R = $CH(C_5H_{11})_2$).

Li, Xiao, Ng, Steigerwald, Nuckolls and coworkers reported the synthesis of a coral-shaped molecule containing perylene diimide (PDI) by using Yamago's method (Figure 3.37) [55]. Distannylated 1,7-diarylPDI **126a**, prepared from dibromoPDI by the Stille coupling reaction, was converted into diplatinum complex **126b**. Complexation of **126b** with 5,5′-bis(trimethylstannyl)-2,2′-bithiophene and subsequent reductive elimination accelerated by triphenylphosphine yielded PDI-containing ring **127**. Rotational isomers of **127** were also isolated.

Stępień synthesized π-extended [6]CPP by using carbazole or benzocarbazole as building blocks (Figure 3.38) [56]. Reaction of dibromocarbazole **128** and 1,3,5-tris(bromomethyl)benzene produced intermediate **130**. Nickel(0)-mediated homocoupling reaction of **130** proceeded intramolecularly to yield [3]cyclo-2,7-carbazolylene **132**. When benzocarbazole derivative **129** was used instead [3],cyclo-2,6-benzo[*def*]carbazolylene **133** can be obtained through intermediate **131**.

Osuka reported the synthesis of macrocyclic molecule consisting solely of porphyrin rings via platinum complexes (Figure 3.39) [57]. Diarylporphyrin **134** was subjected to iridium-catalyzed C–H borylation to yield 2,12-diborylporphyrin **135a**. The incorporated metal (zinc) was then replaced by nickel (**135b**). Complexation of **135b** with platinum was carried out to obtain a mixture of cyclic platinum complex **136**, which was converted into cyclo-2,12-porphyrinylenes by treatment with triphenylphosphine to yield [3]–[5]cyclo-2,12-porphyrinylene **137–139**.

3.8 Synthesis of Carbon Nanocages

The [*n.n.n*]carbon nanocages with C_3 symmetry consist of a pair of trisubstituted benzene rings and three [*n*]paraphenylene moieties, each linking two benzene rings. In 2013, Itami, Segawa and Kamada reported the first synthesis of [6.6.6]

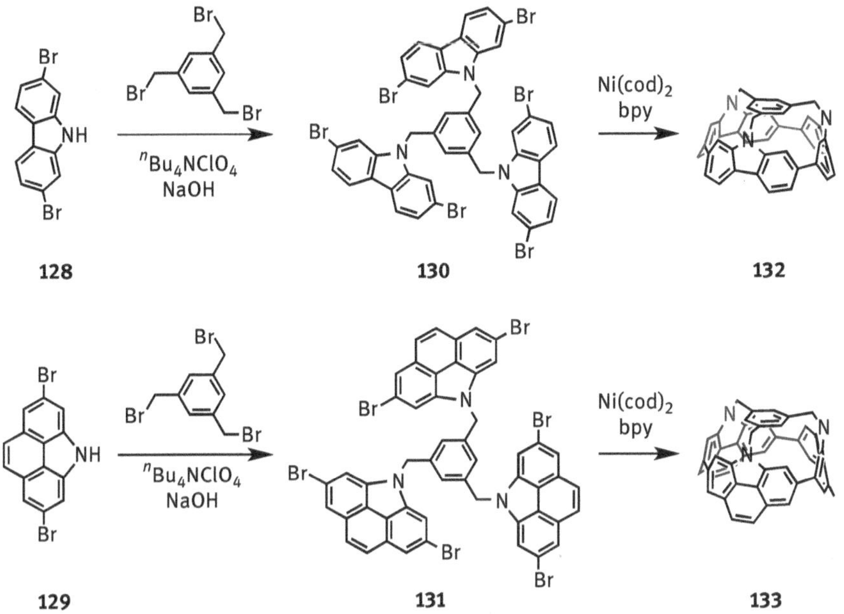

Figure 3.38: Synthesis of [3]cyclo-2,7-carbazolylene **132** and [3]cyclo-2,6-benzo[*def*] carbazolylene **133**.

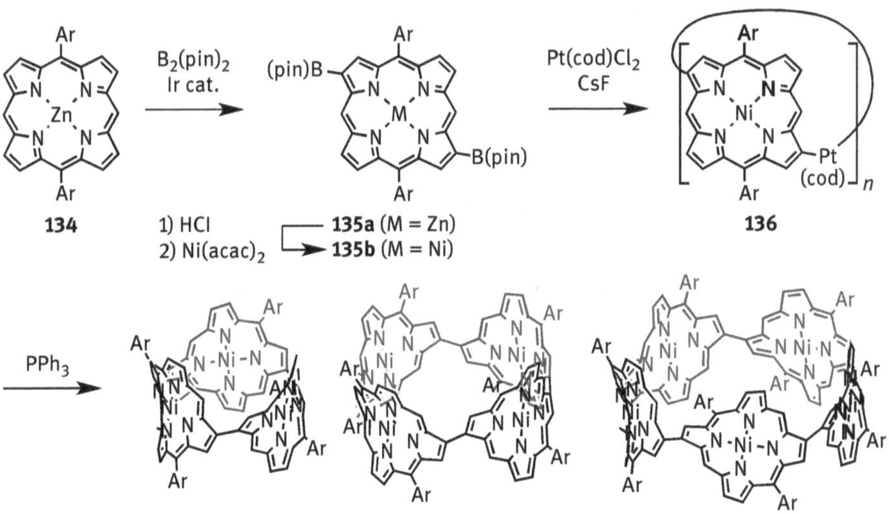

Figure 3.39: Synthesis of [3]–[5]cyclo-2,12-porphyrinylene **137–139** (Ar = 3,5-di-*tert*-butylphenyl).

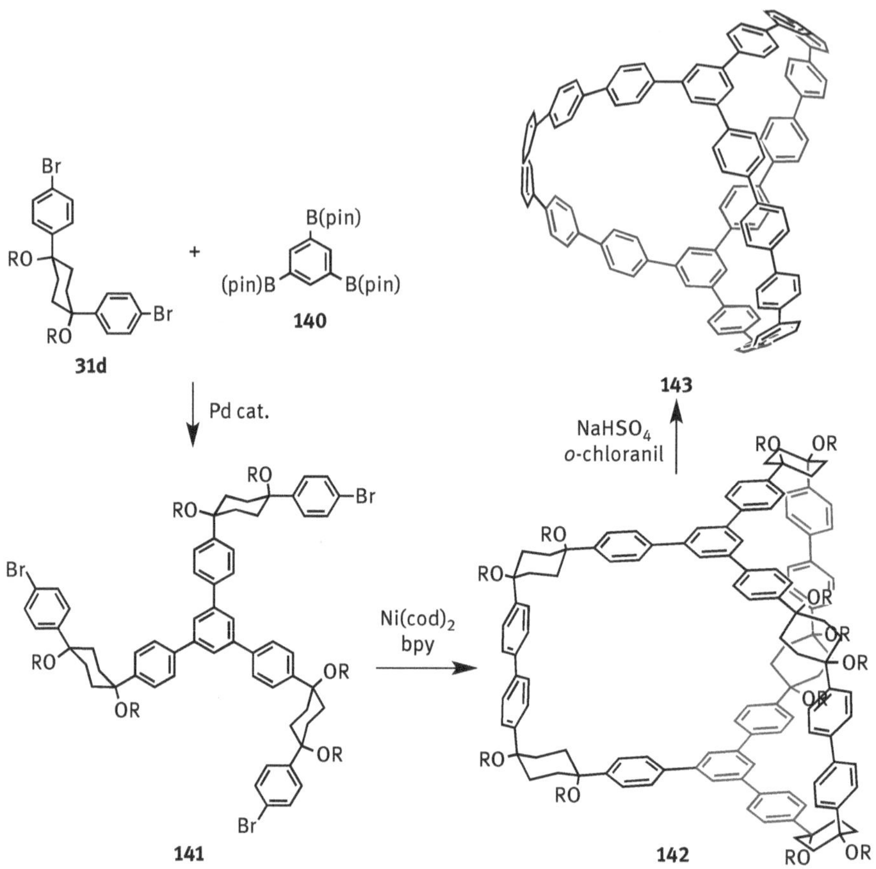

Figure 3.40: Synthesis of [6.6.6]carbon nanocage **143** (R = MOM).

carbon nanocage **143** (Figure 3.40) [58]. Similar to the synthesis of [14]–[16]CPP by Itami, trifurcated unit **141** was obtained by a three-fold Suzuki–Miyaura coupling of 1,3,5-triborylbenzene **140** with L-shaped unit **31b.** Nickel(0)-mediated dimerization of **141** provided a nonstrained bicyclic precursor **142,** and its six cyclohexane moieties were aromatized to afford [6.6.6]carbon nanocage **143.** Itami and Segawa then expanded this synthetic methodology to obtain carbon nanocages of varying sizes; use of appropriately modified starting materials resulted in the synthesis of [4.4.4] carbon nanocage **147** and [5.5.5]carbon nanocage **148** shown in Figure 3.41 [59]. Based on Itami's synthesis of [7]CPP and [8]CPP, the smaller trifurcated unit **144a** was prepared in several steps from 1,3,5-tribromobenzene and small L-shaped unit **47b.** After transforming the chloro groups on **144a** into boryl groups (**144b**), two bicyclic macrocycles **145** and **146** were synthesized under the influence of Ni(cod)$_2$/phen (phen = 1,10-phenanthroline) or palladium catalyst, depending on the reaction conditions and starting materials used. Precursors **145** and **146** were subsequently subjected to

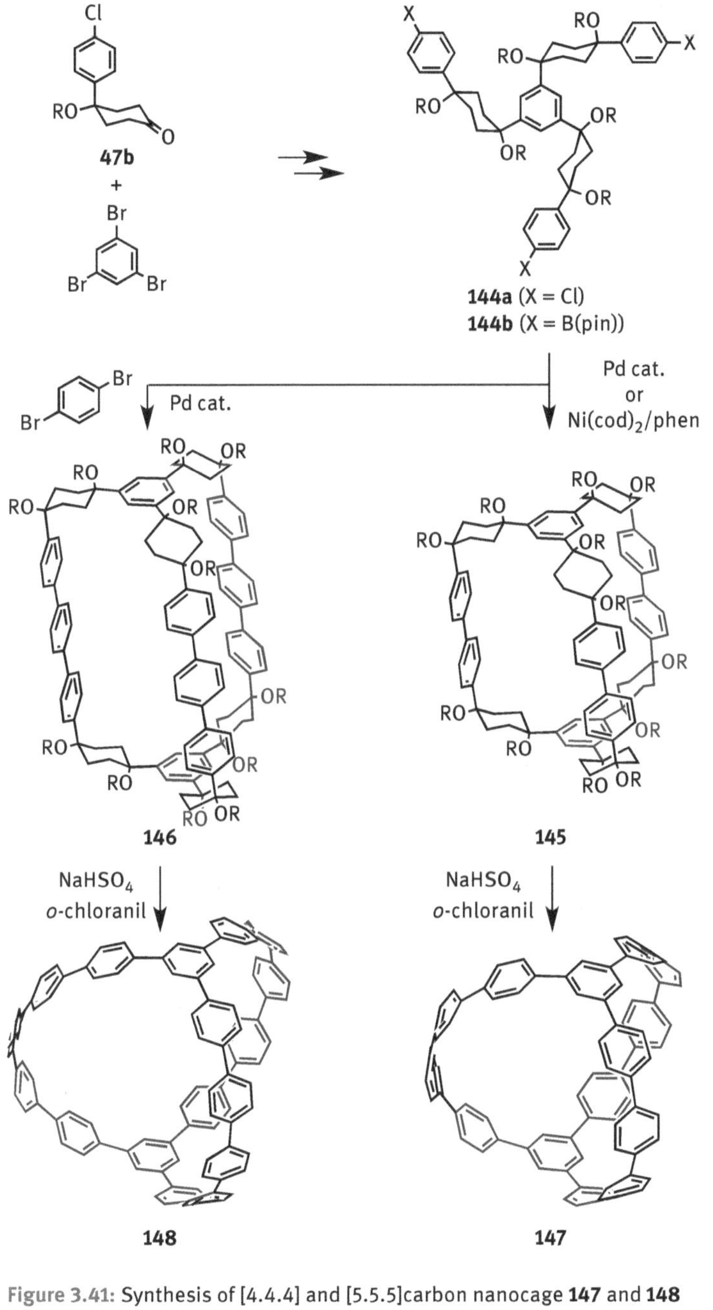

Figure 3.41: Synthesis of [4.4.4] and [5.5.5]carbon nanocage **147** and **148** (R = MOM).

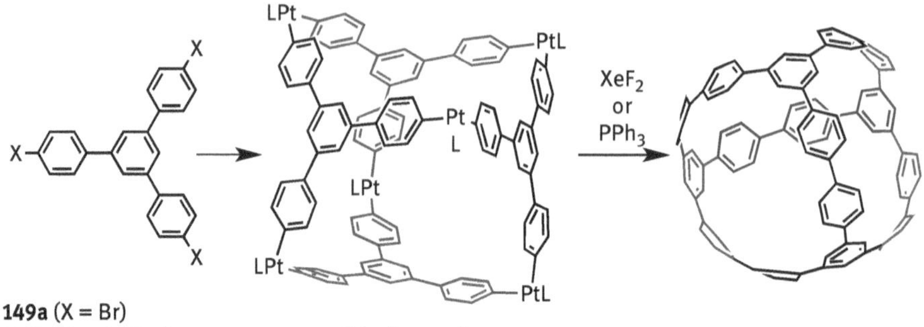

149a (X = Br)
149b (X = $SnMe_3$)
149c (X = Pt(cod)Cl)

150a (L = cod)
150b (L = dppf)

151

Figure 3.42: Synthesis of cage-shaped molecule **151**.

acid-mediated aromatization reactions, affording the corresponding carbon nanocages (**147** and **148**).

Yamago and Kim reported cage-shaped molecule (**151**) consisting of 16 benzene rings (Figure 3.42) [60]. First, 1,3,5-tris(4-bromophenyl)benzene **149a** was converted into stannyl derivative **149b** and then to platinum complex **149c**. Mixing of **149b** and **149c** generated hexaplatinum complex **150a**, which was an analogue of self-assembled molecular capsules developed by Fujita [61]. After exchanging the ligand cod with dppf, reductive elimination was performed to obtain cage-shaped molecule **151**.

3.9 Summary

In this chapter, syntheses of CPPs and derivatives are reviewed. Development of the synthetic methods of CPPs allowed to uncover interesting size-dependent photophysical and redox properties of CPPs [62–79]. Characteristic CPP properties, such as host–guest behavior [80–91] and the complexation with transition metal [92], will provide a better understanding of the fundamental structure–property relationships and allow the application of CPPs in materials science. Growth of CNT by using CPP as a template will open the new field of selective synthesis of CNT [93]. For further details about the properties and applications of CPPs and their derivatives, see also accounts and review articles [94–100].

References

[1] Iijima S. Helical microtubules of graphitic carbon. Nature 1991;354:56–8.
[2] Avouris P, Chen Z, Perebeinos V. Carbon-based electronics. Nat Nanotech 2007;2:605–15.
[3] Avouris P, Freitag M, Perebeinos V. Carbon-nanotube photonics and optoelectronics. Nat Photon 2008;2:341–50.

[4] Sgobba V, Guldi DM. Carbon nanotubes-electronic/electrochemical properties and application for nanoelectronics and photonics. Chem Soc Rev 2009;38:165–84.

[5] Heller DA, Baik S, Eurell TE, Strano MS. Single-walled carbon nanotube spectroscopy in live cells: towards long-term labels and optical sensors. Adv Mater 2005;17:2793–9.

[6] Zhang M, et al. Strong, transparent, multifunctional, carbon nanotube sheets. Science 2005;309:1215–19.

[7] Wu Z, et al. Transparent, conductive carbon nanotube films. Science 2004;305:1273–6.

[8] Dresselhaus MS, Dresselhaus G, Saito R. Physics of carbon nanotubes. Carbon 1995;33: 883–91.

[9] Dai H. Carbon nanotubes: synthesis, integration, and properties. Acc Chem Res 2002;35: 1035–44.

[10] Parekh VC, Guha PC. Synthesis of *pp'*-diphenylenedimonosulphide. J Indian Chem Soc 1934;11:95–100.

[11] Franke J, Vögtle F. Cyclische para-phenylensulfide: selektive synthese, dotierung und elektrische leitfähigkeit. Tetrahedron Lett 1984;25:3445–8.

[12] Friederich R, Nieger M, Vögtle F. On the way to macrocyclic para-phenylenes. Chem Ber Recl 1993;126:1723–32.

[13] McMurry JE, Haley GJ, Matz JR, Clardy JC, Mitchell J. Pentacyclo[12.2.2.22,5.26,9.210,13]-1,5, 9,13-tetracosatetraene and its reaction with silver trifluoromethanesulfonate. Synthesis of a square-planar D_{10} organometallic complex. J Am Chem Soc 1986;108:515–16.

[14] Jasti R, Bhattacharjee J, Neaton JB, Bertozzi CR. Synthesis, characterization, and theory of [9]-, [12]-, and [18]cycloparaphenylene: carbon nanohoop structures. J Am Chem Soc 2008;130:17646–7.

[15] Takaba H, Omachi H, Yamamoto Y, Bouffard J, Itami K. Selective synthesis of [12]cycloparaphenylene. Angew Chem Int Ed 2009;48:6112–16.

[16] Yamago S, Watanabe Y, Iwamoto T. Synthesis of [8]cycloparaphenylene from a square-shaped tetranuclear platinum complex. Angew Chem Int Ed 2010;49:757–9.

[17] Fuhrmann G, Debaerdemaeker T, Bäuerle P. C–C bond formation through oxidatively induced elimination of platinum complexes – a novel approach towards conjugated macrocycles. Chem Commun 2003;948–9.

[18] Sisto TJ, Golder MR, Hirst ES, Jasti R. Selective synthesis of strained [7]cycloparaphenylene: an orange-emitting fluorophore. J Am Chem Soc 2011;133:15800–2.

[19] Darzi ER, Sisto TJ, Jasti R. Selective syntheses of [7]–[12]cycloparaphenylenes using orthogonal Suzuki–Miyaura cross-coupling reactions. J Org Chem 2012;77:6624–8.

[20] Xia J, Jasti R. Synthesis, characterization, and crystal structure of [6]cycloparaphenylene. Angew Chem Int Ed 2012;51:2474–6.

[21] Xia J, Bacon JW, Jasti R. Gram-scale synthesis and crystal structures of [8]- and [10]CPP, and the solid-state structure of C_{60}@[10]CPP. Chem Sci 2012;3:3018–21.

[22] Evans PJ, Darzi ER, Jasti R. Efficient room-temperature synthesis of a highly strained carbon nanohoop fragment of buckminsterfullerene. Nat Chem 2014;6:404–8.

[23] Segawa Y, et al. Concise synthesis and crystal structure of [12]cycloparaphenylene. Angew Chem Int Ed 2011;50:3244–8.

[24] Segawa Y, Senel P, Matsuura S, Omachi H, Itami K. [9]cycloparaphenylene: nickel-mediated synthesis and crystal structure. Chem Lett 2011;40:423–5.

[25] Omachi H, Matsuura S, Segawa Y, Itami K. A modular and size-selective synthesis of [n] cycloparaphenylenes: a step toward the selective synthesis of [n, n] single-walled carbon nanotubes. Angew Chem Int Ed 2010;49:10202–5.

[26] Ishii Y, et al. Size-selective synthesis of [9]–[11] and [13]cycloparaphenylenes. Chem Sci 2012;3:2340–5.

[27] Sibbel F, Matsui K, Segawa Y, Studer A, Itami K. Selective synthesis of [7]- and [8]cycloparaphenylenes. Chem Commun 2014;50:954–6.
[28] Segawa Y, Kuwabara T, Matsui K, Kawai S, Itami K. Palladium-free synthesis of [10]cycloparaphenylene. Tetrahedron 2015;71:4500–3.
[29] Iwamoto T, Watanabe Y, Sakamoto Y, Suzuki T, Yamago S. Selective and random syntheses of [n]cycloparaphenylenes (n = 8–13) and size dependence of their electronic properties. J Am Chem Soc 2011;133:8354–61.
[30] Kayahara E, Sakamoto Y, Suzuki T, Yamago S. Selective synthesis and crystal structure of [10]cycloparaphenylene. Org Lett 2012;14:3284–7.
[31] Kayahara E, Iwamoto T, Suzuki T, Yamago S. Selective synthesis of [6]-, [8]-, and [10]cycloparaphenylenes. Chem Lett 2013;42:621–3.
[32] Kayahara E, Patel VK, Yamago S. Synthesis and characterization of [5]cycloparaphenylene. J Am Chem Soc 2014;136:2284–7.
[33] Patel VK, Kayahara E, Yamago S. Practical synthesis of [n]cycloparaphenylenes (n = 5, 7–12) by H_2SnCl_4-mediated aromatization of 1,4-dihydroxycyclo-2,5-diene precursors. Chem Eur J 2015;21:5742–9.
[34] Kayahara E, Patel VK, Xia J, Jasti R, Yamago S. Selective and gram-scale synthesis of [6]cycloparaphenylene. Synlett 2015;26:1615–19.
[35] Sisto TJ, Tian X, Jasti R. Synthesis of tetraphenyl-substituted [12]cycloparaphenylene: toward a rationally designed ultrashort carbon nanotube. J Org Chem 2012;77:5857–60.
[36] Nishiuchi T, Feng X, Enkelmann V, Wagner M, Müllen K. Three-dimensionally arranged cyclic *p*-hexaphenylbenzene: toward a bottom-up synthesis of size-defined carbon nanotubes. Chem Eur J 2012;18:16621–5.
[37] Golling FE, Quernheim M, Wagner M, Nishiuchi T, Müllen K. Concise synthesis of 3D π-extended polyphenylene cylinders. Angew Chem Int Ed 2014;53:1525–8.
[38] Quernheim M, et al. The precise synthesis of phenylene-extended cyclic hexa-peri-hexabenzocoronenes from polyarylated [n]cycloparaphenylenes by the scholl reaction. Angew Chem Int Ed 2015;54:10341–6.
[39] Tran-Van A-F, et al. Synthesis of substituted [8]cycloparaphenylenes by [2+2+2] cycloaddition. Org Lett 2014;16:1594–7.
[40] Xia J, Golder MR, Foster ME, Wong BM, Jasti R. Synthesis, characterization, and computational studies of cycloparaphenylene dimers. J Am Chem Soc 2012;134:19709–15.
[41] Ishii Y, Matsuura S, Segawa Y, Itami K. Synthesis and dimerization of chloro[10]cycloparaphenylene: a directly connected cycloparaphenylene dimer. Org Lett 2014;16:2174–6.
[42] Batson JM, Swager TM. Towards a perylene-containing nanohoop. Synlett 2013;24:2545–9.
[43] Huang C, Huang Y, Akhmedov NG, Popp BV, Petersen JL, Wang KK. Functionalized carbon nanohoops: synthesis and structure of a [9]cycloparaphenylene bearing three 5,8-dimethoxynaphth-1,4-diyl units. Org Lett 2014;16:2672–5.
[44] Yagi A, Venkataramana G, Segawa Y, Itami K. Synthesis and properties of cycloparaphenylene-2,7-pyrenylene: a pyrene-containing carbon nanoring. Chem Commun 2014;50:957–9.
[45] Yagi A, Segawa Y, Itami K. Synthesis and properties of [9]cyclo-1,4-naphthylene: a π-extended carbon nanoring. J Am Chem Soc 2012;134:2962–5.
[46] Iwamoto T, Kayahara E, Yasuda N, Suzuki T, Yamago S. Synthesis, characterization, and properties of [4]cyclo-2,7-pyrenylene: effects of cyclic structure on the electronic properties of pyrene oligomers. Angew Chem Int Ed 2014;53:6430–4.
[47] Omachi H, Segawa Y, Itami K. Synthesis and racemization process of chiral carbon nanorings: a step toward the chemical synthesis of chiral carbon nanotubes. Org Lett 2011;13:2480–3.
[48] Hitosugi S, Nakanishi W, Yamasaki T, Isobe H. Bottom-up synthesis of finite models of helical (n,m)-single-wall carbon nanotubes. Nat Commun 2011;2:492.

[49] Matsuno T, Kamata S, Hitosugi S, Isobe H. Bottom-up synthesis and structures of π-lengthened tubular macrocycles. Chem Sci 2013;4:3179–83.
[50] Hitosugi S, Yamasaki T, Isobe H. Bottom-up synthesis and thread-in-bead structures of finite (n,0)-zigzag single-wall carbon nanotubes. J Am Chem Soc 2012;134:12442–5.
[51] Matsui K, Segawa Y, Itami K. Synthesis and properties of cycloparaphenylene-2,5-pyridylidene: a nitrogen-containing carbon nanoring. Org Lett 2012;14:1888–91.
[52] Kuwabara T, Orii J, Segawa Y, Itami K. Curved oligophenylenes as donors in shape-persistent donor–acceptor macrocycles with solvatofluorochromic properties. Angew Chem Int Ed 2015;54:9646–9.
[53] Ito H, Mitamura Y, Segawa Y, Itami K. Thiophene-based radial π-conjugation: synthesis, structure, and photophysical properties of cyclo-1,4-phenylene-2',5'-thienylenes. Angew Chem Int Ed 2015;54:159–63.
[54] Thakellapalli H, et al. Syntheses and structures of thiophene-containing cycloparaphenylenes and related carbon nanohoops. Org Lett 2015;17:3470–3.
[55] Ball M, et al. Chiral conjugated corrals. J Am Chem Soc 2015;137:9982–7.
[56] Myśliwiec D, Kondratowicz M, Lis T, Chmielewski PJ, Stępień M. Highly strained nonclassical nanotube end-caps. A single-step solution synthesis from strain-free, non-macrocyclic precursors. J Am Chem Soc 2015;137:1643–9.
[57] Jiang H-W, Tanaka T, Mori H, Park KH, Kim D, Osuka A. Cyclic 2,12-porphyrinylene nanorings as a porphyrin analogue of cycloparaphenylenes. J Am Chem Soc 2015;137:2219–22.
[58] Matsui K, Segawa Y, Namikawa T, Kamada K, Itami K. Synthesis and properties of all-benzene carbon nanocages: a junction unit of branched carbon nanotubes. Chem Sci 2013;4:84–8.
[59] Matsui K, Segawa Y, Itami K. All-benzene carbon nanocages: size-selective synthesis, photophysical properties, and crystal structure. J Am Chem Soc 2014;136:16452–8.
[60] Kayahara E, et al. Synthesis and physical properties of a ball-like three-dimensional π-conjugated molecule. Nat Commun 2013;4:2694.
[61] Fujita M, Oguro D, Miyazawa M, Oka H, Yamaguchi K, Ogura K. Self-assembly of ten molecules into nanometre-sized organic host frameworks. Nature 1995;378:469–71.
[62] Segawa Y, Omachi H, Itami K. Theoretical studies on the structures and strain energies of cycloparaphenylenes. Org Lett 2010;12:2262–5.
[63] Sundholm D, Taubert S, Pichierri F. Calculation of absorption and emission spectra of [n]cycloparaphenylenes: the reason for the large stokes shift. Phys Chem Chem Phys 2010;12:2751–7.
[64] Wong BM, Lee JW. Anomalous optoelectronic properties of chiral carbon nanorings ... And one ring to rule them all. J Phys Chem Lett 2011;2:2702–6.
[65] Segawa Y, et al. Combined experimental and theoretical studies on the photophysical properties of cycloparaphenylenes. Org Biomol Chem 2012;10:5979–84.
[66] Nishihara T, Segawa Y, Itami K, Kanemitsu Y. Excited states in cycloparaphenylenes: dependence of optical properties on ring length. J Phys Chem Lett 2012;3:3125–8.
[67] Fujitsuka M, Cho DW, Iwamoto T, Yamago S, Majima T. Size-dependent fluorescence properties of [n]cycloparaphenylenes (n = 8–13), hoop-shaped π-conjugated molecules. Phys Chem Chem Phys 2012;14:14585–8.
[68] Camacho C, Niehaus TA, Itami K, Irle S. Origin of the size-dependent fluorescence blueshift in [n]cycloparaphenylenes. Chem Sci 2013;4:187–95.
[69] Fujitsuka M, Iwamoto T, Kayahara E, Yamago S, Majima T. Enhancement of the quinoidal character for smaller [n]cycloparaphenylenes probed by Raman spectroscopy. ChemPhysChem 2013;14:1570–2.
[70] Kayahara E, Kouyama T, Kato T, Takaya H, Yasuda N, Yamago S. Isolation and characterization of the cycloparaphenylene radical cation and dication. Angew Chem Int Ed 2013;52:13722–6.

[71] Golder MR, Wong BM, Jasti R. Photophysical and theoretical investigations of the [8]cycloparaphenylene radical cation and its charge-resonance dimer. Chem Sci 2013;4:4285–91.

[72] Zabula AV, Filatov AS, Xia J, Jasti R, Petrukhina MA. Tightening of the nanobelt upon multielectron reduction. Angew Chem Int Ed 2013;52:5033–6.

[73] Nishihara T, Segawa Y, Itami K, Kanemitsu Y. Exciton recombination dynamics in nanoring cycloparaphenylenes. Chem Sci 2014;5:2293–6.

[74] Fujitsuka M, Lu C, Iwamoto T, Kayahara E, Yamago S, Majima T. Properties of triplet-excited [n]cycloparaphenylenes (n = 8–12): excitation energies lower than those of linear oligomers and polymers. J Phys Chem A 2014;118:4527–32.

[75] Fujitsuka M, Tojo S, Iwamoto T, Kayahara E, Yamago S, Majima T. Radical ions of cycloparaphenylenes: size dependence contrary to the neutral molecules. J Phys Chem Lett 2014;5:2302–5.

[76] Adamska L, et al. Self-trapping of excitons, violation of condon approximation, and efficient fluorescence in conjugated cycloparaphenylenes. Nano Lett 2014;14:6539–46.

[77] Hines DA, Darzi ER, Jasti R, Kamat PV. Carbon nanohoops: excited singlet and triplet behavior of [9]- and [12]-cycloparaphenylene. J Phys Chem A 2014;118:1595–600.

[78] Chen H, Golder MR, Wang F, Jasti R, Swan AK. Raman spectroscopy of carbon nanohoops. Carbon 2014;67:203–13.

[79] Toriumi N, Muranaka A, Kayahara E, Yamago S, Uchiyama M. In-plane aromaticity in cycloparaphenylene dications: a magnetic circular dichroism and theoretical study. J Am Chem Soc 2015;137:82–5.

[80] Iwamoto T, Watanabe Y, Sadahiro T, Haino T, Yamago S. Size-selective encapsulation of C_{60} by [10]cycloparaphenylene: formation of the shortest fullerene-peapod. Angew Chem Int Ed 2011;50:8342–4.

[81] Iwamoto T, Watanabe Y, Takaya H, Haino T, Yasuda N, Yamago S. Size- and orientation-selective encapsulation of C_{70} by cycloparaphenylenes. Chem Eur J 2013;19:14061–8.

[82] Nakanishi Y, et al. Size-selective complexation and extraction of endohedral metallofullerenes with cycloparaphenylene. Angew Chem Int Ed 2014;53:3102–6.

[83] Isobe H, Hitosugi S, Yamasaki T, Iizuka R. Molecular bearings of finite carbon nanotubes and fullerenes in ensemble rolling motion. Chem Sci 2013;4:1293–7.

[84] Hitosugi S, et al. Photoinduced electron transfer in a dynamic supramolecular system with curved π-structures. Org Lett 2014;16:3352–5.

[85] Sato S, Yamasaki T, Isobe H. Solid-state structures of peapod bearings composed of finite single-wall carbon nanotube and fullerene molecules. Proc Natl Acad Sci USA 2014;111:8374–9.

[86] Alvarez MP, et al. Properties of sizeable [n]cycloparaphenylenes as molecular models of single-wall carbon nanotubes elucidated by Raman spectroscopy: structural and electron-transfer responses under mechanical stress. Angew Chem Int Ed 2014;53:7033–7.

[87] Alvarez MP, et al. Chameleon-like behaviour of cyclo[n] paraphenylenes in complexes with C_{70}: on their impressive electronic and structural adaptability as probed by Raman spectroscopy. Faraday Discuss 2014;173:157–71.

[88] Iwamoto T, et al. Partial charge transfer in the shortest possible metallofullerene peapod, La@C_{82}⊂[11]cycloparaphenylene. Chem Eur J 2014;20:14403–9.

[89] Hitosugi S, et al. Modulation of energy conversion processes in carbonaceous molecular bearings. Chem Asian J 2015;10:2404–10.

[90] Matsuno T, Sato S, Iizuka R, Isobe H. Molecular recognition in curved π-systems: effects of π-lengthening of tubular molecules on thermodynamics and structures. Chem Sci 2015;6:909–16.

[91] Isobe H, et al. Theoretical studies on a carbonaceous molecular bearing: association thermodynamics and dual-mode rolling dynamics. Chem Sci 2015;6:2746–53.

[92] Kubota N, Segawa Y, Itami K. η^6-Cycloparaphenylene transition metal complexes: synthesis, structure, photophysical properties, and application to the selective monofunctionalization of cycloparaphenylenes. J Am Chem Soc 2015;137:1356–61.
[93] Omachi H, Nakayama T, Takahashi E, Segawa Y, Itami K. Initiation of carbon nanotube growth by well-defined carbon nanorings. Nat Chem 2013;5:572–6.
[94] Omachi H, Segawa Y, Itami K. Synthesis of cycloparaphenylenes and related carbon nanorings: a step toward the controlled synthesis of carbon nanotubes. Acc Chem Res 2012;45:1378–89.
[95] Hirst ES, Jasti R. Bending benzene: syntheses of [n]cycloparaphenylenes. J Org Chem 2012;77:10473–8.
[96] Yamago S, Kayahara E, Iwamoto T. Organoplatinum-mediated synthesis of cyclic π-conjugated molecules: towards a new era of three-dimensional aromatic compounds. Chem Rec 2014;14:84–100.
[97] Yamago S, Kayahara E, Iwamoto T. New organic chemistry of three-dimensional π-conjugated compounds. J Synth Org Chem Jpn 2014;72:992–1005.
[98] Golder MR, Jasti R. Syntheses of the smallest carbon nanohoops and the emergence of unique physical phenomena. Acc Chem Res 2015;48:557–66.
[99] Darzi ER, Jasti R. The dynamic, size-dependent properties of [5]–[12]cycloparaphenylenes. Chem Soc Rev 2015;44:6401–10.
[100] Lewis SE. Cycloparaphenylenes and related nanohoops. Chem Soc Rev 2015;44: 2221–304.

Hongtao Liu and Yunqi Liu

4 Controlled Chemical Synthesis in CVD Graphene

Abstract: Due to the unique properties of graphene, single layer, bilayer or even few layer graphene peeled off from bulk graphite cannot meet the need of practical applications. Large size graphene with quality comparable to mechanically exfoliated graphene has been synthesized by chemical vapor deposition (CVD). The main development and the key issues in controllable chemical vapor deposition of graphene has been briefly discussed in this chapter. Various strategies for graphene layer number and stacking control, large size single crystal graphene domains on copper, graphene direct growth on dielectric substrates, and doping of graphene have been demonstrated. The methods summarized here will provide guidance on how to synthesize other two-dimensional materials beyond graphene.

4.1 Introduction

Chemical vapor deposition (CVD) is a chemical process widely used to grow high-quality, high-performance thin films. CVD is the most promising method to produce large-area, high-quality graphene, among various methods of graphene synthesis, such as micromechanic exfoliation of graphite using scotch tape, thermal decomposition of silicon carbide (SiC), reduction of graphene oxide, liquid exfoliation of graphite and even bottom-up organic synthesis [1–17]. Graphene can be grown on the surface of many transition metals (e.g., Ni, Cu) and alloys. Various gaseous, liquid or solid precursors can be used as carbon source. Tuning the growth parameters, like temperature, pressure, flow of carrier gas and hydrogen and so on, can also control the growth process. It can produce large-area, high-quality graphene films and large-sized single-crystal graphene flakes with different shapes and layers by choosing appropriate catalytic substrate and finely controlling the growth parameters. At the same time, there is a reservoir of precursors being used to synthesize graphene or doped graphene that has different properties to pristine graphene. CVD method is compatible to standard semiconductor technology, and it will facilitate the commercial application of graphene in integrated circuits (ICs), solar cells, transparent conductive films (TCFs) and so on. The insights gained in CVD graphene will be helpful for manufacturing of other two-dimensional (2D) materials beyond graphene.

For thermal CVD, the growth mechanism of graphene depends on the carbon solubility in transition metal catalysts and the growth parameters. The interaction of graphene and transition metals can generally be categorized as weak, for metals with closed *d*-shells such as Cu, and strong, for metals with open *d*-shells such as Ni [18, 19]. The weak interaction will also result in low carbon solubility in the metal and preserve the linear band structure of graphene near the Fermi level. The strong interaction will lead to high carbon solubility and destroy the linear band structure

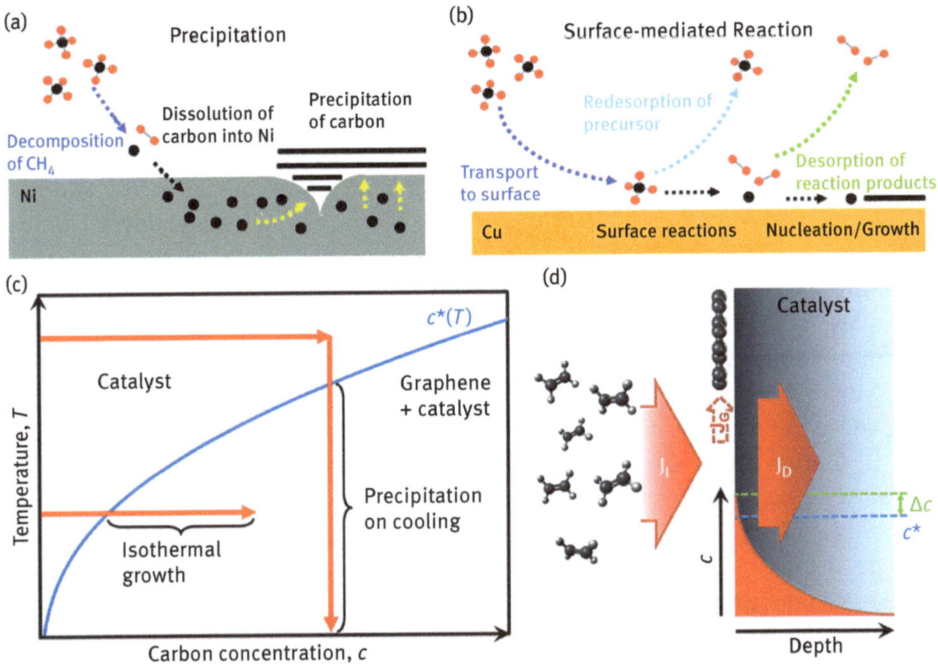

Figure 4.1: Schematic diagrams of graphene growth mechanisms on transition metals. (a) Graphene growth through segregation and precipitation. (b) Graphene growth via surface-mediated reaction [21]. (c) Carbon phase diagram illustrating graphene growth via isothermal growth and precipitation upon cooling. (d) A local carbon supersaturation (Δc) near the catalyst surface leading to graphene growth, which is a competition with carbon diffusion into the bulk metal (J_D) [17].

of graphene. There are two main graphene growth processes corresponding with the different carbon solubility. One is carbon segregation and precipitation (Figure 4.1(a)) [20]. The carbon solubility in those metals is high at high temperature. No stable metal carbide forms or metastable carbide forms and later decomposes to release the carbon. When temperature drops, the carbon solubility decreases and then carbon atoms diffuse out of the metal and bond covalently forming graphite or graphene. In this case, graphene growth on Ni is a typical example. The other growth process is surface-catalyzed growth (Figure 4.1(b)). It applies to metals in which carbon solubility is very low, such as Cu. In this process, carbon precursor adsorbs and decomposes only on the surface of the metal. It is surface mediated and self-limited. However, despite the low carbon solubility of Cu, it has demonstrated that segregation growth of graphene indeed occurs using Cu as substrate [14]. Carbon precursor decomposes on one side of the Cu surface and diffuses through the Cu foil to the other side of the Cu surface forming bilayer or multilayer graphene. Regardless of which process graphene grows, graphene nucleation and growth require carbon concentration near the metal surface to exceed the equilibrium carbon solubility, that is, carbon supersaturation near the metal surface. Carbon supersaturation can be achieved by continued carbon precursor exposure and dissociation at the metal surface. Graphene growth in this process can be considered as isothermal growth because the temperature is usually kept constant (Figure 4.1(c)). Carbon supersaturation can also be achieved by cooling the metal at a specific carbon concentration because the carbon solubility decreases due to falling of temperature (Figure 4.1(c)). Graphene grows via segregation and precipitation in this process. Using metals with low carbon solubility as catalyst, isothermal growth of graphene will be dominant. Whereas using metals with high carbon solubility, graphene can grow isothermally at continued carbon precursor exposure or can also grow during catalyst cooling down. So selecting metals with appropriate carbon solubility is essential to control the layer of CVD graphene.

Hydrocarbons are common precursors to produce graphene, and methane is the most widely used precursor. The growth mechanism can be very briefly described like this: (1) Carbon precursors diffuse to and adsorb on the metal surface. (2) Adsorbed precursors decompose into active carbons, which diffuse on the metal surface or into the bulk metal. (3) Carbon species on the surface or diffused out of the metal nucleate at high-energy sites on the metal surface forming graphene crystallites. (4) Nonactive hydrogen atoms, which desorb from the metal surface, are carried off in the form of hydrogen molecules by carrier gas. Every step is vital to graphene growth. Controllable CVD growth of graphene can be achieved by choosing appropriate metal substrate, substrate pretreatment and finely tuning the growth parameters. Here we will briefly describe the development of the control of layer number, size, direct growth on insulating substrates and doping of graphene. We will mainly focus on the graphene growth using Ni and Cu as catalysts and substrates. For Ni substrate, the layer number control will be discussed, while for Cu substrate, the controllable synthesis of large-sized single-crystal graphene and bilayer graphene with Bernal stacking (AB stacking) will be presented.

4.2 Layer Number Control

4.2.1 Layer Number Control on Ni Substrate

The process of graphene growth on transition metals with high carbon solubility, such as Ni, Co and Fe, involves the following steps: (1) carbon adsorption and decomposition on the metal surface, (2) carbon dissolution and diffusion in the bulk of the metal and (3) graphene nucleation and growth via carbon segregation and precipitation from the bulk to the surface of the metal. Carbon solubility in Ni is high and temperature dependent. Carbon segregation and precipitation from the metal is rapid during the cooling period so that single-layer and few-layer graphene form simultaneously on the metal surface. Figure 4.2 gives the whole graphene growth process on Ni, consisting of catalyst pretreatment, C dissolution, isothermal growth and precipitation. Carbon prefers segregating and precipitating at the step edges, defects and grain

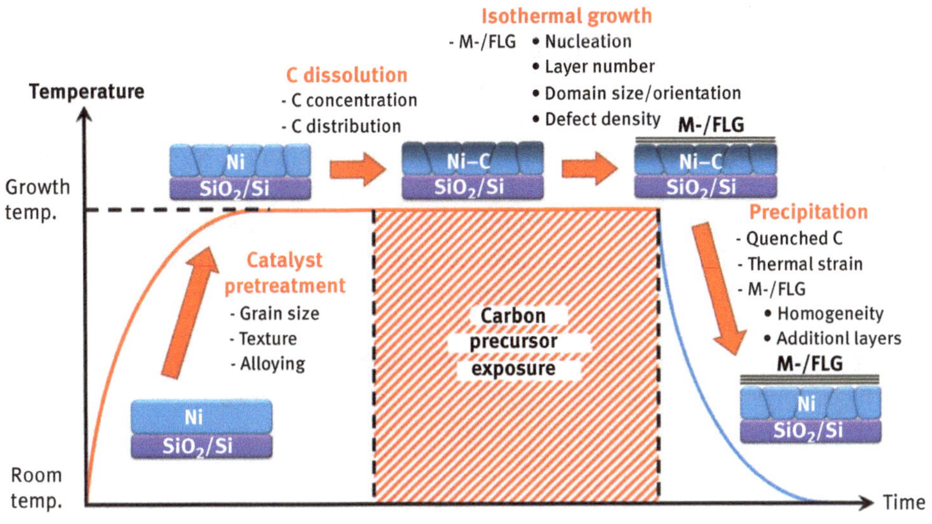

Figure 4.2: Graphene growth process on Ni. Four major phases and their key factors determining the graphene growth behavior are illustrated [22].

boundaries of polycrystalline Ni substrates [23–25]. Single and bilayer graphene prefer growing on the surface of terrace, while few-layer and multilayer graphene will form at the grain boundaries of Ni substrates. As a result, the layer number of graphene at the grain boundaries of Ni substrates is larger than within the grain boundaries. So it is challenging to synthesize uniform single-layer, bilayer or even few-layer graphene on the surfaces of metals with high carbon solubility such as Ni. Controlling the morphology of Ni is important to control the growth behavior of graphene. The diffusion of carbon to the surface of metals can also be kinetically controlled to some extent and thus uniform single-layer, bilayer or few-layer graphene will be formed.

4.2.1.1 Substrate

Substrate Selection

To uniformly grow monolayer (ML)/bilayer graphene on Ni, atomically smooth surfaces with large grain size or in the absence of grain boundaries will be essential [26]. It has been reported that ML/bilayer graphene preferentially grows on bulk Ni(111) single crystal [23, 27] or epitaxial Ni(111) thin films grown on MgO(111) [28] due to the absence of defects and grain boundaries and lattice mismatch. It has been respected that because of the registry between graphene and the Ni(111) lattice, there will be no grain boundaries in continuous graphene films formed by coalescence of neighboring graphene domains on single-crystal Ni. Considering the cost of single-crystal metals is high and the catalyst metals are usually etched away to transfer graphene to target substrate, polycrystalline metals should be chosen as substrates to grow large area graphene films with high productivity [10]. However, graphene growth behaviors on single crystal are quite different from those on polycrystalline metals [29]. Graphene growth under "real-world" conditions, such as ambient pressure and polycrystalline catalysts, should be carefully studied and it should not straightforwardly "extrapolate" from "ideal" conditions with ultra-high vacuum (UHV) on the surface of high-quality single crystals. Graphene cannot grow homogenously on the entire surface of Ni foil or polycrystalline Ni film due to the high carbon solubility at high temperature and nonequilibrium carbon segregation and precipitation upon cooling. Single-layer, bilayer, few-layer and/or multilayer graphene always appear simultaneously. As a result, most approaches to grow uniform graphene with controllable layers are trying to suppress the nonequilibrium carbon precipitation at highly active Ni surface sites, such as step edges, defects and grain boundaries.

Film Thickness

Carbon solubility in Ni is high especially at high temperature; therefore, the metal acts as a carbon reservoir. The film thickness of Ni determines the maximum carbon content dissolved in the metal and also the grain size of the metal. The thicker the metal film is, the more carbon will be dissolved at high temperature [31]. Few-layer graphene or even graphite will form on the surface of thick Ni film via carbon segregation and precipitation upon cooling. Kim et al. [30] found that Ni film deposited on SiO_2/Si by electron beam evaporation with thickness about 100 nm can be used to decrease the carbon in the metal and grow single and bilayer graphene films with a short growth time (Figure 4.3(b)). However, as shown in Figure 4.3(a), when graphene grows at high temperature for a long time, the Ni thin film tends to agglomerate forming a rough surface, which will facilitate the formation of multilayer

(a)
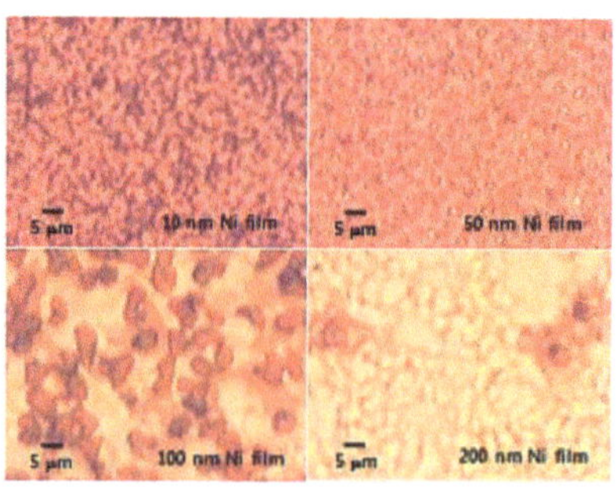

(b)
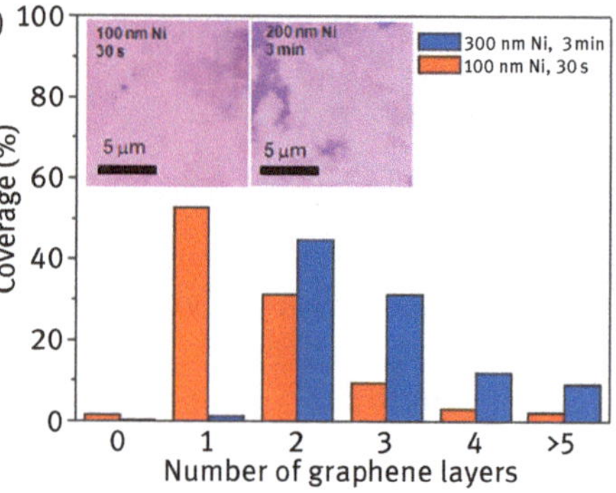

Figure 4.3: (a) Optical images of graphene grown on different thickness Ni film. As the film thickness decreases, it tends to agglomerate forming a rough surface during the growth process at high temperature and thus giving inhomogeneous graphene. (b) Number of graphene layers distributed on Ni thin films with different thickness [30].

graphene and thus give rise to an inhomogeneous graphene film [31, 32]. With the aid of plasma, ultra-thin Ni film (~50 nm or less) and low temperature can be used in a hydrogen-free plasma-enhanced CVD (PECVD) to synthesize graphene with controllable layers [33]. Low temperature determines the low carbon solubility in Ni and a thin Ni film gives a finite dissolution of carbon atoms in the Ni film, which segregate and precipitate on the surface of Ni forming few-layer graphene and suppress the formation of multilayer graphene.

Uniform single-layer graphene or graphene with controllable layers can also be grown on the surface of thick Ni films, foils or even slabs by kinetically controlled local surface saturation of carbon, without fully saturating the bulk throughout its thickness [34]. Graphene grows isothermally on the catalyst surface and the carbon diffused into the bulk will not precipitate upon cooling due to only the near-surface saturation of carbon. Carbon diffusion into the bulk metal mediates the graphene growth. Please note that this graphene growth process is implemented at low temperature (~600°C) and low pressure.

Pretreatment

Both the topography and grain orientation of Ni surface have important bearing on the layer number and domain size of CVD graphene. So, before graphene growth, it is necessary to pretreat the metal substrate. Thermal annealing will smooth the surface, increase the grain size and reduce the grain boundary of the metal, which makes the growth of graphene with controlled layer easier and decrease the thickness variation of graphene films. Annealing in H_2 can further eliminate impurities and reduce the film roughness. As a result, a thinner and more uniform graphene can be obtained on Ni surface [35]. In order to increase the percentage of Ni(111), the annealing rate of polycrystalline Ni should be deliberately controlled. It has been reported that thermal annealing of polycrystalline Ni at a slow rate favors the formation of large Ni(111) grains with less grain boundaries [23]. Larger domain size and predominated Ni(111) grains can be obtained with a prolonged annealing time. Because the morphology of graphene correlates with the metal substrate and multilayer graphene grows at the grain boundary of the metal, graphene grown on the surface of metal with large grain size will be more uniform. Moreover, graphene can also grow across the grain boundaries of the metal forming a continuous film [36–38]. All of those factors make the growing large area, uniform and thin graphene possible.

A flat Ni surface with a roughness less than 5 nm can be obtained by mechanical polish and subsequent chemical etching [39]. There are an extremely small number of graphene nucleation sites on such flat surface, and graphene can grow continuously without changing the crystal orientation over the steps and grain boundaries of the substrate. ML graphene grows at the temperature between 1,100 and 1,200 K. No graphene forms above 1,200 K and multilayer graphene grows below 1,100 K.

Alloys

In order to tune and especially to decrease the carbon content dissolved in Ni bulk and thus to suppress the formation of multilayer graphene, binary metal alloys are selected as catalysts, which preserve the catalytic properties of Ni but impart new properties to Ni [40]. However, different metals or elements added to Ni play different and important roles during graphene growth. Metals with strong metal–carbon interaction or low carbon solubility or low catalytic activity toward carbon precursor decomposition, such as Mo, Cu and Au, can be used to be alloyed with Ni. The component other than Ni in the alloy will decrease the graphene nucleation density [41] compared with the pure Ni catalyst alone by tuning the carbon content and subsequent segregation [31], or inhibiting high active sites, such as grain boundary [42]. To a certain extent, the layer number of graphene is correlated with the Ni mole fraction in the alloy, which can be modulated by the film thickness of the constituent before alloyed by thermal annealing. As a result, uniform graphene with large grain size and controllable layer number will be acquired.

Mo–Ni Alloy

Dai et al. [43] used a Mo–Ni alloy to grow uniform high-quality single-layer graphene. Ni is the catalyst to decompose carbon precursors, while Mo acts as carbon sink to trap dissolved carbons in Ni via forming stable metal carbides and to suppress the carbon segregation and precipitation (Figure 4.4). Graphene growth on this alloy catalyst is surface limited, like the growth behavior of single-layer graphene on the surface

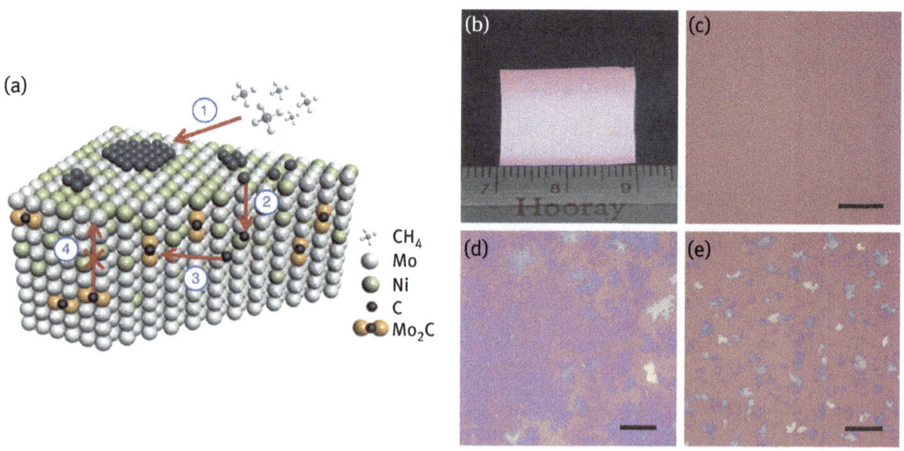

Figure 4.4: (a) Schematic of graphene growth on Mo–Ni alloy [43]. (b) and (c) Optical images of graphene on Mo–Ni alloy. Graphene uniformly covers the entire surface on the alloy. (d) and (e) Optical images of graphene on Ni surface grown under different cooling rates, which distribute inhomogeneously on the Ni surface. All scale bars are 20 μm.

of Cu. Uniform single-layer graphene films with a coverage of 100 % can be obtained over a wide range of growth conditions (different concentrations of carbon precursor, Mo film thickness, growth temperature and cooling rate), as long as there is enough Mo remaining in the alloy to react with carbon species and trap them. The catalytic metal Ni can be replaced with other catalytic metals such as Co or Fe, and other metals like W and V, which can form stable carbides, can be chosen as carbon trap metals.

Graphene nucleation and growth on this kind of Mo–Ni alloy with different Mo/Ni ratio achieved by varying the Ni film thickness deposited over Mo foils have been studied by Rümmeli et al. [44]. The study of the surface structure of the Mo–Ni binary substrates shows that during graphene growth, Ni thin films diffuse into the underlying Mo foils and the Ni films break up forming Ni islands embedded in the Mo foil at the surface (Figure 4.5). This will take place during the annealing pretreatment. When thinner Ni thin film is deposited on Mo foil, the penetration depth and lateral density of Ni islands are smaller than that of thicker one. Spatially resolved Auger spectroscopy studies reveal that graphene nucleates only on Ni islands surface rather than on Mo islands surface. This is because stable molybdenum carbide forms on Mo surface and it cannot transform into graphene. Carbon will be depleted over and near the Mo surface and carbon is also deficient on Ni surface surrounded by Mo. So the growth of graphene on this Mo–Ni binary substrate is self-terminated. The growth of multilayer graphene is suppressed and single-layer graphene with high uniformity grows over the substrate. The density of graphene nucleation sites can be tuned by altering the Ni film thickness deposited on Mo foil. When thinner Ni film is deposited on Mo foil, the density of Ni islands at the surface decreases and there will be fewer

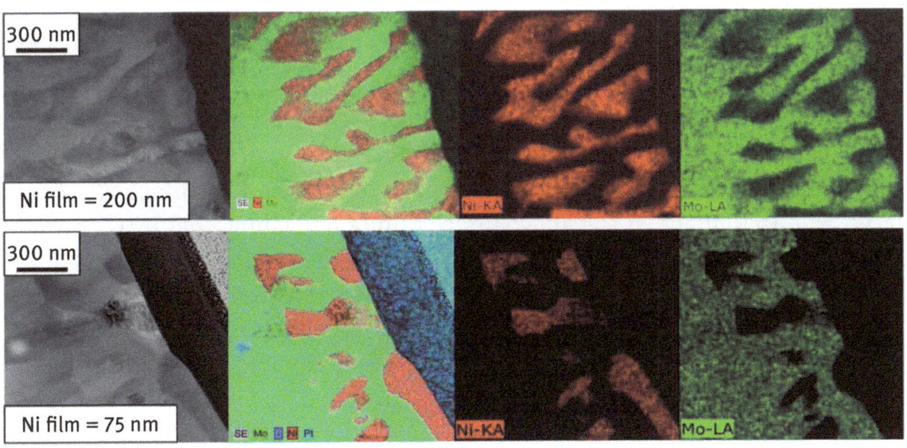

Figure 4.5: Scanning transmission electron microscopy (STEM) and energy-dispersive X-ray spectroscopy (EDS) of cross sections of Mo–Ni alloy. Ni film breaks up and diffuses into the underlying Mo foil. Ni islands form at the alloy surface. When the Ni film is thicker, the number of Ni islands and the diffusion depth are larger [44].

graphene nucleation sites, thus leading to larger graphene domain size. As carbon precursor supply continues, graphene flakes grow larger and merge with each other into a continuous single-layer graphene film.

Cu–Ni Alloy

Commercial Cu–Ni foils with a fixed Cu/Ni ratio, which have moderate carbon solubility at high growth temperature and large grain size after annealing, can be used as catalysts to deposit highly uniform graphene films with controllable layer number, for example, single layer, bilayer and few layer, through carbon segregation and precipitation by tuning the growth temperature, carbon precursor exposure time and cooling rate [42, 45]. The moderate carbon solubility in Cu–Ni alloy foil makes controlling the graphene thickness much easier. Not only the graphene layer number but also the substrate cover level can be controlled by varying the growth parameters. The surface coverage and uniformity of graphene films increase as the growth temperature increases, due to the enhanced carbon solubility and diffusivity at high temperature. The surface coverage can be attributed mainly to two factors: density of graphene nucleation sites, which is mainly temperature dependent, and the size of individual graphene sheet, which is mainly sensitive to the carbon precursor exposure time. For example, uniform ML, bilayer (about 70 % coverage and mainly AB stack) and multilayer graphene films (2–5 ML) were obtained at the growth temperature of 975°C, 1,000°C and 1,030°C, respectively, at the same cooling rate (100°C/s) for "70–30" commercial Cu–Ni foils (weight percent: 67.80 % Cu, 31.00 % Ni) [42]. Slow cooling rate favors the AB-stacked bilayer graphene. And a low-carbon precursor flow rate and a long exposure time will enhance the uniformity of bilayer graphene. Multilayer graphene or thick graphite will deposit when more carbon atoms dissolved into the alloy foil and when they have enough time to segregate and precipitate during the cooling period.

Cu–Ni ratio can be tuned by adjusting the thickness of the deposited Cu and Ni film before pre-annealing to form an alloy [46]. As the concentration of Cu increases in Cu–Ni alloy, graphene growth mode changes from segregation mode on pure Ni and Ni-rich surface to surface catalytic mode on pure Cu and Cu-rich surface [47]. Moreover, Cu will accumulate at the alloy surface during annealing, that is, Cu concentration at the top surface is higher than that in bulk. The Cu-rich surface suppresses the carbon diffusion into the alloy bulk. As a result, graphene with controllable layer number (ML, bilayer and trilayer) can be obtained through this surface catalytic mode at different growth pressures and carbon precursor (methane) partial pressures. And the yield of AB-stacked bilayer graphene is strongly influenced by the Cu/Ni ratio and growth temperature as shown in Figure 4.6, where both are achieved by changing the carbon concentration and graphene growth mode. However, surface concentration enhancement of Ni at Cu–Ni surface is also confirmed by angle-resolved X-ray photoelectron spectroscopy after graphene growth at high temperature (1,050°C), which probably results from the high vapor pressure of Cu (about three times higher than Ni)

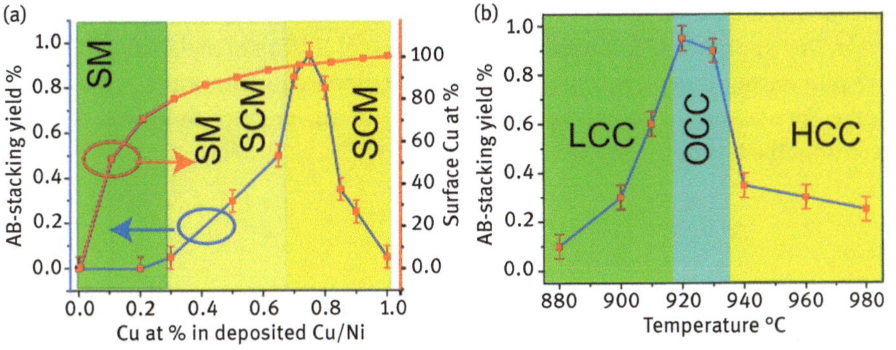

Figure 4.6: Graphene growth mode and the yield of AB stacking bilayer graphene with the relationship of (a) the ratio of Cu in Cu–Ni alloy and (b) growth temperature [47]. SM, segregation mode; SCM, surface catalytic mode; HCC, high carbon concentration; LCC, low carbon concentration; OCC, optimal carbon concentration.

at high temperature [48]. In a word, the growth of large area, uniform graphene with controllable layer number and stacking order (>1 L) needs to optimize the metal substrate and growth conditions.

Large-area single-crystal graphene has grown by locally feeding carbon precursor on an optimized Cu–Ni alloy [46]. Local precursor feeding through a quartz nozzle makes it possible to grow large-sized single-crystal graphene from only one nucleus and even at designated locations. In contrast, multiple site nucleation and thus polycrystalline graphene usually occur in a conventional CVD process. The optimized ratio of Cu/Ni is 85/15, which enables a nickel-mediated isothermal segregation growth of graphene at a fast rate due to the Ni-enhanced catalytic power and abundant carbon atoms dissolved in the alloy. This method avoids long-time growth of graphene and ~1.5-inch hexagonally shaped single-crystal ML graphene can be synthesized in 2.5 h at 1,100°C.

Cu–Ni alloy can also be used as substrate to grow wafer size uniform ML and bilayer graphene through segregation without introducing extraneous carbon precursors [31, 49]. Cu in the alloy was used as segregation controller, while Ni was used as carbon source and graphene thickness controller. By tuning the Cu/Ni ratio (Ni film thickness), graphene with controllable layer number was obtained at appropriate annealing temperature and segregation time, while the bilayer graphene here obtained is non-AB stacking. Carbon content in metal substrate and layer number of subsequently precipitated graphene can also be controlled precisely by varying the fluence of the carbon ion implantation combined with Cu–Ni binary metal substrate [50]. All the carbon ions implanted in Ni-coated Cu foil are expelled toward the substrate surface and precipitate forming graphene during annealing, which is enhanced by inward diffusion of Ni and outward diffusion of Cu leading to a decreased carbon solubility in Cu-rich surface. This approach also applies to Co–Cu alloy but using an exogenous carbon source in an atmospheric pressure CVD (APCVD) system [51]. Moreover, this approach is less sensitive to the thermal history of metal substrate and cooling rate during the growth of graphene and compatible with large-scale microelectronics processing. It might be a great practical convenience to grow graphene with controllable layer number.

Au–Ni Alloy

Graphene growth using Au–Ni alloy catalyst can be performed at low temperature (~450°C) [34, 41]. Due to the low carbon solubility in Ni at low temperature, graphene was grown isothermally on the surface of alloy catalyst and it was not a segregation and precipitation process. Au decorates the majority active surface sites of Ni, such as step edges, and the presence of Au decreases the graphene nucleation density and thus reduces the graphene thickness variation yielding more uniform graphene films and larger domains. Being different from Mo, Au and Cu cannot react with carbon and no stable carbide is formed. As a result, there are no carbon traps in Au-Ni alloy, and multilayer graphene or even graphite will grow when more carbon is supplied.

Calculations on Alloys for Graphene Growth

Density functional theory (DFT) calculations of Au alloying on Ni(111) are consistent with the experimental works. It is energetically favorable for Au atoms to substitute Ni atoms at the step edges on Ni(111) due to the least coordination numbers of Ni atoms at the step edges, which will passivate the highly active step edges and makes the step edges much more like the Ni terrace in catalyzing graphene growth [52]. The uniformity of Au-modified Ni(111) step edges and clean Ni(111) terrace in carbon adsorption, diffusion and segregation will reduce the density of the nucleation centers and facilitate the growth of large area and uniform graphene film. Nevertheless, before all the Ni atoms at the step edges are replaced by Au atoms, the Au atoms begin to substitute Ni atoms on the terrace of Ni(111), because high density of Au atoms at the step edges will lead to a high formation energy which is equivalent to the formation energy of Au atoms deposited on Ni(111) terrace [52]. Au atoms modified on Ni(111) terrace can lower the energy barrier of carbon atoms incorporation onto graphene growth front and thus facilitate the growth of armchair/zigzag (AC/ZZ) graphene edges [53]. As a consequence, high-quality graphene will be obtained at a lower growth temperature on Au-alloyed Ni(111).

Calculations have also demonstrated that other metals, such as Mn, with strong metal–carbon interaction and repulsive with each other atoms on Cu(111) surface, can be alloyed in Cu forming a $\sqrt{3} \times \sqrt{3}$ R30° superstructure [54]. Carbon clusters with only one orientation will seed on the surface of this Mn–Cu(111) superstructure due to the strong Mn–carbon interaction. Those carbon clusters subsequently grow larger and coalesce with each other into a graphene sheet with no grain boundaries. Not only the alloyed metals on the uppermost metal surface will affect the nucleation and growth of graphene, but also the subsurface alloy will also affect the graphene growth behavior by modulating the topmost metal–carbon coupling and carbon–carbon interaction. For example [55], for metal Rh(111), when it is alloyed with relatively inert transition metals right side of Rh in the periodic table, such as Pd and Ag, carbon atoms repel with each other when they get closer. They prefer separation due to the repulsive effect and graphene nucleation will be suppressed. However, alloying with those late transition metals will suppress the orientation disordering of graphene domains. When Rh(111) is alloyed with more active transition metals on the left side of Rh, such as Tc and Ru, the carbon–substrate interaction will weaken and two carbon atoms energetically and kinetically tend to form a dimer on the topmost host Rh(111), which will facilitate the nucleation of graphene, but the orientations of graphene domains will be disordering.

4.2.1.2 Growth Parameters

The correlation of hydrocarbon pressure with the epitaxial growth of graphene on Ni(111) surface will shed light on CVD growth of few-layer graphene on polycrystalline Ni catalyst [56]. There is a strong interaction between epitaxial graphene and Ni(111), which leads to a depletion of dissolved carbon closed to the catalyst surface and prevents the formation of additional layer of graphene. Single-layer graphene is thus commonly achieved in a wide range of low-exposure pressures. A further increase of carbon precursor pressure, dissolved carbon accumulates in the bulk catalyst, and the concentration of dissolved carbon near the catalyst surface increases, which weakens the graphene–Ni interaction causing a formation of rotated graphene and enabling the nucleation and growth of additional graphene layers.

To minimize the carbon dissolved in Ni and thus reducing the graphene thickness variation, highly diluted hydrocarbon should be introduced into the growth system [34, 36]. Hydrogen (H_2) is also important to the graphene growth. As high H_2/CH_4 ratio is used, the coverage of multilayer graphene decreases and the percentage of single-layer graphene increases [57]. However, no graphene grows without H_2 and too much H_2 will etch the graphene and suppress the formation of graphene. The amount of H_2 also plays a key role in determining the layer number of graphene.

Growth temperature determines the activity of the metal catalyst, the sorption/desorption and decomposition of carbon precursors, the carbon solubility and diffusion rate of carbon in the bulk and on the surface [31, 58]. At high temperature (>900°C), the carbon solubility in Ni is high. When the pressure of carbon precursor is high, more carbon will be dissolved in the bulk of Ni and there is no graphene formed on the surface of Ni at such high temperature, but only dilute carbon atoms [59]. Graphene will form upon cooling by segregating and precipitating dissolved carbon as a consequence of decreased solubility of carbon in Ni. The majority of the graphene is rotated or mismatched with Ni(111) surface at high temperature, which, due to the weak interaction of the Ni(111) and mismatched graphene, favors the precipitation and segregation of carbon from bulk to the surface forming a second layer of surface carbide or graphene [60]. This process depends on the concentration of carbon in the bulk Ni. It is challenging to suppress carbon diffusion to the surface upon cooling when the carbon precursor pressure is high. Multilayer graphene will difficult to be avoided. However, at a relatively low temperature (~600°C or ~550°C), carbon dissolution into the bulk is limited. Self-limiting surface growth or isothermal growth of graphene on polycrystalline Ni film dominates and single-layer graphene will be obtained [22]. At a much lower temperature (<500°C), a single layer of nickel carbide forms, which inhibits the nucleation and growth of high-quality graphene [27, 61]. However, the role of the surface carbide is open to discussion. It has been reported that a single layer of nickel carbide could transform into graphene [56, 60, 62]. Graphene grown at very low temperature is usually not continuous and defective, and the majority of the graphene is in registry with Ni(111) surface. Most of the above-mentioned processes to grow graphene on the surface of Ni(111) were carried out in UHV, in which the pressure of the carbon precursors is very low and it is easy to keep the Ni surface clean. In ambient pressure growth systems, it is difficult to grow large-area graphene with controllable layers at high temperature due to the high pressure of carbon precursors and oxidation of Ni. Other metals with low carbon solubility and weak metal–graphene interaction such as Cu and Pt will be alternatives to Ni to be used as substrate to grow uniformly large-area high-quality graphene with controllable layer.

The layer number of graphene grown on the surface of Ni depends not only on the carbon content dissolved in the metal but also on the cooling rate of the metal. When low concentration of carbon precursor exposes to Ni, thus low carbon concentration in the bulk metal, the coverage of single-layer and bilayer graphene increases at a slow cooling rate (<25°C/min) [63]. The low carbon content in Ni will suppress the carbon segregation on the Ni surface and the slow cooling rate will promote the carbon segregation under conditions close to equilibrium. Both of those conditions favor the growth of single-layer and bilayer graphene. The density of nucleation sites of multilayer graphene decreases but the thickness of multilayer graphene increases at a slow cooling rate. The single-layer and bilayer graphene will grow across the grain boundary of the metal substrate and the size of the graphene is limited only by the size of metal film and the CVD chamber. However, when higher concentration of carbon precursor is introduced into the growth system, multilayer graphene with fewer single-layer and bilayer graphene grows on the metal surface independent of the cooling rate. Although the uniformity of CVD graphene improves using a slow cooling rate, there are still thicker-layer graphene appears among single-layer, bilayer and few-layer graphene [43].

It has also reported that at an extremely fast cooling rate, the dissolved carbon atoms do not have enough time to diffuse to the Ni surface (Figure 4.7) [35]. The quench effect allows no graphene to form. Graphite will form at a fast cooling rate (~20°C/s)

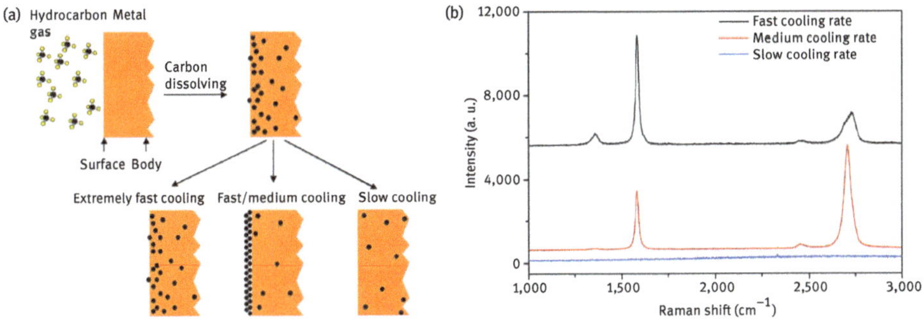

Figure 4.7: The relationship between cooling rate of Ni substrate and (a) the related carbon segregation model and (b) the Raman spectra of as-grown graphene [35].

and there are many defects in those graphite due to not enough time to crystallization at the high cooling rate. The layer number of graphite is reduced and the quality is improved with a slower cooling rate. Few-layer (less than 4) graphene can be obtained at a medium cooling rate of ~10°C/s. However, at a slow cooling rate (0.1°C/s), there is also no graphene formation because of the dissolved carbon atoms diffusing into the bulk of the metal and no carbon segregating and precipitating on the Ni surface. Those results have been verified by Kim et al. [30], in which a fast cooling rate (~10°C/s) can suppress the formation of multilayer graphene. So controlling the cooling rate controls not only the layer number of graphene but also the graphene quality. Because of fast cooling, the formation of graphene is also very fast. Bilayer or few-layer graphene with AB stack is energetically more favorable, and turbostratically arranged layers are usually obtained due to the nonequilibrium graphene process [64]. However, besides cooling rate, other factors such as carbon content in the Ni film, Ni film thickness and surface roughness are also critical to control the layer number of grown graphene.

The atmosphere in which the sample is cooled plays a key role in graphene growth too. Before cooling, exposing the Ni film to H_2, carbon layers formed on the Ni surface will be removed by the etching effect of H_2 with the assistance of Ni at high temperature, and carbon content in bulk Ni is also reduced. Carbon segregation and precipitation from Ni bulk to the surface is strongly suppressed due to the reduced carbon content in Ni. Wafer-scale uniform layer-controlled graphene can be obtained by rapidly cooling and finely tuning the Ni film thickness, carburization and H_2 exposure at ambient pressure [65]. A coverage of ~90 % single-layer and ~6 % bilayer graphene can be obtained. The sheet resistance of a 3–4 layer graphene film is as low as ~100 Ω per square and ~90 % transmittance at 550 nm. The use of H_2 during the cooling process can also etch away the weakly bounded carbon atoms by forming hydrocarbons like methane to obtain graphene free of deposits or flakes. Nevertheless, the hydrocarbon in the system will balance the etching effect.

4.2.2 Bilayer Graphene and Stacking on Cu Substrate

Graphene growth behavior varies on the surface of different transition metals. Graphene growth on the surface of Cu is surface limited and is predominantly single layer [66]. When the (entire) surface is covered by graphene, the second-layer graphene cannot easily grow above (in the absence of the catalyst) or below (deficient or no carbon precursor feeding) the first graphene layer. And the stacking disorder appears when the second layer grows across the grain boundary of the first-layer graphene. Multilayer graphene always precipitated "simultaneously" on the grain boundaries and steps of Ni leading to disordered stacking of graphene [30, 36]. Therefore, controllable growth of large-area single-crystal ML graphene, which will be discussed in the next section, is the prerequisite to synthesize large-area, uniform, orderly stacked bilayer graphene.

Although consensus on the growth of the second-layer graphene on top or below the first-layer graphene has not been reached, the location of the second-layer graphene will depend on the technique and growth conditions [21]. The possible growth mechanisms of bilayer graphene are illustrated in Figure 4.8. It is challenging

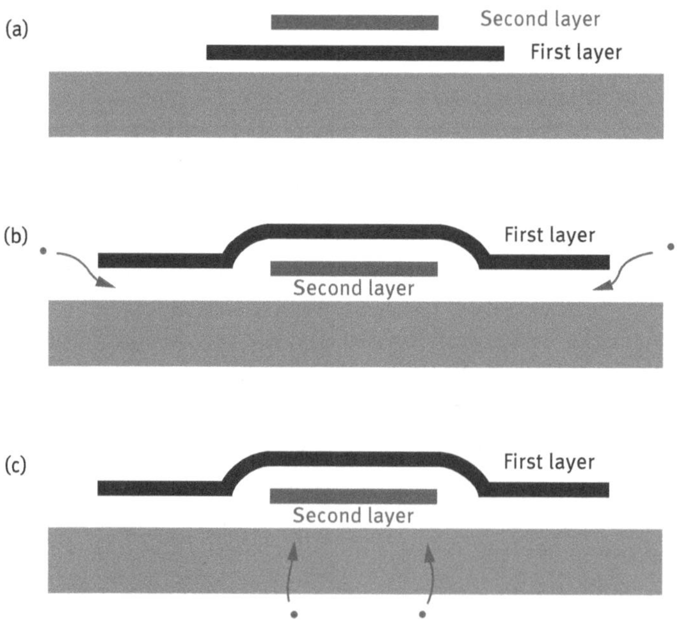

Figure 4.8: Possible bilayer graphene growth mechanisms. (a) The second-layer graphene growth on top of the first-layer graphene. The second-layer graphene growth beneath the first-layer graphene feeding by carbon species diffused from the graphene–metal interface (b) and the bulk metal (c).

to synthesize bilayer graphene in large scale, especially for large-area uniform AB-stacked (Bernal) bilayer graphene, in which tunable band gap opens by applying a perpendicular electric field for breaking the inversion symmetry [67–71] or by chemical doping [72, 73].

Bilayer graphene has been usually synthesized by catalyst engineering, kinetic control [74–77] or a combination of both [34]. For example, as mentioned earlier, alloy substrates can be used to grow graphene with controllable layer number by carefully tuning the alloy composition and the growth conditions [31, 42, 45, 47, 49]. However, the resulting bilayer graphene often consists of a mixture of AB-stacked and randomly stacked bilayer graphene.

In a classical CVD system using Cu as a catalyst, the self-limited growth can be broken and the second-layer graphene typically grows beneath the first-layer graphene by surface reaction, that is between the Cu surface and the top graphene layer (Figure 4.8(b)), at a low pressure [76], high H_2/CH_4 ratio or low flux of methane with a decreased growth rate [78]. As a result, in order to obtain a large-sized bilayer graphene, a long growth time is sometimes necessary. Both graphene layers share the same nucleation center and most bilayer graphene is AB stacked due to the strong interaction of the two layers. The second-layer graphene stops growing when the first-layer graphene coalesce into a continuous film, because the carbon diffusion channel at the edges of the first graphene layer will be shut [76]. Although AB-stacked bilayer and trilayer graphene has been synthesized at atmospheric pressure by finely tuning the growth parameters, the procedures is complicated and other simple and effective methods should be developed to grow bilayer and trilayer graphene in large scale [79].

Epitaxial growth of the second-layer graphene will be the most straightforward approach to synthesize bilayer layer graphene. However, for Cu catalyst, the catalyst is passivated by the covered graphene. In order to grow the second-layer graphene, additional catalyst or process must be introduced to decompose the carbon precursor and generate active carbon species.

A two-step growth process has been developed to grow bilayer graphene by Yan et al. [80]. Uniform ML graphene was first grown on the surface of Cu foil serving as the substrate for the following epitaxial growth of the second-layer graphene. A fresh Cu foil was placed upstream to catalyze the decomposition of methane to feed the second-layer graphene epitaxial growth on the surface of the first-layer graphene. As the growth time increases, the lateral size of the second-layer graphene reaches several tens of micrometers and the shape changes from hexagons to hexagrams or round risks. Most bilayer graphene is AB stacked with a coverage as high as 67 % confirmed by optical microscope, transmission electron microscopy (TEM), selected area electron diffraction (SAED) and Raman. The epitaxial growth of the second-layer graphene and the low coverage can be attributed to the fact that carbon radicals or fragments generated by Cu catalytic decomposition of methane prefer to attach to the graphene edges with dangling bonds rather than on the graphene surface forming an additional layer. Liu et al. [81] have found that a relatively high H_2/CH_4

ratio, high temperature and very low pressure are required to improve the bilayer graphene coverage and AB-stacking ratio in a similar approach to epitaxially grow bilayer graphene. A relatively high H_2/CH_4 ratio will keep the balance of graphene growth and etching, which leaves the upstream Cu foil uncovered by graphene and keeps the catalytic activity. Moreover, the high ratio of H_2 will terminate the graphene edges by hydrogen, which allows the carbon diffuse easily and fast into the region beneath the top graphene layer and thus favors the growth of bilayer or few-layer graphene [82]. Using large-area single-crystal ML graphene as template, large-sized (up to 300 um width) AB-stacked bilayer graphene can be achieved by using a high H_2/CH_4 ratio (4,400) [83]. The growth rate is very slow as it is a diffusion-limited growth. However, the very low pressure is not beneficial for the growth of large-sized bilayer graphene. To remedy this issue, a low-pressure (1 mbar) nucleation and high-pressure (5 mbar) growth procedure are employed to obtain a high coverage (~99 %) and a high AB-stacking ratio (~90 %) bilayer graphene [81]. The introduction of high ratio/concentration of H_2 will lower the growth rate of the second layer and it takes long time to obtain large-sized bilayer graphene. Using ethanol as carbon precursor, epitaxial growth of AB-stacked bilayer layer graphene can be achieved without introducing extra Cu as catalyst due to the catalytic effect of ethanol but require careful control of the growth parameters [84].

The catalytic transition metal thin film can be used as carbon delay to grow the second-layer graphene. That is, the time of carbon reaches to one surface of Cu is different. The early arrived carbon forms the first-layer graphene and the lately arrived carbon that must diffuse through the bulk metal thin film gives the second-layer graphene. In other words, one side of the Cu foil acts as the first-layer graphene growing substrate and the other side acts as catalyst to decompose carbon precursor to diffuse through the bulk foil to feed the growth of the second-layer graphene. Because of the exposed metal surface acting as catalyst to decompose the carbon precursor, it should make sure that the catalytic surface of the metal should not be fully covered or passivated by graphene before the growth of the bilayer graphene on the other side is completed. As mentioned earlier, it is very important to maintain the activity of the catalyst and control the carbon content and the precipitation time to grow large-area bilayer layer graphene.

Fang et al. [85, 86] have shown that bilayer, trilayer or quadlayer graphene can be grown on the outside surface of a Cu enclosure at low pressure. The newly formed graphene grows by capturing carbon species decomposed on the inside surface of the Cu enclosure and diffused through the Cu foil. So it is a carbon diffusion-limited growth model. The growth time, Cu foil thickness and the time of full coverage of the inside surface will determine the size, and coverage of the bilayer graphene on the outside surface. Graphene growth behavior on both sides of a Cu pocket has demonstrated once again that graphene growth on inside and outside of Cu surface is correlated, that is single layer inside versus single layer outside, bare Cu inside versus bilayer or multilayer outside. The bare inside surface of Cu allows the active carbon species to diffuse from the inside to the outside and then nucleates and grows the second-layer graphene underneath the top graphene layer [87]. The full coverage of outside Cu surface by graphene will not terminate the growth of the second-layer graphene. Therefore, graphene with different layer numbers can be obtained by this approach.

It has been demonstrated that oxygen on Cu surface plays a critical role in the growth of bilayer graphene [88]. A Cu pocket was designed to grow bilayer graphene on the outside surface (Figure 4.9(a)). The second-layer graphene nucleates and grows by feeding carbon atoms that are formed by complete dehydrogenation of methane with the assistance of inside surface oxygen, diffusion through the bulk Cu and segregation to the outside Cu–first graphene interface. Both the first-layer and the second-layer graphene grow isothermally rather than by precipitation as verified by isotope labeling. The weak interaction between the two-layer graphene enables the AB stacking, while grain boundary in the first graphene layer is the cause of the

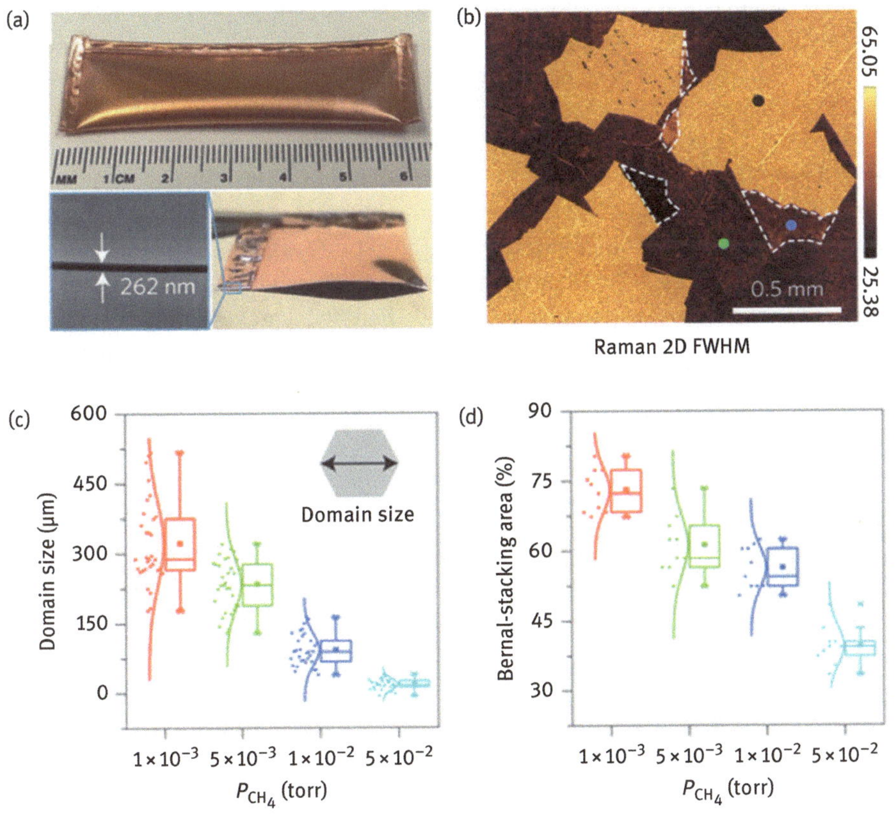

Figure 4.9: Bilayer graphene growth on the outside surface of an oxygen-rich Cu foil pocket [88]. (a) Optical image of the pocket. (b) Raman 2D mapping of as-grown bilayer layer graphene, in which bright areas are Bernal stacking. Bilayer graphene domain size (c) and Bernal-stacking ratio (d) as a function of methane partial pressure, P_{CH4}.

stacking disorder (Figure 4.9(b)). Analogous to the case of single-layer graphene growth, decreasing the pressure of methane will reduce the density of graphene nucleation and thus increase the domain size of the second layer graphene (Figure 4.9(c)). As a result, the ratio of AB stacking increases with the decrease in methane pressure (Figure 4.9(d)). High methane pressure will lead quickly to full coverage of graphene on the inside surface and thus sparse bilayer graphene on the outside surface. Long-time exposure of methane will grow graphene >2 layers. This work points out the key role of surface oxygen in the growth of bilayer graphene and provides the evidence of underneath growth of the second-layer graphene.

Bilayer graphene has been synthesized directly on the surface of silicon oxide using deposited Ni thin film on top of it as a catalyst (Figure 4.10) [89]. Various carbon precursors can be used to grow bilayer graphene. They decompose, diffuse, segregate and precipitate at the constrained Ni–silicon oxide interface, forming bilayer graphene. This growth process resembles the graphene growth behavior on the top surface of Ni, but here the constrained space plays a vital role in determining the layer number. Most of the bilayer graphene (~95 %) are non-AB stacked with twist angles from 0° to 30°. Using polymers or self-assembled MLs (SAMs) on a dielectric substrate as carbon precursors and deposited Ni atop of them as catalyst, bilayer and multilayer graphene can be synthesized by tuning the film thickness or the alkyl length of the SAM [90, 91]. A similar method but using 300 nm Cu thin film as catalyst has also been developed to grow few-layer graphene [92]. AB-stacked bilayer graphene are randomly observed and other stacking configurations also exist.

In contrast to the large number of reports on CVD growing AB-stacked bilayer graphene, twisted bilayer graphene with specified twist angles has been rarely reported. Twisted bilayer graphene has been synthesized by combining surface reaction and segregation growth on the surface of Ni(111) [93]. But this kind of bilayer graphene is a mixture of twisted bilayer graphene with different twisted angles. It has

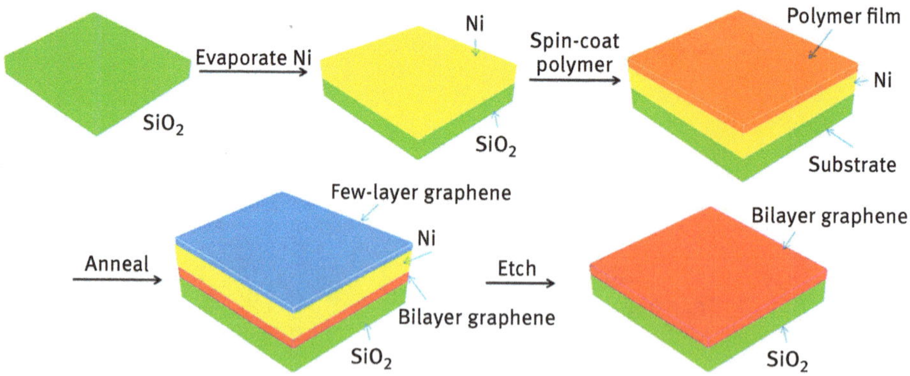

Figure 4.10: Bilayer graphene growth on silicon oxide using solid carbon precursors and a Ni film as a catalytic buffer layer [89].

been reported that twisted bilayer graphene prefers to grow underneath the rotated graphene domains relative to a Ni(111) substrate rather than the no-rotated ones [94]. The formation of twisted bilayer graphene may be attributed to the strong interaction of graphene and Ni, which gives the carbon atoms less mobility and cannot form the thermodynamic stable configurations. Twisted bilayer graphene has also been synthesized using APCVD by optimizing growth parameters, such as temperature, H_2 pressure and CH_4 concentration [77].

4.3 Large-Sized Single-Crystal Graphene on Cu Substrate

Due to the surface-mediated and thus self-limited graphene growth process, as well as its low cost, availability, reliability and scalability, Cu has been considered as an ideal substrate to synthesize large-sized uniform single-layer graphene. However, graphene prepared by this CVD method is typically polycrystalline [95]. Graphene grain boundaries form when graphene flakes nucleated and grown with different lattice orientations coalesce to a continuous film. Although graphene grain boundaries have distinct properties and can be used to tailor the properties of graphene, they will deteriorate graphene's mechanical, electrical and other physical properties [96–98]. It is important to grow large-sized single-crystal graphene without grain boundaries and it has been demonstrated that the quality of large-sized CVD graphene can be comparable to mechanically exfoliated graphene [99, 100].

In order to obtain large-sized single-crystal graphene, decreasing the graphene nucleation density, which makes graphene flakes have enough space to grow to large size, and at the same time ensuring the continuous growth of graphene without forming additional nuclei is one of the key principles [101]. Lowering the nuclei density can be achieved by pretreating the Cu substrate and carefully controlling the growth parameters, such as temperature, pressure, flow rate of carbon precursor and ratio of hydrogen and carbon precursor. Another approach is to grow graphene flakes at different nucleation sites but with same lattice orientation, suitable shape and arrangement, and then coalesce seamlessly to a continuous film free of grain boundaries [15, 102]. It should be noted that the formation and structure of graphene grain boundaries will depend on the growth model. And it should be carefully characterized before identifying the single-crystal graphene flakes, especially for graphene flakes with irregular shapes [103, 104].

4.3.1 Substrate

4.3.1.1 Facets

Cu–graphene interaction is relatively weak. Graphene growth on Cu surface generally cannot be considered as epitaxy because there is no obvious or weak correlation between the lattice of graphene and Cu. Graphene can grow continuously across the imperfections such as grain boundaries, steps, dislocations, clusters and vacancies of Cu and maintain the lattice orientation. On least mismatched Cu(111) (3.9 %), there are still rotations of lattice orientation and grain boundaries in graphene [105, 106]. The grain size, crystallographic orientation, surface roughness, purity, surface oxygen and the geometry of the Cu substrate will affect the size, shape and layer number of graphene sheet. In general, Cu substrate with large grain size, high purity and smooth surface will favor the growth of large-sized single-crystal graphene domains.

The weak correlation between the shape and orientation of graphene domains and the crystallographic orientations are simply shown in Figure 4.11 [107]. Zigzag edges of graphene grown on the surface of Cu(101), Cu(001) and Cu(111) tend to align parallel to the direction of Cu<101>. For Cu(101) and Cu(111), there is only one direction, that is Cu[−101] (Figure 4.11(c) and 4.11(e)). While for Cu(001), there are two directions, that is Cu[−110] and Cu[110], which are perpendicular

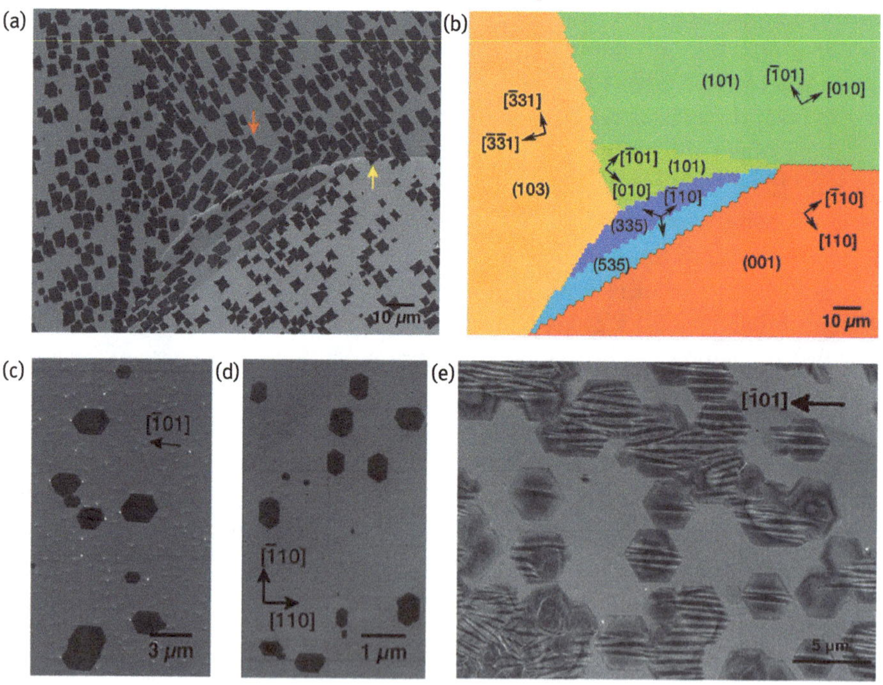

Figure 4.11: Graphene domain shapes and orientations on polycrystalline Cu foil [107]. The arrows in (a) show that graphene domains grow across the Cu grain boundaries, and their shapes change. (b) Electron backscatter diffraction (EBSD) map of the polycrystalline Cu in (a). Graphene domain orientations on (c) and (e) Cu{111}, and (d) Cu{001}. One edge of hexagonal graphene domain is aligned with Cu[−110] and Cu[110] on Cu{001}, which are perpendicular to each other.

to each other (Figure 4.11(d)). The shapes of graphene domains on different Cu facets are also different. For example, the shape of graphene domains on the surface of Cu(111) or the surface with orientations close to Cu(111) is hexagonal. Four-lobed graphene domains on Cu(101) and Cu(001) and six-lobed and star-shaped graphene domains appear on high-index Cu facets. Other experiments have also verified those findings [108–110]. Graphene nucleates and grows more easily on Cu(111) and the quality of graphene grown on Cu(111) is higher than that grown on other surface due to the high carbon diffusion rate and improved carbon adsorption [111]. Cu foil is the most popular substrate to grow graphene. However, it is polycrystalline and contains many structures, such as different facets, grain boundaries and annealing twins. Cu(111) is the lowest energy Cu surface and the (111) surface is enriched in Cu foil after high-temperature annealing [111]. Inspired by those results, we can grow graphene domains with different shapes and orientations by selecting or engineering the Cu orientation. Pretreatment of Cu foil by thermal annealing to product Cu(111) facets will be helpful for the production of large-sized graphene. Graphene domains grown on Cu(111) with alignment in one direction will merge seamlessly and without grain boundaries into a large sheet [101, 112–114]. Nevertheless, the shape of graphene domains also largely depends on the growth parameters.

4.3.1.2 Roughness

The roughness of Cu foil is another factor for the growth of large-sized graphene. On the one hand, graphene should be transferred onto target substrate to realize its various applications. Wrinkles, discontinuity and damage, such as cracks and tears, occur when graphene grown on roughness surface is transferred onto a substrate. On the other hand, nuclei density decreases on smooth Cu surface and the homogeneity of graphene enhances. For example, the roughness of Cu foil reduced by a factor of 10–30 after electropolishing, leading to that uniform high-quality >95 % single-layer graphene with room temperature hole mobility enhanced by a factor of 2–5 was obtained [115]. Due to the reduction of nuclei density (from, e.g., 2.4×10^4 ea/mm^2 on unpolished Cu foil to 5.5×10^3 ea/mm^2 on polished Cu foil), the size of graphene domains increases with a higher growth rate on polished Cu foil [116]. The size of single graphene domains further extends to millimeter size (~2.3 and ~4.5 mm^2) after electrochemical polishing and then high-pressure high-temperature annealing of Cu foil by eliminating sharp wrinkles, steps and defects to improve the quality of Cu foil [117]. The quality of as-prepared graphene is comparable to that of exfoliated graphene showing a room temperature hole mobility as high as 11,000 cm^2/Vs. Seamlessly stitching oriented hexagonal graphene domains into large-area (6 cm × 3 cm) single-crystal graphene without grain boundary was achieved on Cu(111) foil pretreated by high-temperature annealing and subsequent chemical–mechanical polishing [112].

In order to grow large-sized graphene, low pressure and high temperature are typically required. At such high temperature, Cu evaporates and sublimates, which are elevated at low pressure. As a result of this, the Cu surface becomes rough, resulting in a decreased growth rate and degraded quality of graphene. To suppress the evaporation, Cu foil enclosure or tube was designed and large-sized graphene domains grow on the inside surface with a low nuclei density due to the elimination of Cu surface evolution caused by evaporation and sublimation and thus smooth surface, absence of SiO_2 particles and low pressure of carbon precursor [118, 119].

4.3.1.3 Liquid Metal

Liquid metal will be another selection to grow large-sized graphene [40]. The processing history as well as the imperfections (grain boundaries, steps, etc.) of Cu are eliminated forming a quasi-atomic smooth homogeneous surface in liquid state. Geng et al. [120] demonstrated that uniform, single-layer, large-sized and self-aligned hexagonal graphene flakes can be synthesized on liquid Cu surface (Figure 4.12). Graphene domain size larger than 200 μm has also been produced by using liquid Cu as catalytic substrate and combining controlling the amount of hydrogen and the H_2/CH_4 ratio

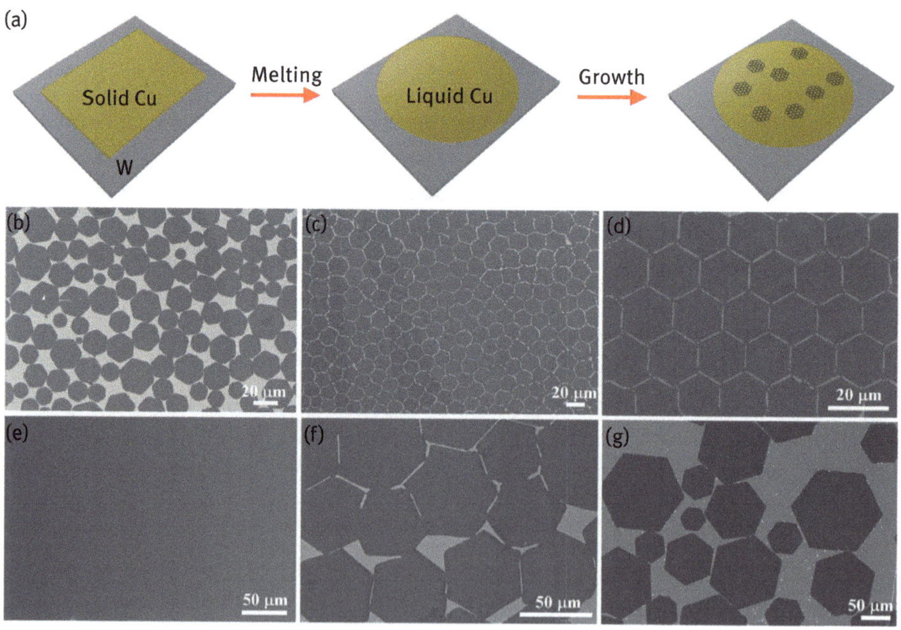

Figure 4.12: Hexagonal graphene domains growth and self-assembly on liquid Cu [120]. (a) Schematic of the growth process. (b)–(d) Surface coverage of graphene domains with an increasing density. At high density, graphene domains self-align into a compact ordered structure. (e) Graphene domains extend to a continuous graphene film. (f) and (g) Typical large-sized graphene domains.

during growth [121]. The alignment of hexagonal graphene domains can be attributed to the epitaxial relationship with the Cu lattice. Because of the high mobility of the liquid metal, large-area continuous single-crystal graphene sheet would be obtained by seamlessly merging the aligned graphene flakes. Resolidified Cu can also be used as substrate to synthesize large-sized graphene and ~1 mm sized hexagonal ML graphene grains were obtained due to the smooth surface of the resolidified Cu [122].

4.3.1.4 Surface Passivation

Annealing the Cu foil in the presence of H_2 will remove the surface oxide, impurities, contaminants and defects to suppress graphene nucleation [123]. However, passivating the active sites on Cu surface, for example, by oxidation, is an effective method to suppress the graphene nucleation density. Whereas because of the weak catalytic activity of the passivated Cu, growth time should be prolonged to obtain large-sized graphene domains. Due to the high nucleation barrier of graphene on Cu foil with native surface oxide by annealing it without H_2, graphene nucleation density greatly reduced from ~10^6 to 4 nuclei/cm and as large as 5 mm sized single-crystal graphene was achieved by further carefully tuning the growth parameters, for example, low pressure, high H_2/CH_4 ratio [83]. Cu surface active sites can also be effectively passivated by exposing Cu foil to O_2 during annealing pretreatment [100]. Graphene nucleation density decreases with increasing O_2 exposure, for example from 2×10^3 mm^{-2} on oxygen-free Cu foil to ~0.01 mm^{-2} on Cu foil with 5 min O_2 exposure. However, it has also been reported that surface oxygen accelerates the graphene domain growth through enhancing adsorption and dissociation of hydrocarbons and thus reducing the edge attachment barrier [124]. Not only the surface oxygen but also the trace amount of gaseous O_2 introduced along with the hydrocarbon into the CVD chamber will reduce the graphene nucleation density and centimeter-sized (1 cm) single-crystalline graphene domains were produced [125]. Here, suppressing nucleation density is realized by oxide etching the metastable carbon islands and enhancing the desorption of carbon species. Other methods such as melamine pretreatment can also be used to passivate the active sites to reduce the nuclei density and grow centimeter-sized (1 cm) single-crystal graphene on Cu foil surface [126].

4.3.1.5 Seeded Growth

Seeding is another approach to control the graphene nucleation density. Graphene domains and graphene arrays with predetermined location and/or controlled layer number can be achieved by using this seeded growth. Patterned multilayer graphene [113], graphene oxide flakes [127], patterned highly oriented pyrolytic graphite (HOPG), Poly(methyl methacrylate) (PMMA) [128], coronene [129], and nanoparticles [130] can be used as the seeds. Occasionally, multiple nucleations at one seed and random newly formed nuclei are also observed. Graphene lattice orientation might be determined by the seed and uniformly aligned graphene domains could merge into a large graphene sheet without grain boundary.

4.3.2 Growth Parameters

Growth parameters, such as H_2/CH_4 ratio, temperature, pressure, hydrocarbon flow rate and partial pressure, can also affect the graphene growth behavior. Figure 4.13 gives a typical example of graphene nucleation and size changing under different growth conditions [131]. It indicates that graphene nucleation density decreases and thus domain size increases as the temperature (T) increases, or methane flow rate (J_{Me}) decreases, or methane partial pressure (P_{Me}) decreases. CVD growth graphene is a complicated process. A large number of reports on graphene growth kinetics and thermodynamics can be referred to those excellent articles [5, 17, 132–136]. Some examples to produce large-sized single-crystal graphene domains are listed in Table 4.1.

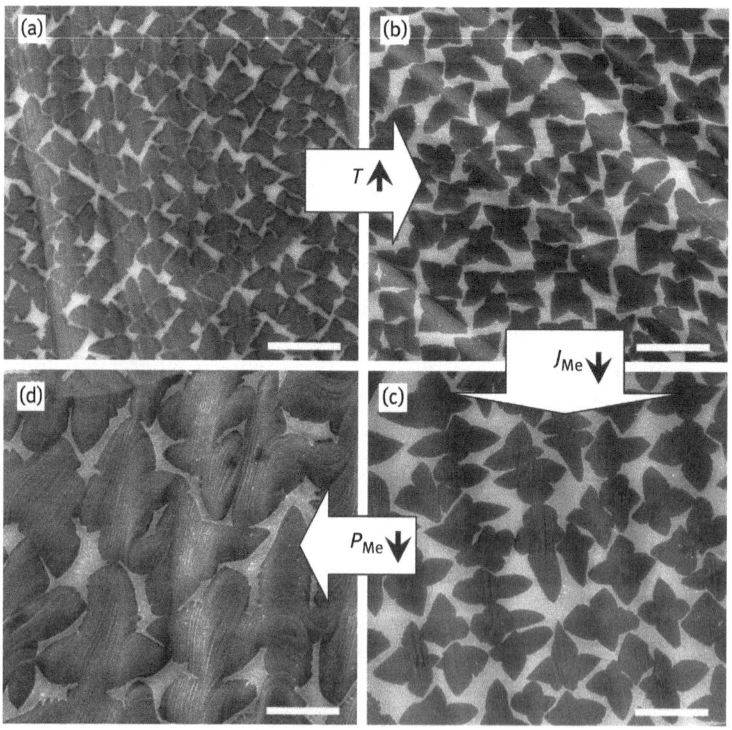

Figure 4.13: Scanning electron microscope (SEM) images of graphene domains grown under different conditions [131]. T (°C)/J_{Me} (sccm)/P_{Me} (mTorr): (a) 985/35/460, (b) 1,035/35/460, (c) 1,035/7/460, (d) 1,035/7/160. Scale bars are 10 µm.

Table 4.1: Large size single crystal graphene produced by CVD.

	Substrate	H_2/CH_4(/Ar) / sccm	Pressure[1]	Temperature / °C	Size and shape	Mobility[2]/cm^2/Vs	Methods or Annealing Pretreatment	Growth time	Ref and Year
1	Cu Foil	2/7–35	LP	1035	hundreds of μm^2 starlike	16000	2 sccm H_2, 1035 °C, 20 min annealing; two-step growth	2.5+1 min	[131] 2010
2	Cu Foil	H_2/Ar, 10/300 CH_4 in Ar 8 ppm	AP	1050	~15 μm hexagonal	$<10^3$–$\sim10^4$	H_2/Ar, 10/300 sccm, 1050 °C, 30 min annealing	~10 min	[113] 2011
3	Cu Film	300/10	AP	1000	~5 μm equiangular hexagonal	1900	tuning H_2/CH_4 ratio, smooth Cu surface, 200 sccm H_2, 1000 °C, 30 min annealing	1.5–60 min	[135] 2011
4	Cu Foil	2/1.3	LP	~1035	0.5 mm, dendrites	4000 (e)	inside surface of copper-foil enclosures		[118] 2011
5	Ni(111)	Propylene gas	UHV	600–680	millimeter size		Ni(111) heteroepitaxially grown on MgO(111)	5 min	[28] 2011
6	Pt Foil	700/4	AP	1040	~1.3 mm hexagonal	7100	700 sccm H_2 1040 °C, 10 min	180 min	[103] 2012
7	Cu Foil	500/0.5	AP	1045	submillimeter, 0.4 × 0.4 mm^2, square		H_2/Ar, 50/300 sccm, 1045 °C, 3 h	15.5 min	[123] 2012
8	Cu Foil	12.5/1	LP	1000	100 μm six-lobed flower	4200; 20000 (hBN)	7 sccm H_2, 1000 °C, 40 min; vapour trapping	30 min	[138] 2012
9	Liquid Cu	300/6	AP	1160	>100 μm, hexagonal	1000–2500	200 sccm H_2, 1100 °C, 30 min	~30 min 10–50 μm/min	[120] 2012
10	Liquid Cu	80/10, CH_4:Ar, 1:99	AP	1090	>200 μm, hexagonal		100 sccm(1:3 H_2/Ar mix) 1090 °C, 30 min	~15 min	[121] 2012
11	Cu Foil	70/0.15	LP	1077	~2.3 mm, ~4.5 mm^2 hexagonal	~11000	electrochemical polishing and high-pressure annealing (1500 Torr, 500 sccm H_2, 1077 °C, 7 h)	125 min	[117] 2012

(Continued)

Table 4.1: (*Continued*)

	Substrate	H_2/CH_4(/Ar) / sccm	Pressure[1]	Temperature / °C	Size and shape	Mobility[2]/cm^2/Vs	Methods or Annealing Pretreatment	Growth time	Ref and Year
12	Cu Foil	10/0.1	LP	1035	centimeter-size	40000–65000 (1.7 K); 15000–30000 (r. t.)	0.1–10 Torr H_2, 1035 °C, 30 min; 1 × 10^{-3} Torr O_2 5 min, oxygen rich surface	12 h	[100] 2013
13	Cu Foil	Polystyrene	AP	1050	~1.2 mm hexagonal	5000–8000	mechanical and electrochemical polishing, H_2/Ar (1:10), 1050 °C, 60 min, pulse heating polystyrene	30–80 min	[139] 2013
14	Cu Foil	10/0.1	LP	1035	~2 mm	5200	inside surface of Cu tube electropolishing	6 h	[140] 2013
15	Resolidified Cu	100/46/854 (CH_4 diluted by Ar, 0.1%)	AP	1075	~1 mm hexagonal		H_2/Ar, 60/940 sccm 1100 °C, 30 min	5 h	[122] 2013
16	Cu Foil	H_2/CH_4 (1320–8800)	LP	1070	~ 5 mm, hexagonal	16000	300 sccm Ar, 1070 °C, 25 min pure Ar, maintaining native oxide, high H_2/CH_4	48 h	[83] 2013
17	Cu Foil	21/15/320(N_2) (CH_4 diluted by Ar, 500 ppm)	AP	1050	5.9 mm, hexagonal		mechanically polishing; pure Ar, 320 sccm Ar; Ar/H_2, 30 min; seeded growth	hours to tens of hours	[130] 2013
18	Ge(110)	CH_4 (Diluted by H_2 to 1–2%)	LP	900–930	wafer size	7250 ± 1390 max. 10620	hydrogen-terminated Ge(111) aligned growth and coalescence	5–120 min	[102] 2014
19	$Cu_{85}Ni_{15}$	15/80–120/300 (0.05% CH_4 diluted by Ar)	AP	1050–1100	~1.5 inch	10,000 ~ 20,000	local carbon precursor feeding Cu-Ni alloy substrate, H_2/Ar 50/1000 sccm, 1050 °C, 2h	2.5 h	[46] 2016

[1] LP stands for low pressure; AP stands for ambient pressure.
[2] e stands for electron mobility.

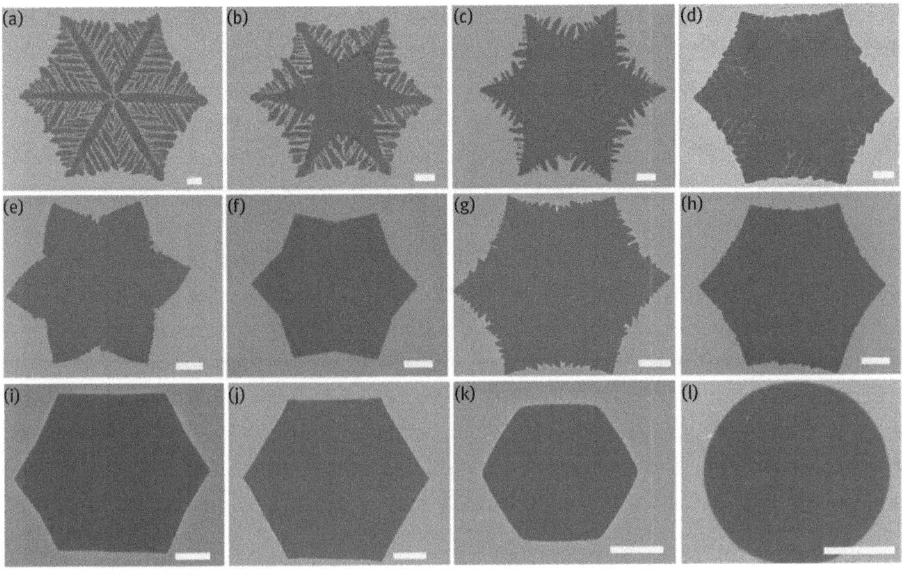

Figure 4.14: Graphene domain shape evolution under different H_2/CH_4 ratio [137]. From (a) to (h), the H_2/CH_4 ratios are 20, 30, 40, 60, 70, 80, 100, 120, respectively and the CH_4 flow rate maintains at 0.5 sccm. The flow rate of H_2/CH_4 for (i)–(l) are 200/2, 200/4, 300/5 and 300/22 sccm, respectively. All scale bars are 5 μm.

Low hydrocarbon concentration and high H_2/CH_4 ratio are typically used to reduce graphene nucleation density with or without the nucleation suppressing methods mentioned earlier. In addition, H_2/CH_4 ratio can also be used to tailor the size, shape or layer number of graphene domains due to the etching effect of H_2 [16, 119, 133, 137]. Generally, dendritic graphene domains are obtained at low H_2/CH_4 ratio due to diffusion (mass transport)-limited growth, while compact hexagonal graphene domains grow at high H_2/CH_4 ratio due to the edge-attachment-limited growth. The evolution of graphene domain shape is illustrated in Figure 4.14.

4.4 Direct Growth on Insulating Substrates

Graphene grown on transition metal substrates must be transferred onto insulating substrates to measure its optoelectronic properties and realize its various applications [2, 141]. During the transfer process, the catalytic substrates typically etched away, which will cause waste and pollution. The metal and support polymer materials such as PMMA cannot be completely removed. As a result, unintentional contaminants, cracks and wrinkles, introduced during transfer process, will degrade the electrical and mechanical properties of graphene [141]. Transferring graphene is heavily manual and time-consuming and not compatible with industrial batch processing.

Although a face-to-face transfer method has been developed by Gao et al. [142], much effort has to be made to find novel transfer methods.

To overcome those challenges, direct growth graphene on dielectric substrate free of transfer is highly desirable [143, 144]. Due to the no or weak catalytic activity of insulating substrate, additional procedures, such as gaseous catalyst (e.g., metal vapor) assisted catalyzation or plasma enhancement, have to be applied to promote the carbon precursor decomposition. To date, various substrates such as SiO_2 [145–150], hexagonal boron nitride (hBN) [151–155], high-k materials (Al_2O_3 [148, 156–158], $SrTiO_3$ [159] and Si_3N_4 [160, 161]), quartz [162] glass [163–165], oxide powders [166, 167] or even water-soluble NaCl [168] powders have been used as substrate to grow graphene. As-grown graphene always conforms to the morphology of the substrate and thus graphene with various structures can be produced. It should be noted that hBN is an ideal substrate for high-performance graphene devices, due to the atomically flat surface free of dangling bonds and charge traps [169, 170]. Graphene band structure can also be modulated by varying the crystallographic alignment with the hBN substrate [171, 172]. However, for conventional CVD, during the growth of graphene, etching of the pregrown hBN is unavoidable. Because of atomic flatness and small lattice mismatch (1.8 %), van der Waals epitaxial growth of single-domain graphene on exfoliated hBN has been achieved using PECVD [151]. Graphene lattice orientation locked into a fixed orientation with respect to the hBN substrate and the size of the well-stitched continuous single-crystalline graphene free of grain boundaries will be limited only by the size of the hBN substrate.

Cu vapor [147, 149], Ni vapor [173] and Ga vapor [165, 174] have been used as catalysts to decompose carbon precursors into active carbon species and then nucleate and grow graphene. Besides growth parameters (e.g., temperature and pressure), metal location and the distance of metal and the growth substrate also play a critical role in determining the size, quality and layer number of graphene. The graphene quality is not very well and nucleation seeds such as graphite debris have been introduced to improve the graphene quality [147]. Although no detectable metal residues were confirmed in as-grown graphene, it could not exclude the possibility of metal contamination. Silane and germane are another highly effective gaseous catalysts to directly grow high-quality graphene on exfoliated hBN with a high growth rate and an alignment with hBN [155]. Silicon atoms attach to the edge of graphene, thus leading to a much reduced reaction barrier. Graphene domain size as large as 20 μm with quality comparable to mechanical exfoliated graphene is obtained, which is evidenced by high hole and electron mobility of 19,000 and 23,000 cm^2/Vs, respectively, and two peaks owing to the secondary Dirac cones of the graphene/hBN superlattice in transport measurement.

Catalyst-free growth of graphene on insulating substrate at low temperature can be achieved using PECVD [148, 175–177]. The plasma provides the energy for bond breaking of carbon precursors and decomposes them into highly reactive radicals and atoms, leading to the possibility of low temperature and rapid synthesis of graphene. Atomic hydrogen generated by decomposing carbon precursor or introduced H_2 acts as important etchant to remove amorphous carbon and grow high-quality graphene [178]. The growth behavior is not self-limiting. Graphene layer number or film thickness can be modulated by varying the growth time or flow rate of carbon precursor.

Other metal-free methods to promote carbon precursor decomposition or graphene nucleation include high-temperature thermal decomposition of carbon precursors [163] and surface oxygen-aided nucleation [145]. Additionally, a two-stage growth (first nucleation and then growth under different conditions) is sometimes adopted to expand the lateral size [156, 160].

No matter what method is applied, most graphene directly grown on insulating substrate typically suffers a low growth rate and long growth time, but poor quality with nonnegligible defects and grain boundaries, and the domains size are limited to several microns. Direct synthesis of high-quality, large-sized graphene on insulating substrate especially at low temperature compatible with flexible plastics still needs further study. However, graphene with inferior quality could find its suitable places for various applications, for example, defoggers [163, 165] and cell culture [164].

4.5 Doping

The electrical properties of graphene can be tailored by doping, rendering graphene suitable for various applications, such as n- or p-type transistors, band gap opening, sensors, work function tuning, catalyst, energy conversion and storage [179–182]. Here, we aim to briefly discuss the direct CVD synthesis of nitrogen (N) substitutional-doped graphene [180, 183–186].

Various heteroatoms, like single heteroatom, such as N [187], B [188–193], S [194–196], Si [197, 198], and binary heteroatoms, such as B:N [199, 200], S:N [201], B:P [202], N:P [203], and N:F [204], have been substitutionally doped graphene by direct CVD synthesis, among which, N-doped graphene has been extensively studied. Several nitrogen-containing compounds, such as NH_3 [177, 205–209], pyridine [210–212], s-triazine [213, 214], melamine [215, 216], acetonitrile [217], urea [218], N, N-dimethylformamide (DMF) [219] and methylamines [220], and other compounds [221] can be used as nitrogen precursors to synthesize N-doped graphene. The nitrogen-containing precursors have a significant influence on the graphene growth mechanism and thus the N configuration in as-grown graphene [220, 222]. There are three common bonding configurations of N in graphene sp^2 carbon network, including pyridinic, pyrrolic and graphitic N (or quaternary N) [205]. A mixture of those N configurations usually appears in as-grown N-doped graphene. The doping level can be tuned by varying the ratio of nitrogen precursor and carbon precursor. As high as 8.9 % of N and graphitic N dominated few-layer graphene was produced on Cu thin film by Wei et al. using NH_3 and CH_4 as nitrogen precursor and carbon precursor, respectively [205].

Temperature has a profound effect on the doping level, N configuration and defect density of as-grown graphene [212, 216]. Typically, doping level decreases as temperature increases [214]. Low temperature creates conditions favorable to the synthesis of N-doped graphene dominated by graphitic N but more defects, while high temperature usually leads to pyridinic N and less defects. A temperature-dependent N configuration change from pyrrolic N (~880°C) to pyridinic N (~1,050°C) has been observed [209]. Single-layer pure pyridinic N-doped graphene with atomic percentage of N up to 16 % has been synthesized on Cu foil by low-pressure CVD at 900°C using ethylene, H_2 and NH_3 as gas source [206]. N-doped graphene with two graphitic N atoms locating in the same sublattice separated by a carbon atom was grown by carefully controlling the growth parameters (growth temperature 850°C, NH_3 reaction time) [208]. Highly N-doped single-layer single-crystal graphene arrays have been grown at temperature as low as 300°C by self-assembling of pyridine on Cu foil [211]. Almost 100 % of the as-grown graphenes are of single layer. Nitrogen content reaches as high as 16.7 % and graphitic N is dominant. Direct growth N-doped graphene on dielectric substrates by PECVD at low temperature (435°C, using C_2H_2 and NH_3 as carbon precursor and nitrogen precursor, respectively) has also been developed by Wei et al. [177].

Nitrogen atoms randomly distribute in the graphene matrix. However, a lower N concentration or N depletion near the grain boundaries and nucleation region than that in the interior was observed by Zhao et al. (Figure 4.15) [223]. Different N configurations will lead to different doping types and carrier concentration [224]. Generally, N doping of graphene leads to n-type doping of graphene and enhancement of state density near the Fermi level because of N delocalizing extra electrons into the graphene lattice and shifting the Fermi level above the Dirac point [225].

The extensive studies of nitrogen doping of graphene will provide important guidance for synthesizing other heteroatoms doped graphene, which we do not discuss here due to space limitations.

4.6 Conclusions and Perspectives

In this chapter, we briefly summarize the controllable CVD synthesis of graphene. It is challenging to grow uniform single-layer, bilayer and few-layer graphene on Ni substrates, due to the high carbon solubility in Ni and the nonequilibrium carbon precipitation. Various methods to produce uniform single-layer, bilayer and/or few-layer graphene on Ni substrate are discussed. The methods consist of substrate selection, thin-film thickness, pretreatment, alloys, growth parameters and heat/cool rate, and the overall aim of which is to obtain a large grain size, smooth Ni surface and suppress the nonequilibrium carbon precipitation at highly active Ni surface sites. For Cu substrate, the surface-mediated reaction and thus self-limited growth of graphene make it suitable to grow single-layer graphene. However, the Cu surface

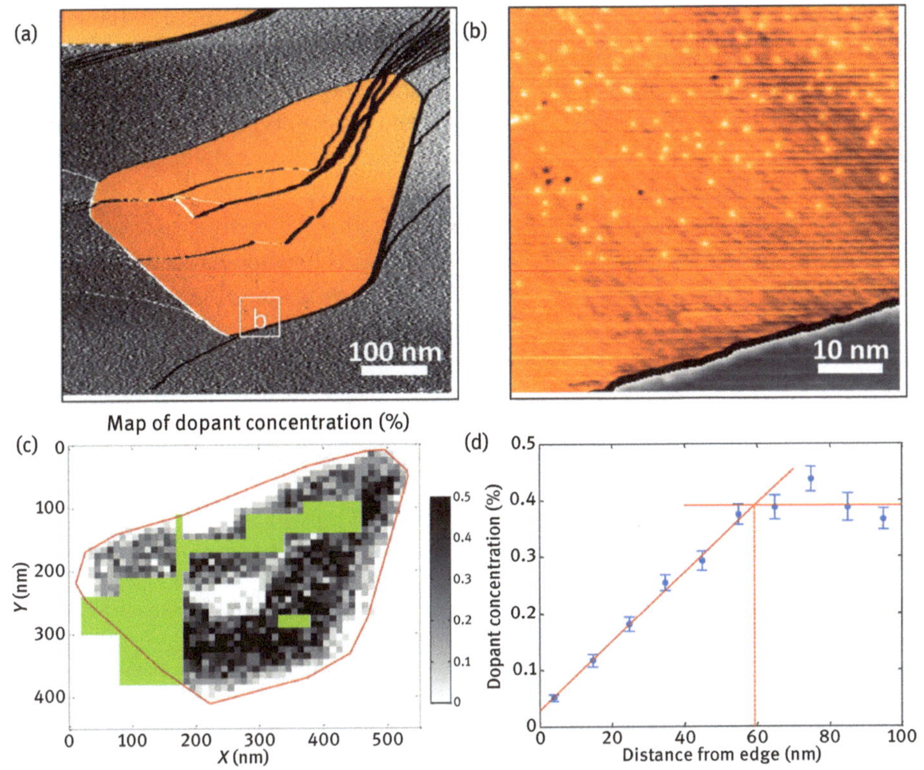

Figure 4.15: Nitrogen dopants distribution in graphene island [223]. (a) Scanning tunneling microscopy (STM) image of a graphene domain. (b) Enlarged STM image of the area boxed in (a). (c) Nitrogen dopants distribution over the entire graphene island. Red line is the graphene edge. There are no data for the shaded area. (d) Nitrogen concentration as a function of distance from the graphene edge.

imperfections such as grain boundaries and steps will hinder the growth of large-sized single-crystal graphene. The approaches to synthesize bilayer graphene and large-sized single-crystal graphene domains on Cu substrate are discussed then. To grow large-sized single-crystal graphene, reducing the graphene nucleation density or (seeded) growth of oriented graphene domains and then seamlessly coalescing into a large graphene sheet free of boundaries are the fundamental principles. Bilayer layer graphene will be obtained by breaking the self-limited growth of graphene on Cu by epitaxial growth or asymmetric growth on two sides of Cu enclosure or Cu pocket. Direct growth graphene on insulating substrate through catalytically decomposing carbon precursors by gaseous catalyst and PECVD is briefly addressed. At last, doping of graphene by direct CVD synthesis using substitutional nitrogen doping as an example is presented. The relationships of doping level and nitrogen configurations with the nitrogen source and temperature are demonstrated.

Although progress has been achieved in the growth of uniform large-sized single-crystal graphene, bilayer graphene, doping as well as direct synthesis on insulating substrates, there are still various issues and challenges remaining, and much effort has to be taken to grow uniform large-sized single-crystal graphene on polycrystalline transition metals such as Ni and Cu. Wafer-scale single-crystal graphene sheet as well as bilayer or few-layer graphene with controllable stacking order will be required for various applications. The quality of graphene directly grown on insulating substrate needs further improvement, especially at low temperature compatible with flexible electronics. CVD growth doped graphene with controllable doping level, configurations and heteroatoms scattering will be helpful to further understand the structure–property relationships. Addressing those issues may offer bright future for graphene applications.

References

[1] Wei D, Liu Y. Controllable synthesis of graphene and its applications. Adv Mater 2010;22: 3225–41.
[2] Mattevi C, Kim H, Chhowalla M. A review of chemical vapour deposition of graphene on copper. J Mater Chem 2011;21:3324–34.
[3] Ruemmeli MH, Rocha CG, Ortmann F, et al. Graphene: piecing it together. Adv Mater 2011;23:4471–90.
[4] Kang J, Shin D, Bae S, Hong BH. Graphene transfer: key for applications. Nanoscale 2012;4:5527–37.
[5] Munoz R, Gomez-Aleixandre C. Review of CVD synthesis of graphene. Chem Vapor Depos 2013;19:297–322.
[6] Yan K, Fu L, Peng H, Liu Z. Designed CVD growth of graphene via process engineering. Acc Chem Res 2013;46:2263–74.
[7] Wu H-Q, Linghu C-Y, Lu H-M, Qian H. Graphene applications in electronic and optoelectronic devices and circuits. Chin Phys B 2013;22:098106–9.
[8] Sutter P, Sutter E. Microscopy of graphene growth, processing, and properties. Adv Funct Mater 2013;23:2617–34.
[9] Iski EV, Yitamben EN, Gao L, Guisinger NP. Graphene at the atomic-scale: synthesis, characterization, and modification. Adv Funct Mater 2013;23:2554–64.
[10] Edwards RS, Coleman KS. Graphene film growth on polycrystalline metals. Acc Chem Res 2013;46:23–30.
[11] Zhang Y, Zhang L, Zhou C. Review of chemical vapor deposition of graphene and related applications. Acc Chem Res 2013;46:2329–39.
[12] Wei D, Wu B, Guo Y, Yu G, Liu Y. Controllable chemical vapor deposition growth of few layer graphene for electronic devices. Acc Chem Res 2013;46:106–15.
[13] Yan Z, Peng Z, Tour JM. Chemical vapor deposition of graphene single crystals. Acc Chem Res 2014;47:1327–37.
[14] Chen X, Zhang L, Chen S. Large area CVD growth of graphene. Synthetic Met 2015;210: 95–108.
[15] Nguyen VL, Lee YH. Towards wafer-scale monocrystalline graphene growth and characterization. Small 2015;11:3512–28.
[16] Geng D, Wang H, Yu G. Graphene single crystals: size and morphology engineering. Adv Mater 2015;27:2821–37.

[17] Hofmann S, Braeuninger-Weimer P, Weatherup RS. CVD-enabled graphene manufacture and technology. J Phys Chem Lett 2015;6:2714–21.

[18] Giovannetti G, Khomyakov PA, Brocks G, Karpan VM, van den Brink J, Kelly PJ. Doping graphene with metal contacts. Phys Rev Lett 2008;101:026803.

[19] Khomyakov PA, Giovannetti G, Rusu PC, Brocks G, van den Brink J, Kelly PJ. First-principles study of the interaction and charge transfer between graphene and metals. Phys Rev B 2009;79:195425.

[20] Li X, Cai W, Colombo L, Ruoff RS. Evolution of graphene growth on Ni and Cu by carbon isotope labeling. Nano Lett 2009;9:4268–72.

[21] Fang W, Hsu AL, Song Y, Kong J. A review of large-area bilayer graphene synthesis by chemical vapor deposition. Nanoscale 2015;7:20335–51.

[22] Weatherup RS, Bayer BC, Blume R, et al. On the mechanisms of Ni-catalysed graphene chemical vapour deposition. ChemPhysChem 2012;13:2544–9.

[23] Zhang Y, Gomez L, Ishikawa FN, et al. Comparison of graphene growth on single-crystalline and polycrystalline Ni by chemical vapor deposition. J Phys Chem Lett 2010;1:3101–7.

[24] Yoshii S, Nozawa K, Toyoda K, Matsukawa N, Odagawa A, Tsujimura A. Suppression of inhomogeneous segregation in graphene growth on epitaxial metal films. Nano Lett 2011;11:2628–33.

[25] Gao J, Yip J, Zhao J, Yakobson BI, Ding F. Graphene nucleation on transition metal surface: structure transformation and role of the metal step edge. J Am Chem Soc 2011;133:5009–15.

[26] Thiele S, Reina A, Healey P, et al. Engineering polycrystalline Ni films to improve thickness uniformity of the chemical-vapor-deposition-grown graphene films. Nanotechnology 2010;21:015601.

[27] Addou R, Dahal A, Sutter P, Batzill M. Monolayer graphene growth on Ni(111) by low temperature chemical vapor deposition. Appl Phys Lett 1601;2012:100, 02.

[28] Iwasaki T, Park HJ, Konuma M, Lee DS, Smet JH, Starke U. Long-range ordered single-crystal graphene on high-quality heteroepitaxial Ni thin films grown on MgO(111). Nano Lett 2011;11:79–84.

[29] Zhang Y, Gao T, Xie S, et al. Different growth behaviors of ambient pressure chemical vapor deposition graphene on Ni(111) and Ni films: a scanning tunneling microscopy study. Nano Res 2012;5:402–11.

[30] Kim KS, Zhao Y, Jang H, et al. Large-scale pattern growth of graphene films for stretchable transparent electrodes. Nature 2009;457:706–10.

[31] Liu N, Fu L, Dai B, et al. Universal segregation growth approach to wafer-size graphene from non-noble metals. Nano Lett 2011;11:297–303.

[32] Kim E, Lee W-G, Jung J. Agglomeration effects of thin metal catalyst on graphene film synthesized by chemical vapor deposition. Electron Mater Lett 2011;7:261–4.

[33] Peng K-J, Wu C-L, Lin Y-H, et al. Hydrogen-free PECVD growth of few-layer graphene on an ultra-thin nickel film at the threshold dissolution temperature. J Mater Chem C 2013;1:3862–70.

[34] Weatherup RS, Dlubak B, Hofmann S. Kinetic control of catalytic CVD for high-quality graphene at low temperatures. ACS Nano 2012;6:9996–10003.

[35] Yu Q, Lian J, Siriponglert S, Li H, Chen YP, Pei S-S. Graphene segregated on Ni surfaces and transferred to insulators. Appl Phys Lett 2008;93:113103.

[36] Reina A, Jia XT, Ho J, et al. Large area, few-layer graphene films on arbitrary substrates by chemical vapor deposition. Nano Lett 2009;9:30–5.

[37] De Arco LG, Zhang Y, Schlenker CW, Ryu K, Thompson ME, Zhou C. Continuous, highly flexible, and transparent graphene films by chemical vapor deposition for organic photovoltaics. ACS Nano 2010;4:2865–73.

[38] Kahng YH, Lee S, Choe M, et al. A study of graphene films synthesized on nickel substrates: existence and origin of small-base-area peaks. Nanotechnology 2011;22:045706.

[39] Odahara G, Hibino H, Nakayama N, et al. Macroscopic single-domain graphene growth on polycrystalline nickel surface. Appl Phys Express 2012;5:035501.
[40] Tan L, Zeng M, Zhang T, Fu L. Design of catalytic substrates for uniform graphene films: from solid-metal to liquid-metal. Nanoscale 2015;7:9105–21.
[41] Weatherup RS, Bayer BC, Blume R, et al. In situ characterization of alloy catalysts for low-temperature graphene growth. Nano Lett 2011;11:4154–60.
[42] Chen S, Cai W, Piner RD, et al. Synthesis and characterization of large-area graphene and graphite films on commercial Cu-Ni alloy foils. Nano Lett 2011;11:3519–25.
[43] Dai B, Fu L, Zou Z, et al. Rational design of a binary metal alloy for chemical vapour deposition growth of uniform single-layer graphene. Nat Commun 2011;2:522.
[44] Rümmeli MH, Zeng M, Melkhanova S, et al. Insights into the early growth of homogeneous single-layer graphene over Ni-Mo binary substrates. Chem Mater 2013;25:3880–7.
[45] Wu Y, Chou H, Ji H, et al. Growth mechanism and controlled synthesis of AB-stacked bilayer graphene on Cu-Ni alloy foils. ACS Nano 2012;6:7731–8.
[46] Wu T, Zhang X, Yuan Q, et al. Fast growth of inch-sized single-crystalline graphene from a controlled single nucleus on Cu-Ni alloys. Nat Mater 2016;15:43–7.
[47] Liu W, Kraemer S, Sarkar D, Li H, Ajayan PM, Banerjeet K. Controllable and rapid synthesis of high-quality and large-area Bernal stacked bilayer graphene using chemical vapor deposition. Chem Mater 2014;26:907–15.
[48] Tyagi P, Robinson ZR, Munson A, et al. Characterization of graphene films grown on CuNi foil substrates. Surf Sci 2015;634:16–24.
[49] Liu X, Fu L, Liu N, et al. Segregation growth of graphene on Cu-Ni alloy for precise layer control. J Phys Chem C 2011;115:11976–82.
[50] Wang G, Zhang M, Liu S, et al. Synthesis of layer-tunable graphene: a combined kinetic implantation and thermal ejection approach. Adv Funct Mater 2015;25:3666–75.
[51] Lin T, Huang F, Wan D, Bi H, Xie X, Jiang M. Self-regulating homogenous growth of high-quality graphene on Co-Cu composite substrate for layer control. Nanoscale 2013;5:5847–53.
[52] Feng Y, Yao X, Hu Z, Xu J-J, Zhang L. Passivating a transition-metal surface for more uniform growth of graphene: effect of Au alloying on Ni(111). Phys Rev B 2013;87:195421.
[53] Huang Y, Du J, Zhou T, Ling C, Wang S, Geng B. Role of Au in graphene growth on a Ni surface. ACS Catal 2014;4:892–902.
[54] Chen W, Chen H, Lan H, et al. Suppression of grain boundaries in graphene growth on superstructured Mn-Cu(111) surface. Phys Rev Lett 2012;109:265507.
[55] Zhang L, Zhao X, Xue X, et al. Sub-surface alloying largely influences graphene nucleation and growth over transition metal substrates. Phys Chem Chem Phys 2015;17:30270–8.
[56] Weatherup RS, Amara H, Blume R, et al. Interdependency of subsurface carbon distribution and graphene-catalyst interaction. J Am Chem Soc 2014;136:13698–708.
[57] Chae SJ, Guenes F, Kim KK, et al. Synthesis of large-area graphene layers on poly-nickel substrate by chemical vapor deposition: wrinkle formation. Adv Mater 2009;21:2328–33.
[58] Dahal A, Batzill M. Graphene-nickel interfaces: a review. Nanoscale 2014;6:2548–62.
[59] Shelton JC, Patil HR, Blakely JM. Equilibrium segregation of carbon to a nickel (111) surface: a surface phase transition. Surf Sci 1974;43:493–520.
[60] Patera LL, Africh C, Weatherup RS, et al. In situ observations of the atomistic mechanisms of Ni catalyzed low temperature graphene growth. ACS Nano 2013;7:7901–12.
[61] Seah C-M, Chai S-P, Mohamed AR. Mechanisms of graphene growth by chemical vapour deposition on transition metals. Carbon 2014;70:1–21.
[62] Lahiri J, Miller T, Adamska L, Oleynik II, Batzill M. Graphene growth on Ni(111) by transformation of a surface carbide. Nano Lett 2011;11:518–22.
[63] Reina A, Thiele S, Jia X, et al. Growth of large-area single- and bi-layer graphene by controlled carbon precipitation on polycrystalline Ni surfaces. Nano Res 2009;2:509–16.

[64] Park HJ, Meyer J, Roth S, Skakalova V. Growth and properties of few-layer graphene prepared by chemical vapor deposition. Carbon 2010;48:1088–94.
[65] Gong Y, Zhang X, Liu G, et al. Layer-controlled and wafer-scale synthesis of uniform and high-quality graphene films on a polycrystalline nickel catalyst. Adv Funct Mater 2012;22: 3153–9.
[66] Li XS, Cai WW, An JH, et al. Large-area synthesis of high-quality and uniform graphene films on copper foils. Science 2009;324:1312–14.
[67] McCann E. Asymmetry gap in the electronic band structure of bilayer graphene. Phys Rev B 2006;74:161403.
[68] Castro EV, Novoselov KS, Morozov SV, et al. Biased bilayer graphene: Semiconductor with a gap tunable by the electric field effect. Phys Rev Lett 2007;99:216802.
[69] Min HK, Sahu B, Banerjee SK, MacDonald AH. Ab initio theory of gate induced gaps in graphene bilayers. Phys Rev B 2007;75:155115.
[70] Zhang YB, Tang TT, Girit C, et al. Direct observation of a widely tunable bandgap in bilayer graphene. Nature 2009;459:820–3.
[71] Xia FN, Farmer DB, Lin YM, Avouris P. Graphene field-effect transistors with high on/off current ratio and large transport band gap at room temperature. Nano Lett 2010;10:715–18.
[72] Ohta T, Bostwick A, Seyller T, Horn K, Rotenberg E. Controlling the electronic structure of bilayer graphene. Science 2006;313:951–4.
[73] Samuels AJ, Carey JD. Molecular doping and band-gap opening of bilayer graphene. ACS Nano 2013;7:2790–9.
[74] Lee S, Lee K, Zhong Z. Wafer scale homogeneous bilayer graphene films by chemical vapor deposition. Nano Lett 2010;10:4702–7.
[75] Wassei JK, Mecklenburg M, Torres JA, et al. Chemical vapor deposition of graphene on copper from methane, ethane and propane: evidence for bilayer selectivity. Small 2012;8:1415–22.
[76] Sun Z, Abdul-Rahman OR, Zhu Y, et al. Large-area Bernal-stacked Bi-, Tr-, and tetralayer graphene. ACS Nano 2012;6:9790–6.
[77] Lu C-C, Lin Y-C, Liu Z, Yeh C-H, Suenaga K, Chiu P-W. Twisting bilayer graphene superlattices. ACS Nano 2013;7:2587–94.
[78] Li Q, Chou H, Zhong J-H, et al. Growth of adlayer graphene on Cu studied by carbon isotope labeling. Nano Lett 2013;13:486–90.
[79] Li J, Ji H, Zhang X, et al. Controllable atmospheric pressure growth of mono-layer, bi-layer and tri-layer graphene. Chem Commun 2014;50:11012–15.
[80] Yan K, Peng H, Zhou Y, Li H, Liu Z. Formation of bilayer Bernal graphene: layer-by-layer epitaxy via chemical vapor deposition. Nano Lett 2011;11:1106–10.
[81] Liu L, Zhou H, Cheng R, et al. High-yield chemical vapor deposition growth of high-quality large-area AB-stacked bilayer graphene. ACS Nano 2012;6:8241–9.
[82] Zhang X, Wang L, Xin J, Yakobson BI, Ding F. Role of hydrogen in graphene chemical vapor deposition growth on a copper surface. J Am Chem Soc 2014;136:3040–7.
[83] Zhou H, Yu WJ, Liu L, et al. Chemical vapour deposition growth of large single crystals of monolayer and bilayer graphene. Nat Commun 2013;4:2096.
[84] Zhao P, Kim S, Chen X, et al. Equilibrium chemical vapor deposition growth of Bernal-stacked bilayer graphene. ACS Nano 2014;8:11631–8.
[85] Fang W, Hsu AL, Caudillo R, et al. Rapid identification of stacking orientation in isotopically labeled chemical-vapor grown bilayer graphene by Raman spectroscopy. Nano Lett 2013;13:1541–8.
[86] Fang W, Hsu AL, Song Y, et al. Asymmetric growth of bilayer graphene on copper enclosures using low-pressure chemical vapor deposition. ACS Nano 2014;8:6491–9.
[87] Zhao Z, Shan Z, Zhang C, et al. Study on the diffusion mechanism of graphene grown on copper pockets. Small 2015;11:1418–22.

[88] Hao Y, Wang L, Liu Y, et al. Oxygen-activated growth and bandgap tunability of large single-crystal bilayer graphene. Nat Nanotechnol 2016;11:426–31.
[89] Peng Z, Yan Z, Sun Z, Tour JM. direct growth of bilayer graphene on SiO_2 substrates by carbon diffusion through nickel. ACS Nano 2011;5:8241–7.
[90] Shin H-J, Choi WM, Yoon S-M, et al. Transfer-free growth of few-layer graphene by self-assembled monolayers. Adv Mater 2011;23:4392–7.
[91] Yan Z, Peng Z, Sun Z, et al. Growth of bilayer graphene on insulating substrates. ACS Nano 2011;5:8187–92.
[92] Su C-Y, Lu A-Y, Wu C-Y, et al. Direct formation of wafer scale graphene thin layers on insulating substrates by chemical vapor deposition. Nano Lett 2011;11:3612–16.
[93] Iwasaki T, Zakharov AA, Eelbo T, et al. Formation and structural analysis of twisted bilayer graphene on Ni(111) thin films. Surf Sci 2014;625:44–9.
[94] Dahal A, Addou R, Sutter P, Batzill M. Graphene monolayer rotation on Ni(111) facilitates bilayer graphene growth. Appl Phys Lett 2012;100, 241602.
[95] Huang PY, Ruiz-Vargas CS, van der Zande AM, et al. Grains and grain boundaries in single-layer graphene atomic patchwork quilts. Nature 2011;469:389–92.
[96] Yazyev OV, Louie SG. Electronic transport in polycrystalline graphene. Nat Mater 2010;9:806–9.
[97] Yazyev OV, Chen YP. Polycrystalline graphene and other two-dimensional materials. Nat Nanotechnol 2014;9:755–67.
[98] Cummings AW, Dinh Loc D, Van Luan N, et al. Charge transport in polycrystalline graphene: challenges and opportunities. Adv Mater 2014;26:5079–94.
[99] Petrone N, Dean CR, Meric I, et al. Chemical vapor deposition-derived graphene with electrical performance of exfoliated graphene. Nano Lett 2012;12:2751–6.
[100] Hao Y, Bharathi MS, Wang L, et al. The role of surface oxygen in the growth of large single-crystal graphene on copper. Science 2013;342:720–3.
[101] Ago H, Ogawa Y, Tsuji M, Mizuno S, Hibino H. catalytic growth of graphene: toward large-area single-crystalline graphene. J Phys Chem Lett 2012;3:2228–36.
[102] Lee J-H, Lee EK, Joo W-J, et al. Wafer-scale growth of single-crystal monolayer graphene on reusable hydrogen-terminated germanium. Science 2014;344:286–9.
[103] Gao L, Ren W, Xu H, et al. Repeated growth and bubbling transfer of graphene with millimetre-size single-crystal grains using platinum. Nat Commun 2012;3:699.
[104] Guo W, Wu B, Li Y, et al. Governing rule for dynamic formation of grain boundaries in grown graphene. ACS Nano 2015;9:5792–8.
[105] Gao L, Guest JR, Guisinger NP. Epitaxial graphene on Cu(111). Nano Lett 2010;10:3512–16.
[106] Kim K, Lee Z, Regan W, Kisielowski C, Crommie MF, Zettl A. Grain boundary mapping in polycrystalline graphene. ACS Nano 2011;5:2142–6.
[107] Murdock AT, Koos A, Ben Britton T, et al. Controlling the orientation, edge geometry, and thickness of chemical vapor deposition graphene. ACS Nano 2013;7:1351–9.
[108] Ogawa Y, Hu B, Orofeo CM, et al. Domain structure and boundary in single-layer graphene grown on Cu(111) and Cu(100) films. J Phys Chem Lett 2012;3:219–26.
[109] Meca E, Lowengrub J, Kim H, Mattevi C, Shenoy VB. Epitaxial graphene growth and shape dynamics on copper: phase-field modeling and experiments. Nano Lett 2013;13:5692–7.
[110] Hayashi K, Yamada A, Sato S, Yokoyama N. Crystallographic characterization and control of domain structure within individual graphene islands. J Phys Chem C 2015;119:4286–93.
[111] Wood JD, Schmucker SW, Lyons AS, Pop E, Lyding JW. Effects of polycrystalline Cu substrate on graphene growth by chemical vapor deposition. Nano Lett 2011;11:4547–54.
[112] Nguyen VL, Shin BG, Duong DL, et al. Seamless stitching of graphene domains on polished copper (111) foil. Adv Mater 2015;27:1376–82.
[113] Yu Q, Jauregui LA, Wu W, et al. Control and characterization of individual grains and grain boundaries in graphene grown by chemical vapour deposition. Nat Mater 2011;10:443–9.

[114] Brown L, Lochocki EB, Avila J, et al. Polycrystalline graphene with single crystalline electronic structure. Nano Lett 2014;14:5706–11.
[115] Luo Z, Lu Y, Singer DW, et al. Effect of substrate roughness and feedstock concentration on growth of wafer-scale graphene at atmospheric pressure. Chem Mater 2011;23:1441–7.
[116] Han GH, Guenes F, Bae JJ, et al. Influence of copper morphology in forming nucleation seeds for graphene growth. Nano Lett 2011;11:4144–8.
[117] Yan Z, Lin J, Peng Z, et al. Toward the synthesis of wafer-scale single-crystal graphene on copper foils. ACS Nano 2012;6:9110–17.
[118] Li X, Magnuson CW, Venugopal A, et al. Large-area graphene single crystals grown by low-pressure chemical vapor deposition of methane on copper. J Am Chem Soc 2011;133: 2816–19.
[119] Jacobberger RM, Arnold MS. Graphene growth dynamics on epitaxial copper thin films. Chem Mater 2013;25:871–7.
[120] Geng D, Wu B, Guo Y, et al. Uniform hexagonal graphene flakes and films grown on liquid copper surface. Proc Natl Acad Sci U S A 2012;109:7992–6.
[121] Wu YA, Fan Y, Speller S, et al. Large single crystals of graphene on melted copper using chemical vapor deposition. ACS Nano 2012;6:5010–17.
[122] Mohsin A, Liu L, Liu P, et al. Synthesis of millimeter-size hexagon-shaped graphene single crystals on resolidified copper. ACS Nano 2013;7:8924–31.
[123] Wang H, Wang G, Bao P, et al. Controllable synthesis of submillimeter single-crystal monolayer graphene domains on copper foils by suppressing nucleation. J Am Chem Soc 2012;134: 3627–30.
[124] Liang T, He G, Huang G, et al. Graphene nucleation preferentially at oxygen-rich Cu sites rather than on pure Cu surface. Adv Mater 2015;27:6404–10.
[125] Guo W, Jing F, Xiao J, Zhou C, Lin Y, Wang S. Oxidative-etching-assisted synthesis of centimeter-sized single-crystalline graphene. Adv Mater 2016;28:3152–8.
[126] Lin L, Li J, Ren H, et al. Surface engineering of copper foils for growing centimeter-sized single-crystalline graphene. ACS Nano 2016;10:2922–9.
[127] Li Q, Zhang C, Lin W, et al. Controllable seeding of single crystal graphene islands from graphene oxide flakes. Carbon 2014;79:406–12.
[128] Wu W, Jauregui LA, Su Z, et al. Growth of single crystal graphene arrays by locally controlling nucleation on polycrystalline Cu using chemical vapor deposition. Adv Mater 2011;23:4898–903.
[129] Wu T, Ding G, Shen H, et al. Continuous graphene films synthesized at low temperatures by introducing coronene as nucleation seeds. Nanoscale 2013;5:5456–61.
[130] Gan L, Luo Z. Turning off hydrogen to realize seeded growth of subcentimeter single-crystal graphene grains on copper. ACS Nano 2013;7:9480–8.
[131] Li X, Magnuson CW, Venugopal A, et al. Graphene films with large domain size by a two-step chemical vapor deposition process. Nano Lett 2010;10:4328–34.
[132] Bhaviripudi S, Jia X, Dresselhaus MS, Kong J. Role of kinetic factors in chemical vapor deposition synthesis of uniform large area graphene using copper catalyst. Nano Lett 2010;10:4128–33.
[133] Vlassiouk I, Regmi M, Fulvio PF, et al. Role of hydrogen in chemical vapor deposition growth of large single-crystal graphene. ACS Nano 2011;5:6069–76.
[134] Lewis AM, Derby B, Kinloch IA. Influence of gas phase equilibria on the chemical vapor deposition of graphene. ACS Nano 2013;7:3104–17.
[135] Wu B, Geng D, Guo Y, et al. Equiangular hexagon-shape-controlled synthesis of graphene on copper surface. Adv Mater 2011;23:3522–5.
[136] Jung DH, Kang C, Kim M, Cheong H, Lee H, Lee JS. Effects of hydrogen partial pressure in the annealing process on graphene growth. J Phys Chem C 2014;118:3574–80.

[137] Wu B, Geng D, Xu Z, et al. Self-organized graphene crystal patterns. NPG Asia Mater 2013;5:e63.
[138] Zhang Y, Zhang L, Kim P, Ge M, Li Z, Zhou C. Vapor trapping growth of single-crystalline graphene flowers: synthesis, morphology, and electronic properties. Nano Lett 2012;12: 2810–16.
[139] Wu T, Ding G, Shen H, et al. Triggering the continuous growth of graphene toward millimeter-sized grains. Adv Funct Mater 2013;23:198–203.
[140] Chen S, Ji H, Chou H, et al. Millimeter-Size Single-Crystal Graphene by Suppressing Evaporative Loss of Cu During Low Pressure Chemical Vapor Deposition. Adv Mater 2013;25:2062–5.
[141] Ferrari AC, Bonaccorso F, Fal'ko V, et al. Science and technology roadmap for graphene, related two-dimensional crystals, and hybrid systems. Nanoscale 2015;7:4598–810.
[142] Gao L, Ni G-X, Liu Y, Liu B, Castro Neto AH, Loh KP. Face-to-face transfer of wafer-scale graphene films. Nature 2014;505:190–4.
[143] Chen X, Wu B, Liu Y. Direct preparation of high quality graphene on dielectric substrates. Chem Soc Rev 2016;45:2057–74.
[144] Sun J, Zhang Y, Liu Z. Direct chemical vapor deposition growth of graphene on insulating substrates. ChemNanoMat 2016;2:9–18.
[145] Chen J, Wen Y, Guo Y, et al. Oxygen-aided synthesis of polycrystalline graphene on silicon dioxide substrates. J Am Chem Soc 2011;133:17548–51.
[146] Bi H, Sun S, Huang F, Xie X, Jiang M. Direct growth of few-layer graphene films on SiO_2 substrates and their photovoltaic applications. J Mater Chem 2012;22:411–16.
[147] Teng P-Y, Lu C-C, Akiyama-Hasegawa K, et al. Remote catalyzation for direct formation of graphene layers on oxides. Nano Lett 2012;12:1379–84.
[148] Wei D, Lu Y, Han C, Niu T, Chen W, Wee ATS. Critical crystal growth of graphene on dielectric substrates at low temperature for electronic devices. Angew Chem Int Edit 2013;52:14121–6.
[149] Kim H, Song I, Park C, et al. Copper-vapor-assisted chemical vapor deposition for high-quality and metal-free single-layer graphene on amorphous SiO_2 substrate. ACS Nano 2013;7:6575–82.
[150] Kim YS, Joo K, Jerng S-K, Lee JH, Yoon E, Chun S-H. Direct growth of patterned graphene on SiO_2 substrates without the use of catalysts or lithography. Nanoscale 2014;6:10100–5.
[151] Yang W, Chen G, Shi Z, et al. Epitaxial growth of single-domain graphene on hexagonal boron nitride. Nat Mater 2013;12:792–7.
[152] Wang M, Jang SK, Jang W-J, et al. A platform for large-scale graphene electronics – CVD growth of single-layer graphene on CVD-grown hexagonal boron nitride. Adv Mater 2013;25: 2746–52.
[153] Zhang C, Zhao S, Jin C, et al. Direct growth of large-area graphene and boron nitride heterostructures by a co-segregation method. Nat Commun 2015;6:6519.
[154] Gao T, Song X, Du H, et al. Temperature-triggered chemical switching growth of in-plane and vertically stacked graphene-boron nitride heterostructures. Nat Commun 2015;6:6835.
[155] Tang S, Wang H, Wang HS, et al. Silane-catalysed fast growth of large single-crystalline graphene on hexagonal boron nitride. Nat Commun 2015;6:6499.
[156] Hwang J, Kim M, Campbell D, et al. van der Waals epitaxial growth of graphene on sapphire by chemical vapor deposition without a metal catalyst. ACS Nano 2013;7:385–95.
[157] Song HJ, Son M, Park C, et al. Large scale metal-free synthesis of graphene on sapphire and transfer-free device fabrication. Nanoscale 2012;4:3050–4.
[158] Fanton MA, Robinson JA, Puls C, et al. Characterization of graphene films and transistors grown on sapphire by metal-free chemical vapor deposition. ACS Nano 2011;5:8062–9.
[159] Sun J, Gao T, Song X, et al. Direct growth of high-quality graphene on high-κ dielectric $SrTiO_3$ substrates. J Am Chem Soc 2014;136:6574–7.

[160] Chen J, Guo Y, Wen Y, et al. Two-stage metal-catalyst-free growth of high-quality polycrystalline graphene films on silicon nitride substrates. Adv Mater 2013;25:992–7.
[161] Chen J, Guo Y, Jiang L, et al. Near-equilibrium chemical vapor deposition of high-quality single-crystal graphene directly on various dielectric substrates. Adv Mater 2014;26:1348–53.
[162] Medina H, Lin Y-C, Jin C, et al. Metal-free growth of nanographene on silicon oxides for transparent conducting applications. Adv Funct Mater 2012;22:2123–8.
[163] Sun J, Chen Y, Priydarshi MK, et al. Direct chemical vapor deposition-derived graphene glasses targeting wide ranged applications. Nano Lett 2015;15:5846–54.
[164] Chen Y, Sun J, Gao J, et al. Growing uniform graphene disks and films on molten glass for heating devices and cell culture. Adv Mater 2015;27:7839–46.
[165] Tan L, Zeng M, Wu Q, et al. Direct growth of ultrafast transparent single-layer graphene defoggers. Small 2015;11:1840–6.
[166] Ruemmeli MH, Kramberger C, Grueneis A, et al. On the graphitization nature of oxides for the formation of carbon nanostructures. Chem Mater 2007;19:4105–7.
[167] Ruemmeli MH, Bachmatiuk A, Scott A, et al. Direct low-temperature nanographene CVD synthesis over a dielectric insulator. ACS Nano 2010;4:4206–10.
[168] Shi L, Chen K, Du R, et al. Direct synthesis of few-layer graphene on NaCl crystals. Small 2015;11:6302–8.
[169] Dean CR, Young AF, Meric I, et al. Boron nitride substrates for high-quality graphene electronics. Nat Nanotechnol 2010;5:722–6.
[170] Xue J, Sanchez-Yamagishi J, Bulmash D, et al. Scanning tunnelling microscopy and spectroscopy of ultra-flat graphene on hexagonal boron nitride. Nat Mater 2011;10:282–5.
[171] Hunt B, Sanchez-Yamagishi JD, Young AF, et al. Massive dirac fermions and hofstadter butterfly in a van der Waals heterostructure. Science 2013;340:1427–30.
[172] Ponomarenko LA, Gorbachev RV, Yu GL, et al. Cloning of dirac fermions in graphene superlattices. Nature 2013;497:594–7.
[173] Yen W-C, Chen Y-Z, Yeh C-H, He J-H, Chiu P-W, Chueh Y-L. Direct growth of self-crystallized graphene and graphite nanoballs with Ni vapor-assisted growth: From controllable growth to material characterization. Sci Rep 2014;4:4739.
[174] Murakami K, Tanaka S, Hirukawa A, et al. Direct synthesis of large area graphene on insulating substrate by gallium vapor-assisted chemical vapor deposition. Appl Phys Lett 2015;106:093112.
[175] Zhang L, Shi Z, Wang Y, Yang R, Shi D, Zhang G. Catalyst-free growth of nanographene films on various substrates. Nano Res 2011;4:315–21.
[176] Chugh S, Mehta R, Lu N, Dios FD, Kim MJ, Chen Z. Comparison of graphene growth on arbitrary non-catalytic substrates using low-temperature PECVD. Carbon 2015;93:393–9.
[177] Wei D, Peng L, Li M, et al. Low temperature critical growth of high quality nitrogen doped graphene on dielectrics by plasma-enhanced chemical vapor deposition. ACS Nano 2015;9:164–71.
[178] Yang R, Zhang L, Wang Y, et al. An anisotropic etching effect in the graphene basal plane. Adv Mater 2010;22:4014–19.
[179] Liu HT, Liu YQ, Zhu DB. Chemical doping of graphene. J Mater Chem 2011;21:3335–45.
[180] Wang X, Sun G, Routh P, Kim D-H, Huang W, Chen P. Heteroatom-doped graphene materials: syntheses, properties and applications. Chem Soc Rev 2014;43:7067–98.
[181] Kong X-K, Chen C-L, Chen Q-W. Doped graphene for metal-free catalysis. Chem Soc Rev 2014;43:2841–57.
[182] Maiti UN, Lee WJ, Lee JM, et al. 25th anniversary article: chemically modified/doped carbon nanotubes & graphene for optimized nanostructures & nanodevices. Adv Mater 2014;26:40–67.

[183] Rao CNR, Gopalakrishnan K, Govindaraj A. Synthesis, properties and applications of graphene doped with boron, nitrogen and other elements. Nano Today 2014;9:324–43.

[184] Wang X, Shi G. An introduction to the chemistry of graphene. Phys Chem Chem Phys 2015;17:28484–504.

[185] Chen N, Huang X, Qu L. Heteroatom substituted and decorated graphene: preparation and applications. Phys Chem Chem Phys 2015;17:32077–98.

[186] Xue Y, Wu B, Bao Q, Liu Y. Controllable synthesis of doped graphene and its applications. Small 2014;10:2975–91.

[187] Wang HB, Maiyalagan T, Wang X. Review on recent progress in nitrogen-doped graphene: synthesis, characterization, and its potential applications. ACS Catal 2012;2:781–94.

[188] Li X, Fan L, Li Z, et al. Boron doping of graphene for graphene-silicon p-n junction solar cells. Adv Energy Mater 2012;2:425–9.

[189] Cattelan M, Agnoli S, Favaro M, et al. Microscopic view on a chemical vapor deposition route to boron-doped graphene nanostructures. Chem Mater 2013;25:1490–5.

[190] Zhao L, Levendorf M, Goncher S, et al. Local atomic and electronic structure of boron chemical doping in monolayer graphene. Nano Lett 2013;13:4659–65.

[191] Lv R, Chen G, Li Q, et al. Ultrasensitive gas detection of large-area boron-doped graphene. P Natl Acad Sci USA 2015;112:14527–32.

[192] Usachov DY, Fedorov AV, Petukhov AE, et al. Epitaxial B-graphene: large-scale growth and atomic structure. ACS Nano 2015;9:7314–22.

[193] Agnoli S, Favaro M. Doping graphene with boron: a review of synthesis methods, physico-chemical characterization, and emerging applications. J Mater Chem A 2016;4:5002–25.

[194] Gao H, Liu Z, Song L, et al. Synthesis of S-doped graphene by liquid precursor. Nanotechnology 2012;23:275605.

[195] Zhang J, Li J, Wang Z, et al. Low-temperature growth of large-area heteroatom-doped graphene film. Chem Mater 2014;26:2460–6.

[196] Xu X, Yang W, Li Y, et al. Heteroatom-doped graphene-like carbon films prepared by chemical vapour deposition for bifacial dye-sensitized solar cells. Chem Eng J 2015;267:289–96.

[197] Wang Z, Li P, Chen Y, et al. Synthesis, characterization and electrical properties of silicon-doped graphene films. J Mater Chem C 2015;3:6301–6.

[198] Zhang SJ, Lin SS, Li XQ, et al. Opening the band gap of graphene through silicon doping for the improved performance of graphene/GaAs heterojunction solar cells. Nanoscale 2016;8:226–32.

[199] Ci L, Song L, Jin CH, et al. Atomic layers of hybridized boron nitride and graphene domains. Nat Mater 2010;9:430–5.

[200] Chang C-K, Kataria S, Kuo C-C, et al. Band gap engineering of chemical vapor deposited graphene by in situ BN doping. ACS Nano 2013;7:1333–41.

[201] Xu J, Dong G, Jin C, Huang M, Guan L. Sulfur and nitrogen co-doped, few-layered graphene oxide as a highly efficient electrocatalyst for the oxygen-reduction reaction. ChemSusChem 2013;6:493–9.

[202] Ovezmyradov M, Magedov IV, Frolova LV, et al. Chemical vapor deposition of phosphorous- and boron-doped graphene using phenyl-containing molecules. J Nanosci Nanotechnol 2015;15:4883–6.

[203] Ma X, Ning G, Qi C, Xu C, Gao J. Phosphorus and nitrogen dual-doped few-layered porous graphene: a high-performance anode material for lithium-ion batteries. ACS Appl Mater Inter 2014;6:14415–22.

[204] Wang Q, Zhang P, Zhuo Q, Lv X, Wang J, Sun X. Direct synthesis of co-doped graphene on dielectric substrates using solid carbon sources. Nano-Micro Lett 2015;7:368–73.

[205] Wei DC, Liu YQ, Wang Y, Zhang HL, Huang LP, Yu G. Synthesis of N-doped graphene by chemical vapor deposition and its electrical properties. Nano Lett 2009;9:1752–8.

[206] Luo Z, Lim S, Tian Z, et al. Pyridinic N doped graphene: synthesis, electronic structure, and electrocatalytic property. J Mater Chem 2011;21:8038–44.

[207] Qu LT, Liu Y, Baek JB, Dai LM. Nitrogen-doped graphene as efficient metal-free electrocatalyst for oxygen reduction in fuel cells. ACS Nano 2010;4:1321–6.

[208] Lv R, Li Q, Botello-Mendez AR, et al. Nitrogen-doped graphene: beyond single substitution and enhanced molecular sensing. Sci Rep 2012;2:586.

[209] Sui Y, Zhu B, Zhang H, et al. Temperature-dependent nitrogen configuration of N-doped graphene by chemical vapor deposition. Carbon 2015;81:814–20.

[210] Jin Z, Yao J, Kittrell C, Tour JM. Large-scale growth and characterizations of nitrogen-doped monolayer graphene sheets. ACS Nano 2011;5:4112–17.

[211] Xue Y, Wu B, Jiang L, et al. Low temperature growth of highly nitrogen-doped single crystal graphene arrays by chemical vapor deposition. J Am Chem Soc 2012;134:11060–3.

[212] Capasso A, Dikonimos T, Sarto F, et al. Nitrogen-doped graphene films from chemical vapor deposition of pyridine: influence of process parameters on the electrical and optical properties. Beilstein J Nanotechnol 2015;6:2028–38.

[213] Usachov D, Vilkov O, Grueneis A, et al. Nitrogen-doped graphene: efficient growth, structure, and electronic properties. Nano Lett 2011;11:5401–7.

[214] Lu Y-F, Lo S-T, Lin J-C, et al. Nitrogen-doped graphene sheets grown by chemical vapor deposition: synthesis and influence of nitrogen impurities on carrier transport. ACS Nano 2013;7:6522–32.

[215] Sun Z, Yan Z, Yao J, Beitler E, Zhu Y, Tour JM. Growth of graphene from solid carbon sources. Nature 2010;468:549–52.

[216] Wang Z, Li P, Chen Y, et al. Synthesis of nitrogen-doped graphene by chemical vapour deposition using melamine as the sole solid source of carbon and nitrogen. J Mater Chem C 2014;2:7396–401.

[217] Reddy ALM, Srivastava A, Gowda SR, Gullapalli H, Dubey M, Ajayan PM. Synthesis of nitrogen-doped graphene films for lithium battery application. ACS Nano 2010;4:6337–42.

[218] Zhang C, Lin W, Zhao Z, et al. CVD synthesis of nitrogen-doped graphene using urea. Sci China-Phys Mech Astron 2015;58:107801.

[219] Gao H, Song L, Guo W, et al. A simple method to synthesize continuous large area nitrogen-doped graphene. Carbon 2012;50:4476–82.

[220] Ito Y, Christodoulou C, Nardi MV, Koch N, Sachdev H, Muellen K. Chemical vapor deposition of N-doped graphene and carbon films: the role of precursors and gas phase. ACS Nano 2014;8:3337–46.

[221] Bao JF, Kishi N, Soga T. Synthesis of nitrogen-doped graphene by the thermal chemical vapor deposition method from a single liquid precursor. Mater Lett 2014;117:199–203.

[222] Katoh T, Imamura G, Obata S, Saiki K. Growth of N-doped graphene from nitrogen containing aromatic compounds: the effect of precursors on the doped site. RSC Adv 2016;6:13392–8.

[223] Zhao L, He R, Zabet-Khosousi A, et al. Dopant segregation in polycrystalline monolayer graphene. Nano Lett 2015;15:1428–36.

[224] Schiros T, Nordlund D, Palova L, et al. Connecting dopant bond type with electronic structure in N-doped graphene. Nano Lett 2012;12:4025–31.

[225] Zhao L, He R, Rim KT, et al. Visualizing individual nitrogen dopants in monolayer graphene. Science 2011;333:999–1003.

Isabella Anna Vacchi, Cécilia Ménard-Moyon and Alberto Bianco

5 Chemical Functionalization of Graphene Family Members

Abstract: Thanks to their outstanding physicochemical properties, graphene and its derivatives are interesting nanomaterials with a high potential in several fields. Graphene, graphene oxide, and reduced graphene oxide, however, differ partially in their characteristics due to their diverse surface composition. Those differences influence the chemical reactivity of these materials. In the following chapter the reactivity and main functionalization reactions performed on graphene, graphene oxide, and reduced graphene oxide are discussed. A part is also dedicated to the main analytical techniques used for characterization of these materials. Functionalization of graphene and its derivatives is highly important to modulate their characteristics and design graphene-based conjugates with novel properties. Functionalization can be covalent by forming strong and stable bonds with the graphene surface, or non-covalent via π–π, electrostatic, hydrophobic, and/or van der Waals interactions. Both types of functionalization are currently exploited.

Keywords: Graphene, graphene oxide, reduced graphene oxide, covalent functionalization, non-covalent functionalization, characterization

Nanomaterials are an important product of nanotechnologies, and carbon nanomaterials are part of this family. They are getting lots of attention from different fields due to their interesting characteristics. Often, to facilitate their employment carbon nanomaterials are modified through covalent or non-covalent functionalization. In this chapter we will focus on the functionalization of graphene and its derivatives, graphene oxide (GO) and reduced graphene oxide (rGO).

5.1 Graphene

Graphene is a nanomaterial made of sp^2-hybridized carbon atoms, one-atom thick in the form of a planar sheet. It is a building block of graphite and was isolated for the first time in 2004 by Novoselov et al. [1]. Earlier, it was theoretically hypothesized as a part of graphite in 1940's [2] but considered thermodynamically unstable under ambient conditions [3].

Graphene presents excellent electrical and mechanical characteristics [1, 4–8]; however, it tends to form stable agglomerates or even restack to form graphite [9]. The functionalization of graphene can prevent this phenomenon because of strong polar–polar interactions of hydrophilic groups or bulky size of hydrophobic functions [10].

Due to its delocalized π-bonding system, graphene is chemically inert. This stable system limits graphene from covalent addition, but it makes complexation reaction possible through π–π, H–π, cation–π, anion–π and metal–π interactions. Compared to ideal graphene, graphene material produced until now contains vacancies, edges, curvatures and chemical impurities. Edge and vacancies in the basal plane of graphene are very reactive. The curvature, due to the occurrence of defects, diminishes instead of the overlap of the p_z atomic orbital. Thus, the reactivity of the carbon atoms is increased [11, 12]. Damages can be introduced in graphene also by the sonication procedure used to disperse it. Thus, it is mandatory not to sonicate it for a long time to preserve the integrity of its structure. To diminish the sonication time it is necessary to use solvents in which graphene is easily dispersible [13]. The defects are mainly oxidized carbon atoms at the edges of the layer. Ultrasound waves break the structure of graphene and form graphitic carbon fragments of variable sizes. This method is used to give the necessary energy to exfoliate graphite and let the solvent intercalate between the layers. Solvent exhibiting surface tension close to 40–50 mJ m^{-2} should be chosen, for example, benzyl benzoate, *N*,*N*-dimethylacetamide, *γ*-butyrolactone, 1,3-dimethyl-2-imidazolidinone, *N*,*N*-dimethylformamide or *N*-methyl-2-pyrrolidone [14, 15]. It is possible to decrease the damage introduced by sonication also by the use of reducing agents that prevent the formation of oxidizing radicals such as ferrocene aldehyde or tiopronin [11].

Graphene can be functionalized on the two faces [16], at the edges and at the defect sides. Compared to carbon nanotubes (CNTs), graphene has a higher exposed surface but is less reactive because there is no bond tension caused by the curvature like on CNTs. Anyway, most of the reactions that were used on CNTs have also been applied to graphene.

Graphene can be covalently or non-covalently functionalized. In the first case, there is the formation of a stable covalent bond; in the second case weak bonds like π–π interactions, hydrogen bonds and van der Waals interactions are formed. During covalent functionalization, defects are introduced and consequently there is a partial loss of the electrical conductivity of graphene; meanwhile, in non-covalent functionalization the sp^2 structure is preserved.

5.1.1 Covalent Functionalization Reactions

Covalent surface modification of graphene induces rehybridization of one or more sp^2 carbon atoms of the aromatic structure, causing the loss of electronic conjugation and a strain in the plane. Indeed, because of the change of hybridization the strictly constrained sp^2 carbon atoms have to protrude out of the plane in a tetrahedral sp^3 geometry. For this reason, edge carbon atoms react easier in covalent addition reaction [12]. The covalent organic functionalization is useful to increase the dispersibility of this material in both organic solvents and water and to combine the characteristics of graphene with other functional materials such as chromophores or polymers [17, 18]. Functionalization can happen on the surface, at the edges and at the defect sites. The defects are caused by the chemical methods used to obtain graphene and include damage in the carbon lattice, structural imperfections, adatoms and solvent molecules randomly adsorbed [13, 19]. The covalent modification can be classified in free radical addition, atomic radical addition, nucleophilic addition, cycloaddition and electrophilic substitution reactions.

5.1.1.1 Free Radical Addition Reactions

Free radicals are highly reactive organic species that interact with the aromatic structure of graphene to form covalent bonds. This functionalization has been achieved either by thermal or by photochemical treatments. However, by employing high-energy reactants there are also unintended side reactions or deficiency in homogeneity in the functionalization reaction [20]. The most common radical addition reaction is performed using aryl diazonium salts. Other reactions include addition of peroxides, the Bergman cyclization [21] and the Kolbe electrosynthesis [14, 22]. In this section we will focus on the reaction with aryl diazonium salts.

The addition of aryl diazonium salts on graphene was first accomplished by Tour and coworkers [23]. In this reaction, the aryl diazonium ion becomes an aryl radical after the elimination of nitrogen (Scheme 5.1). It has been postulated that the radical aryl moiety adds to sp^2 surface of graphene which donates an electron [14]. However, this reaction can lead to organic radicals that covalently react with graphene but also are susceptible to self-polymerization [20]. This reaction is strongly dependent on the number of graphene layers. It has been found by Strano et al. that single graphene sheets have a 10 times higher reactivity compared to bi- or multilayer graphene [24]. Meanwhile, the decrease in conductivity due to the disruption of the aromatic system can be controlled. It has been discovered by Tour and coworkers that the decrease in conductivity is related to the duration of the reaction [25].

Aryl diazonium salts can have in the para position different functional groups, such as chlorine, bromine, iodine, nitro, methoxy, carboxy and cyano groups, making this reaction highly versatile [26].

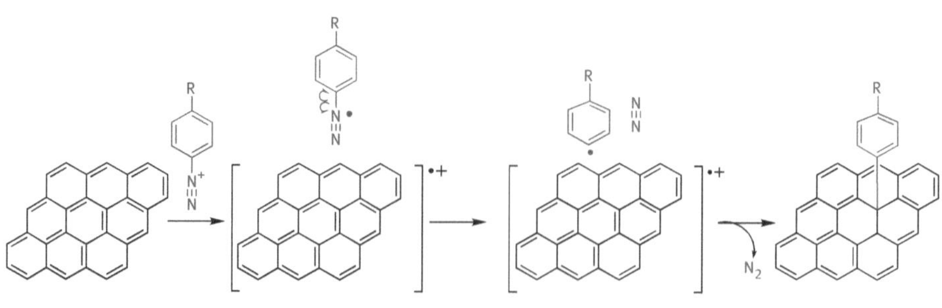

Scheme 5.1: Mechanism of the reaction with aryl diazonium salts.

This reaction, for example, has been used to attach chlorophenyl groups to investigate the antimicrobial activity of the prepared graphene product. Graphene functionalized with chlorophenyl groups has been found to be more than twice effective as antimicrobial agent than pristine graphene [27].

5.1.1.2 Nucleophilic Addition Reactions

In a nucleophilic addition graphene acts as an electron acceptor. An example is the reaction between poly-9,9′-dihexyfluorene carbazole and graphene [28, 29]. The anionic moiety is formed by using a base, leading to the generation of nitrogen anions on carbazole (Scheme 5.2). Subsequently, the nitrogen anion reacts on the surface of graphene with the creation of a covalent bond.

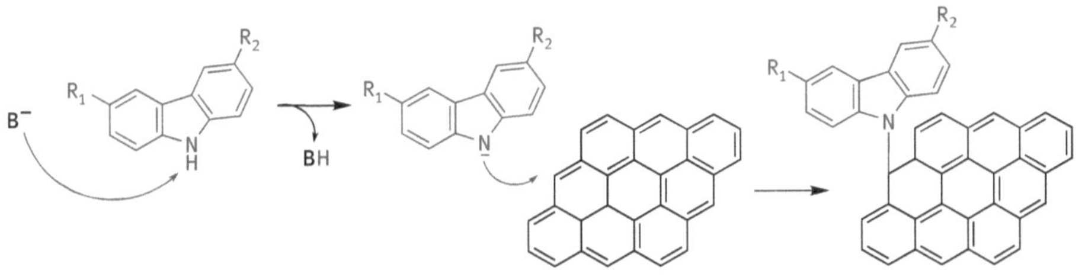

Scheme 5.2: Mechanism of the nucleophilic addition reaction with carbazole.

5.1.1.3 Cycloaddition Reactions

In the cycloaddition reactions the electrons move in a circular way that includes bond cleavage and bond formation processes at the same time without the formation of anions or cations as intermediates. The following cycloadditions have been performed on graphene: [2+1], [2+2], [3+2] and [4+2]. The most famous cycloaddition is the 1,3-dipolar cycloaddition [30], which is a [3+2] cycloaddition reaction [14]. Concerning the other cycloaddition reactions, we can mention the following: (i) [2+1] cycloadditions include reactions with carbenes [31], the Bingel reaction and reactions with nitrenes [32] to form cyclopropane or aziridine adducts, respectively; (ii) [2+2] cycloaddition is the reaction with an aryne or a benzyne to form a four-carbon atoms cycle [33] and (iii) [4+2] cycloaddition is the Diels–Alder cycloaddition [34, 35] between a diene (electron rich) and a dienophile (electron deficient). Graphene is really versatile in the Diels–Alder reaction because it can act as both the diene and the dienophile. On graphene, side reactions often occur. For this reason it is difficult to control the composition of the surface modification. For example nitrenes and carbenes can react with solvents or among themselves as well [36].

In this section we will focus on the 1,3 dipolar cycloaddition reaction and the Bingel reaction. In the 1,3-dipolar cycloaddition reaction the active component is an

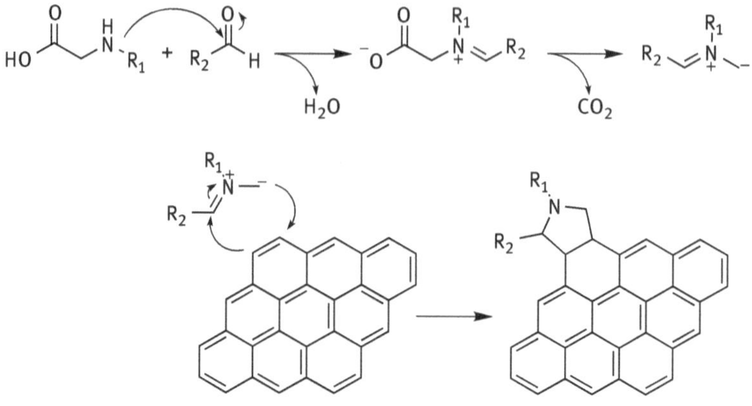

Scheme 5.3: 1,3-dipolar cycloaddition reaction.

azomethine ylide, a very reactive intermediate, obtained from a decarboxylation reaction of an α-amino acid and a carbonyl species performed under heating conditions (Scheme 5.3).

Trapalis and coworkers [37] were the first to perform this reaction on graphene. Meanwhile, Prato and coworkers have discovered that this reaction occurs on both the edges and the basal plane [30]. In this reaction pyrrolidine rings are formed on the surface of graphene. The 1,3-dipolar cycloaddition reaction is very versatile [11]. This is also important because it is possible to choose between several aldehydes or substituted α-amino acids as precursors, yielding in this way a large range of functional groups bound to the graphene surface [17].

With this reaction porphyrin moieties such as tetraphenylporphyrin and palladium tetraphenylporphyrin have been added onto graphene by Feringa and coworkers [38]. The quenching of fluorescence and the reduced excited state lifetimes suggest that there is a photoinduced energy/electron transfer between graphene and the porphyrin units.

The Bingel reaction is one of the most useful reactions for the functionalization of carbon nanomaterials, including graphene, because of the mild conditions. The reaction involves a halide derivative of a malonate ester moiety in the presence of a base (Scheme 5.4). The halide-malonate moiety is usually generated *in situ*. The base subtracts a proton from this derivative giving an enolate that attacks a C=C bond of graphene. The carbanion formed undergoes a nucleophilic substitution with the displacement of the halide to form a cyclopropane ring [14]. Through this reaction a tetrathiafulvalene moiety has been attached onto graphene by Economopoulos et al. [39]. Tetrathiafulvalene and its derivatives are used as biosensors for sugars like fructose, and it has been demonstrated that the association with graphene enhances the sensitivity of the device [40, 41].

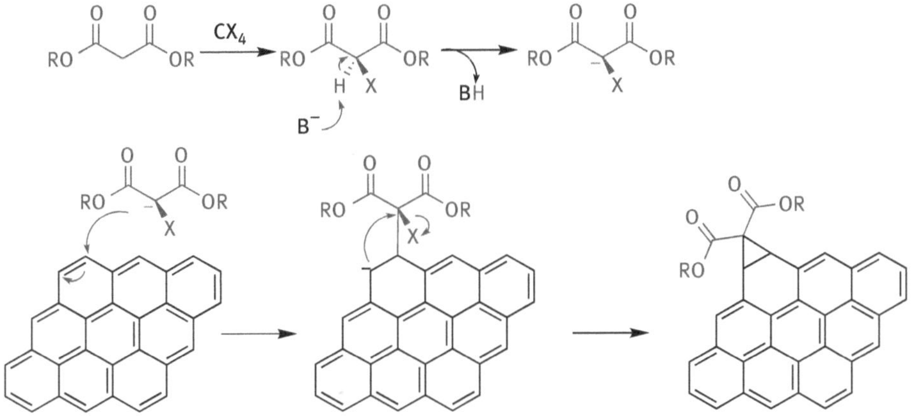

Scheme 5.4: Mechanism of the Bingel reaction.

5.1.1.4 Reaction with Atomic Radicals

If we compare the reactions performed with organic free radicals with those performed with atomic species like hydrogen, fluorine and oxygen (Scheme 5.5), in the last cases there is a minimized possibility of side reactions. Therefore, the functionalization of graphene is uniform and homogeneous [16, 20].

In the hydrogenation, after the attachment of the first atomic hydrogen, the lattice is deformed and can react easier to form the second C–H bond. The total hydrogenated form of graphene is called graphane. The hydrogenation of graphene can be performed by exposition to hydrogen-based plasmas [42].

The fluorination of graphene is similar to the hydrogenation because fluorine also forms a single bond to carbon but with an increased binding strength. As in the case of hydrogenation, fluorination leads to high level of functionalization [43]. Fluorographene is expected to be electrically insulating. There are three main methods to

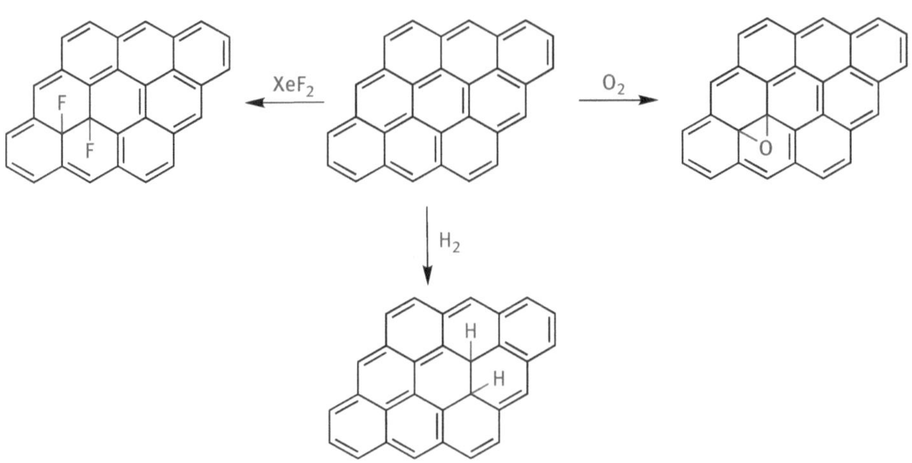

Scheme 5.5: Mechanism of the reactions performed with gas phase atomic radicals.

produce it: (i) by exposition to XeF_2 [44], (ii) by etching using fluorinated compound [45] and (iii) by liquid phase exfoliation of bulk graphite fluoride [20, 46].

Oxygenated graphene obtained through the common Hummers' method [47, 48] is highly inhomogeneous (Section 5.2). In the radical addition of atomic oxygen, oxygen atoms are expected to add on graphene forming epoxides. The formation of epoxide groups has been obtained by exposing graphene at oxygen plasmas and atomic oxygen beams [49, 50].

5.1.1.5 Electrophilic Substitution Reactions

This category of reactions allows to functionalize graphene owing to its electron-rich nature. Examples of substitution reactions are the Friedel–Crafts acylation [51] and hydrogen–lithium exchange [52]. The reactivity of metallated aromatics is higher than simple aromatic moieties. In the hydrogen–lithium exchange there is first the deprotonation and/or the carbometallation of graphene by BuLi (Scheme 5.6). The lithium graphene derivative subsequently reacts with an electrophile. Once the electrophile added, the lithium graphene derivative reacts immediately with the formation of a covalent bond [14, 52].

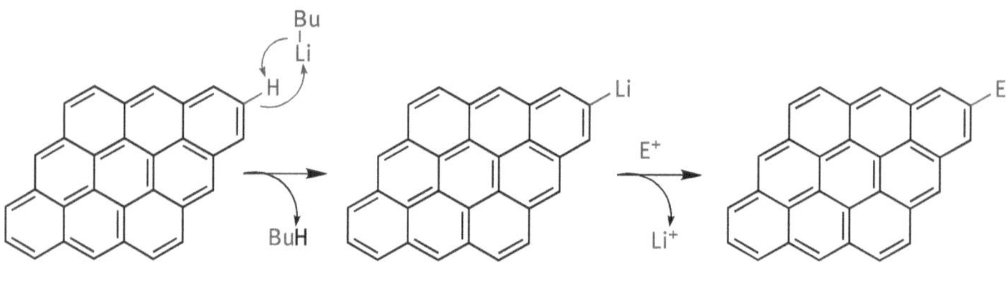

Scheme 5.6: Mechanism of the reaction with BuLi.

The Friedel–Crafts acylation allows to introduce ketone moieties. The reactive intermediate is an acyl cation formed in the presence of a Lewis acid (Scheme 5.7). Brominated flame-retardant high-density polyethylene composites containing graphene nanoplatelets were obtained using this reaction [53].

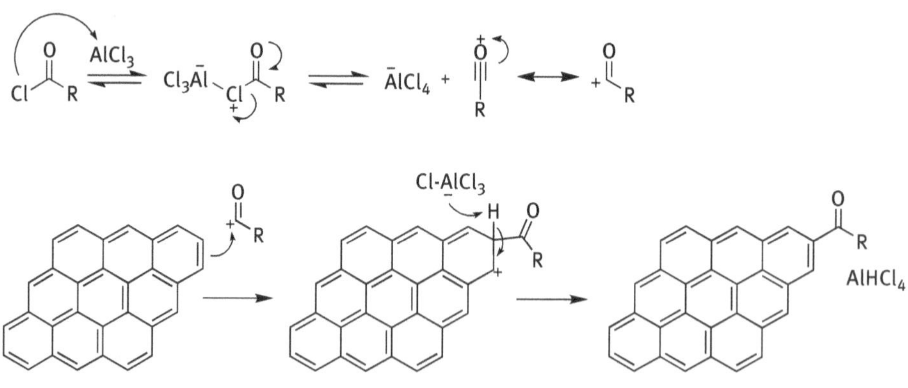

Scheme 5.7: Mechanism of the Friedel–Crafts reaction.

5.1.2 Non-covalent Functionalization Reactions

Graphene has a high specific surface area that helps the physical adsorption of molecules via non-covalent interactions such as hydrophobic, van der Waals and electrostatic [54]. This way of functionalization is an interesting method to immobilize molecules without disturbing the electronic network in an easy and reversible way [55]. The non-covalent functionalization is of fundamental importance for the immobilization of proteins, enzymes, drugs and DNA, mostly in the context of devices as a little modification in the electronic characteristics of the π system can lead to a total change in the structure and properties of the system [17, 56, 57].

π-Complexes based on graphene have been extensively investigated. The interactions have been classified in non-polar gas–π, H–π, π–π, cation–π and anion–π interactions [58, 59]. The strength of these interactions is a combination of different forces that include electrostatic, dispersive and inductive interactions as attractive forces, and exchange repulsion as repulsive forces.

Non-covalently functionalized graphene can be produced from graphite by exfoliation using different types of surfactants, such as sodium cholate [60, 61], cetyltrimethylammonium bromide [62], polyvinylpyrrolidone [63], triphenylene [64] or pyrene derivatives [65]. However, in addition to single-layer graphene a big quantity of few-layer graphene or dispersed graphite is also obtained. Layers with a diameter of a few 100 nm are produced by the use of coronene carboxylate [66]. Instead, to produce larger flakes of graphene it is possible to employ perylene derivatives (Figure 5.1) [67, 68]. Bai et al. [69] dispersed graphene with sulfonated polyaniline and obtained non-covalent functionalized graphene nanoplatelets employed to produce water-soluble composites.

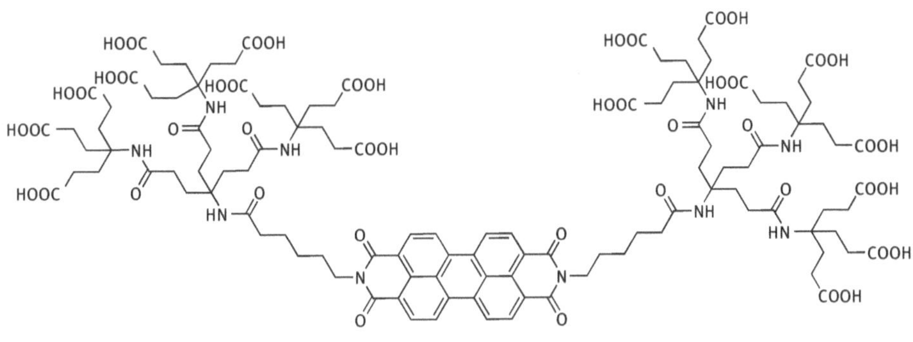

Figure 5.1: Example of a perylene derivative.

Beside graphene stabilization in solution, non-covalent functionalization can be also exploited to immobilize molecules like epirubicin on graphene for targeted chemotherapy and photothermal therapy [70]. Another interesting approach was proposed by Lee et al. which prepared a blood-compatible graphene/heparin conjugate through non-covalent functionalization [71].

5.2 Graphene oxide

GO is commonly obtained through an oxidative exfoliation treatment of graphite. The flakes tend to be predominantly monolayers. This material is made of a single graphitic layer with aromatic and oxygenated aliphatic regions randomly distributed.

This oxidized form of graphene presents different reactive oxygen-bearing functional groups, such as hydroxyl functions and epoxide groups on the basal plane, while carboxyl, carbonyl, phenol, lactone and quinone groups are mainly localized at the edges. The oxygen composition of GO is still debated, in particular regarding the presence or not of carboxylic acids. There are different models in literature [72]. The most recognized is the Lerf–Klinowski model [73], in which GO presents epoxide and hydroxyl groups in the whole structure and carboxylic acids at the edges (Figure 5.2). One important aspect to be considered is that the surface composition of GO is strongly related to the synthesis conditions used to produce it. The presence of polar oxygen-containing species in GO makes it less hydrophobic than pristine graphene. Thus, GO is more dispersible and stable than graphene in water. Thanks to its dispersibility in water, graphene materials have already found applications in the biomedical field for biosensing [74], disease diagnosis [75], cancer targeting [76], photothermal therapy [77], and as antibacterial [78] and antiviral materials [79]. However, GO is characterized by a large quantity of structural defects that can affect its physical properties. On the other hand, the presence of oxygen-containing groups allows further functionalization. The major issue with GO is the disruption of the π-network. A reductive treatment of GO induces the removal of the functional groups and rehybridization of the sp^3 carbon atoms to sp^2, partially restoring aromaticity. GO has been often employed as a starting material for the synthesis of graphene.

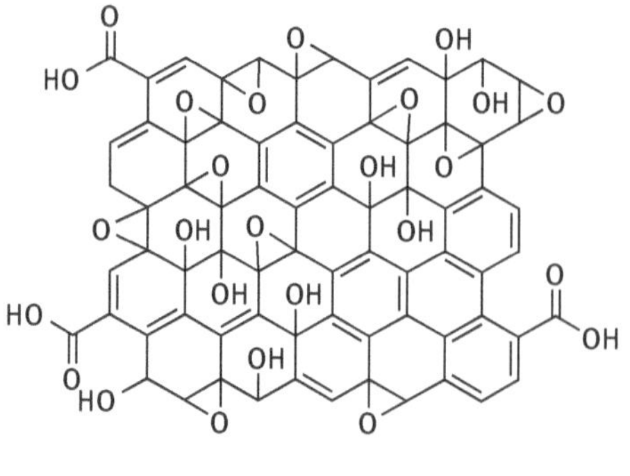

Figure 5.2: Schematic representation of GO based on Lerf–Klinowski model.

Functionalization of GO can be either covalent or non-covalent. Covalent functionalization can be done on both the sp^2-hybridized areas of GO and the oxygenated functional groups present on its surface. Meanwhile, non-covalent functionalization occurs mainly by taking advantage of the aromatic portions through hydrophobic weak interactions.

5.2.1 Covalent Functionalization Reactions

Thanks to the presence of different types of oxygenated groups on GO, such as hydroxyls, epoxides and carboxylic acids, there are many possibilities for derivatization of GO. It is difficult to compare the reactivity of the oxygenated functions of GO with the reactivity of graphene. But, due to the high amount of defects present in GO, we can consider this material more reactive. However, because of the presence of many different functional groups, there is a high risk of secondary reactions. Moreover, the functionalization of GO has to be performed in mild conditions because of the instability of this material at high temperature.

The reactions performed on the sp^2-hybridized portions of GO are mostly the same already mentioned for graphene (see Section 5.1.1) as for example the free radical addition with aryl diazonium salts [80] or the hydrogen lithium exchange. A combination of BuLi and a brominated derivative has been used to functionalize GO, leading to the concomitant alkylation, etherification and esterification (Scheme 5.8) [81].

Scheme 5.8: Reaction with BuLi. The GO model shown is simplified for clarity reasons.

The risk of side reaction during the functionalization of the oxygenated groups on the surface of GO is very high.

The functionalization of GO can be performed by derivatization of the different functional groups present on its surface. But, due to the elevated number of groups, it is difficult to obtain a good degree of selectivity.

5.2.1.1 Functionalization of Hydroxyl Groups

The hydroxyl groups are mainly located on the core surface of the oxidized graphene layer. They can undergo silanization [82] and etherification [83] reactions. The silanization reaction allows to functionalize the hydroxyl groups on the surface of GO. Indeed, R–O–Si bonds in silane are very reactive toward protic groups. For example, the functionalization of GO with *N*-[(3-trimethoxysilyl)propyl]ethylenediamine triacetic acid has been performed to prepare an antimicrobial system. This showed a higher antimicrobial activity compared to GO alone, which could be due to the synergic antimicrobial properties of graphene and ethylenediamine triacetic acid moiety [84]. It is believed that the silanization is a two-step reaction (Scheme 5.9). There is first the hydrolysis of the trialkoxy groups of the silane and then the reaction between the Si–OH groups and the hydroxyl groups of GO forming a Si–O–C bond. This reaction has been done also under microwave irradiation. This procedure permits to fulfill a one-step functionalization of GO under mild conditions in a few minutes.

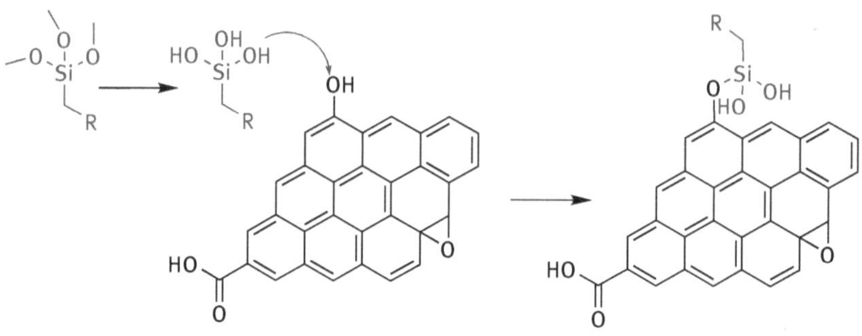

Scheme 5.9: Mechanism of the silanization reaction.

The Johnson–Claisen rearrangement is a modified Claisen reaction in which an allylic alcohol with an orthoester is used instead of the allyl vinyl ether. On GO only the hydroxyl functions that are allylic alcohols are subjected to this reaction. Therefore, not all hydroxyl groups present on the surface of GO will be functionalized. In this reaction there is first the reaction with the triethyl orthoacetate to give an ester, and then the rearrangement and the formation of a C–C bond with the graphitic surface (Scheme 5.10) [85].

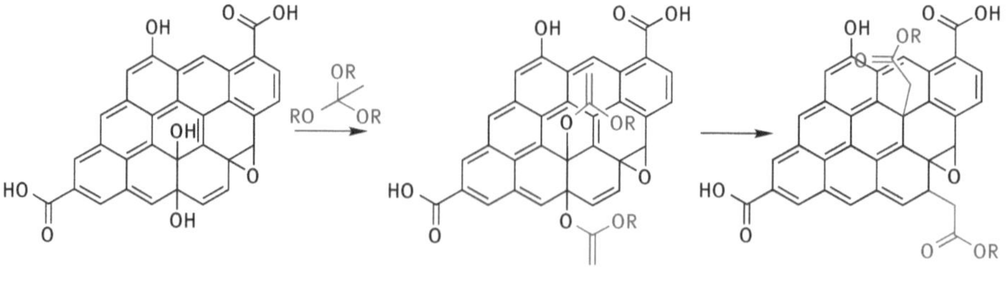

Scheme 5.10: Mechanism of the Johnson–Claisen rearrangement.

5.2.1.2 Functionalization of Carboxylic Acids

The carboxylic acids of GO are located mainly at the edges. These functional groups can react with amines to form amides or with alcohols and phenols to form esters. The amidation reaction on the carboxylic acid groups of GO can be performed in harsh or mild conditions (Scheme 5.11). In the first case the carboxylic acid is activated to form an acyl chloride using thionyl or oxalyl chloride. Then, the amino derivative is added to form the amide. In the second case, coupling agents are employed, like *N,N'*-dicyclohexylcarbodiimide or 1-ethyl-3-(3-dimethylaminopropyl)carbodiimide (EDC) and hydroxybenzotriazole. However, thionyl or oxalyl chlorides are not selective for the carboxylic acid groups as they react with the hydroxyl groups too, causing side reactions. Various amino chains have been attached onto GO [86–89]. For example, Xu et al. prepared a porphyrin-graphene nanohybrid [86].

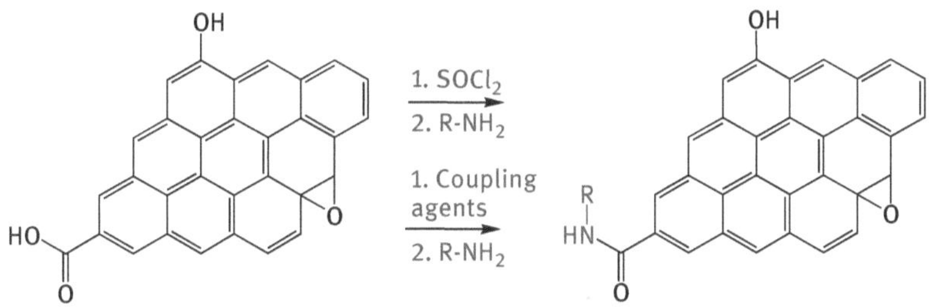

Scheme 5.11: Mechanism of the amidation reaction. In this scheme the side reaction, opening of the epoxide, is not shown.

We would like to highlight that, once we add an amino chain, an important side reaction can happen due to the reaction with the epoxide groups, with a consequent opening of the ring and covalent introduction of the amino chain forming a secondary amine bond (see Section 5.2.1.3 about nucleophilic addition on epoxides). The amidation reaction has often been used to obtain biocompatible GO by introducing polymers like polyethyleneglycol [90], dextran [91] and chitosan [92].

The esterification reaction is another way to functionalize the carboxylic acids on the surface of GO. As in the case of amidation, thionyl (or oxalyl) chloride can be employed to obtain the acyl chlorides. Then, a base has to be used to deprotonate the alcohol, resulting in the formation of an ester bond. Meanwhile, if coupling agents such as EDC and 4-dimethylaminopyridine (DMAP) are instead utilized, a base is not necessary. The conditions using EDC/DMAP are commonly known as the Steglich esterification. The Steglich conditions were first applied on graphene by Mei et al. to obtain a double functionalized GO [93]. The activation of the carboxylic acids via the formation of acyl chlorides has been used to graft a poly(3-hexylthiophene) molecule, to design heterojunction photovoltaic devices [94].

5.2.1.3 Functionalization of Epoxides

The epoxides are mainly located on the basal plane of GO. The epoxy ring can be functionalized with amine-terminated molecules or sodium azide through a nucleophilic ring-opening reaction. The opening of the epoxy ring is mainly performed with amino moieties (Scheme 5.12). Thanks to this nucleophilic addition reaction, chromophores and polymers have been introduced on GO. For example poly-*L*-lysine has been grafted to enhance the dispersibility of GO in water and its biocompatibility [95].

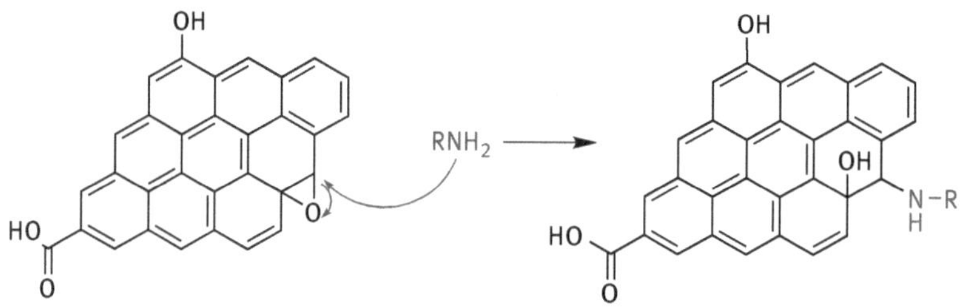

Scheme 5.12: Mechanism of the opening of epoxide with a nucleophile.

5.2.1.4 Reaction with Isocyanate

Often, due to the high amount of functional groups on the surface of GO, the reagents can interact with more than one functional groups, as it is the case for amidation reactions. Organic isocyanate derivatives can react with both carboxyl and hydroxyl groups of GO with the generation of carbamates and amides (Scheme 5.13) [96, 97]. The addition of isocyanate compounds has been used in various GO derivatives, in particular to get a pillared lamellar hybrid porous material [98].

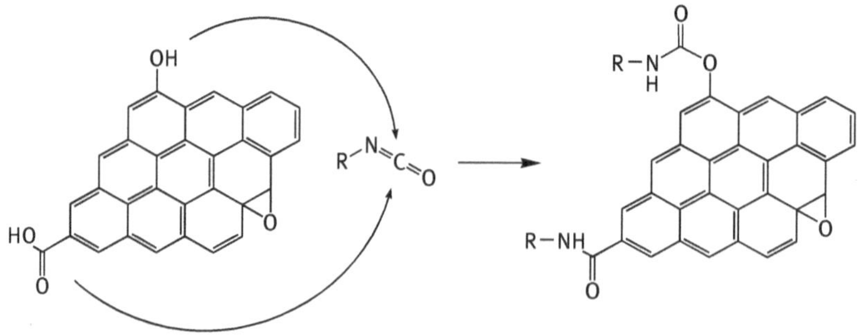

Scheme 5.13: Mechanism of the reaction with isocyanate.

5.2.2 Non-covalent Functionalization Reactions

The non-covalent functionalization of GO occurs through π–π, electrostatic binding, H-bonding, van der Waals, H–π, cation–π and anion–π interactions [99–102]. Non-covalent interactions are usually formed with surfactant molecules or amphiphilic polymers. For example, anticancer drugs were adsorbed onto PEGylated GO nanocarrier [90]. An electrochemical supercapacitor was produced by non-covalent functionalization of GO with sulfanilic acid azocromotrop and consecutive reduction of GO [103].

5.3 Reduced GO

The procedure to produce graphene is expensive, the production yields are low and the graphene has the tendency to reaggregate. For this reason, the research in this field in the last years faced a fast raise to develop methods to obtain cheaper graphene at the industrial scale. Thus, the easiest way to produce graphene is through the oxidation and exfoliation of graphite, followed by the reduction of the resulting GO into graphene [104]. Nevertheless, rGO is constituted of graphene layers that contain a higher number of defects in the basal plane (e.g., holes), compared to graphene (Figure 5.3) [68]. A complete reduction of GO to graphene has not been achieved yet [17]. This material can be obtained starting from GO with different methods: (i) thermal reduction, (ii) chemical reduction and (iii) electrochemical reduction [105–107].

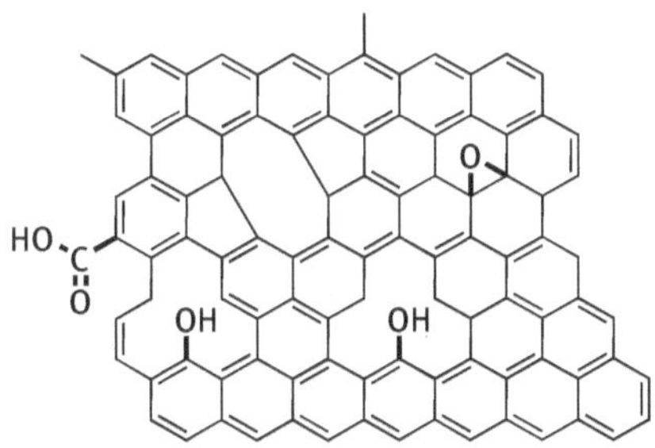

Figure 5.3: Molecular structure of rGO.

Through heat treatment at different temperatures or chemical reduction, not only water but also carbonaceous groups like CO_2 and CO are removed, causing the formation of defects. However, thanks to theoretical and experimental studies, it has been concluded that the elimination of carbonaceous groups is temperature specific. Depending on the temperature there are transformations of various functional groups, starting from the epoxides and carboxylic acids, to thermodynamically stable hydroxyl and carbonyl groups [108, 109]. Like graphene and GO, rGO can also be covalently and non-covalently functionalized. Indeed, rGO presents characteristics of both materials.

It is prevalently aromatic like graphene, but it is more reactive due to the high amount of defects and the presence of remaining oxygen-bearing functional groups.

5.3.1 Covalent Functionalization

Covalent functionalization can happen on both the aromatic structure and the oxygen-containing groups. However, rGO has a lower amount of oxygenated functions. In addition, the quantity of oxygenated functional groups depends on the reduction conditions. Thus, it is necessary to assess which functions are present and to which extent. Anyway, rGO is mainly functionalized on its aromatic surface. All covalent reactions already mentioned for graphene can be applied on rGO (see Section 5.1.1). For example, rGO has already been functionalized with phenyldiazonium derivatives [110, 111] and through carbene addition (Scheme 5.14) [112].

As already mentioned the covalent functionalization of the oxygenated functional groups depends on the type of the groups that are present on the surface. Considering that the main groups found on rGO are the hydroxyls, the functionalization reactions are mainly performed on these functions [108, 109].

The functionalization reactions that can be applied have been already mentioned in the section about the covalent functionalization of the oxygenated groups on the surface of GO (Section 5.2.1). In a rGO sample with residual oxygen-bearing species on the surface, an amine linker was introduced through nucleophilic opening of the epoxide ring [113].

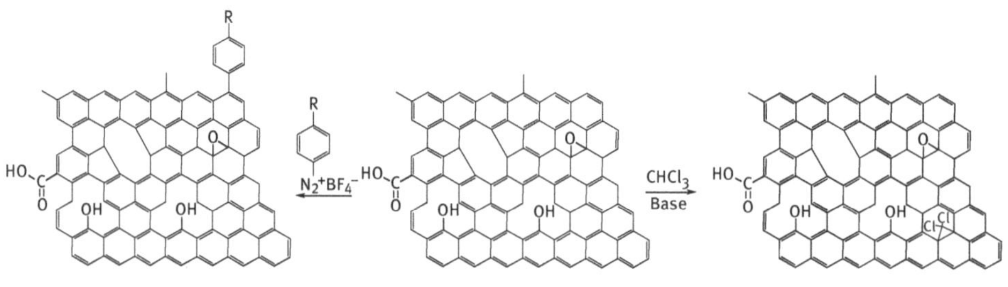

Scheme 5.14: Aryl diazonium salt radical addition and carbene addition reactions on rGO (Possible side reactions are not shown).

5.3.2 Non-covalent Functionalization Reactions

As for GO, rGO can be non-covalently functionalized with different types of polymers and small molecules through π–π stacking or van der Waals interactions [114]. Shen et al. decorated rGO with polystyrene–polyacrylamide copolymer to obtain an amphiphilic graphene that can be dispersed in polar and non-polar solvents [115].

5.4 Characterization of Graphene Family Members

There are many challenges in the nanotechnology field on the identification of the products. A precise assessment of their structure is of fundamental importance. Different characterization techniques for the identification of graphene, GO, rGO and derivatives exist. In this section the UV-visible spectroscopy, Fourier transform infrared (FTIR) spectroscopy, atomic force microscopy, transmission electron microscopy, Raman spectroscopy, X-ray photoelectron spectroscopy (XPS) and the thermogravimetric analysis (TGA) will be discussed in detail.

5.4.1 UV-Visible Spectroscopy

The UV-visible spectroscopy does not give much information on graphene materials. For GO a peak around 224 nm is due to the π–π^* transitions of C=C bonds; meanwhile, a shoulder at 300 nm is attributed to the n–π^* transition of C=O bonds (Figure 5.4). Through this technique it is also possible to estimate the number of layers of graphene because it is related to the transparency of the sample [54, 116].

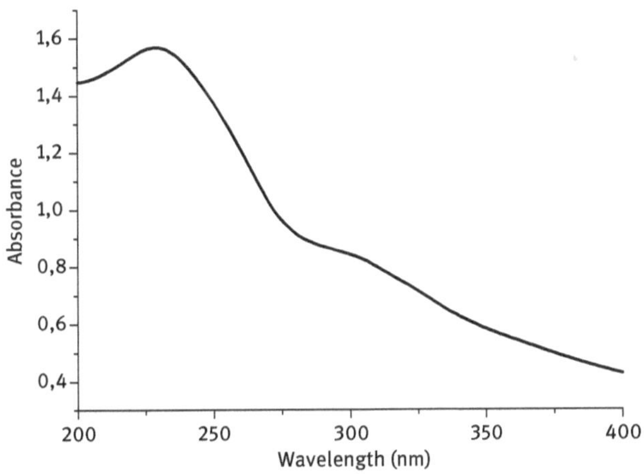

Figure 5.4: UV spectrum of GO.

5.4.2 Fourier transform infrared Spectroscopy

The FTIR spectroscopy is used to determine the functional groups of a molecule. While no significant peaks were found in graphene, for GO there are the characteristic peaks of the oxygenated functional groups (Figure 5.5). However, GO has a high humidity content due to the presence of oxygenated species. The adsorbed water is

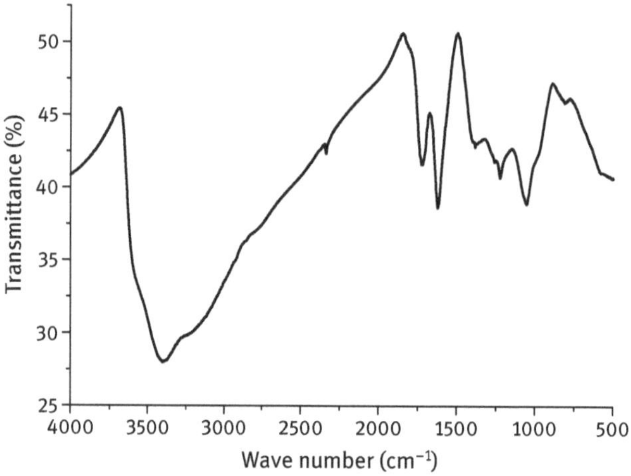

Figure 5.5: FT-IR spectrum of GO.

difficult to eliminate through thermal treatment because of GO thermal instability. For this reason, it is also possible to detect in GO spectra the strong signals of water. In the spectrum of GO, we can identify the following stretchings: the hydroxyl group and the adsorbed water at 3,800–2,200 cm^{-1}, the carbonyl group at 1,714 cm^{-1}, the vibration of the trapped water or of oxygen surface compounds and the skeletal vibration of the sp^2 carbon atoms at 1,620 cm^{-1}, the C–O groups at 1,063 cm^{-1}, and the epoxy and ether functionalities at 1,266 cm^{-1} [116–118].

5.4.3 Atomic Force Microscopy

Atomic force microscopy (AFM) is a high-resolution method used to visualize the surface topology of the materials in three-dimensional detail at the nanometer scale (Figure 5.6). In this technique a tip gathers the surface of the material to scan the surface. It is possible, through this method, to determine the lateral size, the number of layers and the thickness of graphene and GO. In GO, due to the presence of oxygen groups, the distance between layers is increased to 0.7 nm, almost two times more compared to graphene [117].

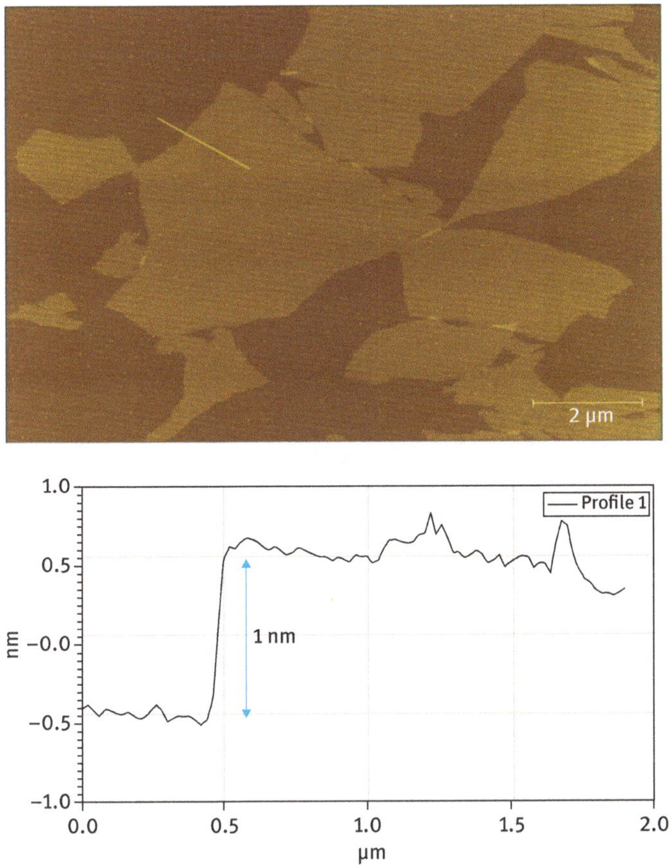

Figure 5.6: AFM image of GO on the top and its high profile on the bottom.

5.4.4 Transmission Electron Microscopy

Transmission electron microscopy (TEM) allows to observe graphene samples and assess their size, thickness, morphology and purity. By analyzing an elevated number of images, it is possible to do flake thickness statistics of graphene materials (Figure 5.7). Meanwhile, the transparency of graphene sheets gives indications about the thickness of the graphene samples. More precisely, looking at the edges it is possible to count the number of layers [54].

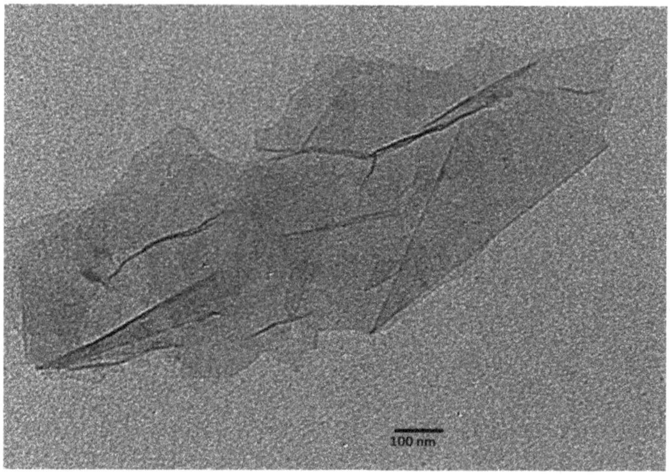

Figure 5.7: TEM image of GO.

5.4.5 Raman spectroscopy

Raman spectroscopy is a technique used for sample identification and quantification that gives information about the molecular vibrations. For carbon nanomaterials it is used to define the disorder and the order in the carbon skeleton of sp^2- and sp^3-hybridized structures. Graphene displays three main peaks at 1,350 cm^{-1} (D band), at 1,580 cm^{-1} (G band) and at 2,700 cm^{-1} (2D band) (Figure 5.8). The G band is caused by planar vibrations of carbon atoms. The D band is due to the amount of defects present in the material. The 2D band gives instead information on the number of layers of the graphite material. In the case of graphite this band has a width of around 45–60 cm^{-1}; meanwhile, for single-layer graphene it is around 24 cm^{-1}.

In GO the G band is larger and shifted to higher values (around 1,590 cm^{-1}), the D band has an increased intensity due to the higher amount of defects and the 2D band decreases and becomes wider (Figure 5.8) [117].

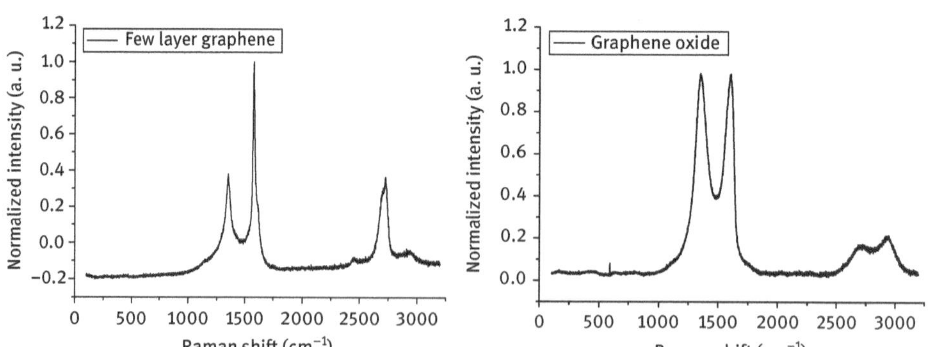

Figure 5.8: Raman spectra of few-layer graphene (left) and GO (right).

5.4.6 X-Ray Photoelectron Spectroscopy

XPS is a semi-quantitative spectroscopic technique giving information on the elemental composition, the empirical formula, the chemical state and the electronic state of the elements in a sample. In this technique the sample is irradiated with a beam of X-rays in ultra-high vacuum conditions exiting the core electrons into unoccupied atomic/molecular orbitals. XPS measures the kinetic energy and the number of electrons that are emitted from the top of the analyzed material. This technique gives information about not only the elemental composition, but also the ratios of the elements detected and the identity of the functional groups. In graphene, mainly one peak can be localized in the C1s spectral region, at 284.5 eV corresponding to the sp^2 C–C. In GO the peaks are more complex, as we can find the sp^2 and sp^3 carbon atoms, the hydroxyl, epoxy, carbonyl and carboxylic acid groups at 284.5, 285.2, 286.4, 287.1, 288.0, 289.2 eV, respectively, in the C1s region (Figure 5.9) [117, 119].

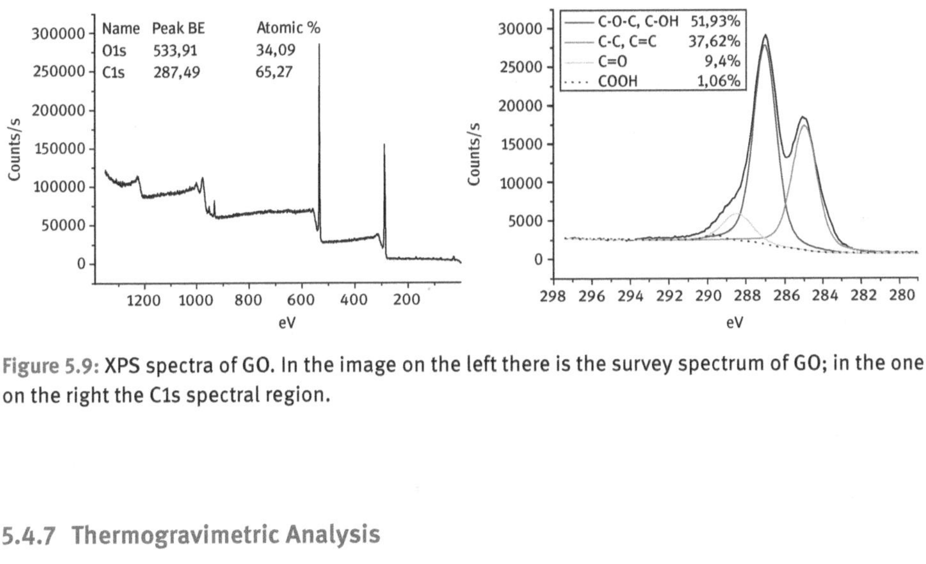

Figure 5.9: XPS spectra of GO. In the image on the left there is the survey spectrum of GO; in the one on the right the C1s spectral region.

5.4.7 Thermogravimetric Analysis

TGA allows to establish the amount of functional groups by calculating the weight loss while increasing the temperature (Figure 5.10). There are different factors that can cause this weight loss. These factors depend on the analyzed compound. Graphene is stable at high temperatures; therefore, the weight loss of the functionalized sample corresponds to the amount of functional groups introduced. Meanwhile, the thermal profile of GO shows a weight loss between 150°C and 400°C that can be assigned to the thermal decomposition of the unreacted oxygenated groups. If we analyze instead functionalized GO, the weight loss in this interval is caused by both, the unreacted oxygenated groups and the molecule covalently introduced. Therefore, to establish

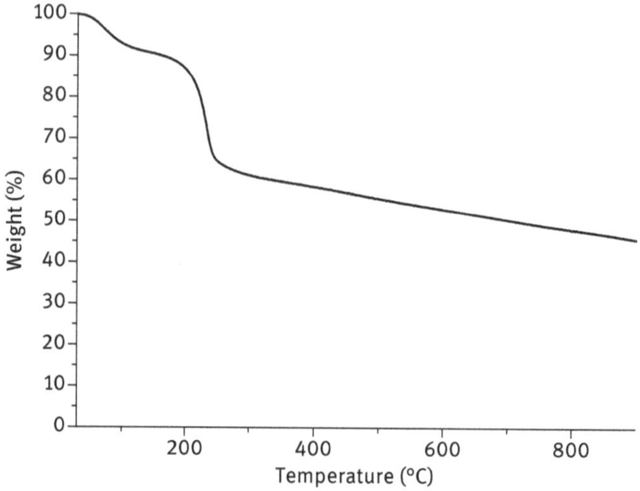

Figure 5.10: Thermogravimetric analysis of GO.

the amount of molecule introduced, the difference between the weight loss of GO and functionalized GO has to be calculated.

5.5 Conclusions

Even if graphene, GO and rGO are part of the same family, they present different characteristics and can undergo different reactions. Graphene can be covalently functionalized directly on the sp^2 carbon atom surface through free radical addition, atomic radical addition, nucleophilic addition, cycloaddition and electrophilic substitution reactions. It can be non-covalently functionalized by the formation of hydrophobic, van der Waals and electrostatic interactions with molecules such as perylene derivatives. GO instead is mainly covalently functionalized on its oxygenated functions, via the amidation and the esterification reactions on the carboxylic acids, the silanization and the Johnson–Claisen rearrangement of the hydroxyl groups and the nucleophilic addition to the epoxides. It can be non-covalently functionalized by π–π, electrostatic binding, H-bonding, van der Waals, H–π, cation–π and anion–π interactions. rGO can be considered between these two materials. It is more similar to graphene or GO, depending on the degree of reduction. The total reduction of GO to graphene has not been achieved yet. However, an important difference between graphene and rGO concerns the presence of defects in the latter one.

Which material to choose between graphene, GO and rGO depends on the purpose. GO, for instance, is mainly explored for composites and biomedical applications because of its high dispersibility. Graphene, instead, is more useful in electronics because of its electrical properties. Graphene and its derivatives are studied only for few years; therefore there is still a lot that has to be understood and discovered. Moreover, because of the complexity of this family of materials, much work remains to be done in developing reliable characterization methods that will aid to obtain precise and unambiguous chemical and structural identification.

Acknowledgments

This work was supported by the Centre National de la Recherche Scientifique (CNRS), by the Agence Nationale de la Recherche (ANR) through the LabEx project Chemistry of Complex Systems (ANR-10-LABX-0026_CSC), and by the International Center for Frontier Research in Chemistry (icFRC). The authors gratefully acknowledge financial support from EU H2020-Adhoc-2014-20 GrapheneCore1 project (No. 696656).

References

[1] Novoselov KS, Geim AK, Morozov SV, et al. Electric field effect in atomically thin carbon films. Science 2004;306(5696):666–9.
[2] Wallace PR. The band theory of graphite. Phys Rev 1947;71(9):622–34.
[3] Mermin ND. Crystalline order in two dimensions. Phys Rev 1968;176(1):250–4.
[4] Dreyer DR, Park S, Bielawski CW, Ruoff RS. The chemistry of graphene oxide. Chem Soc Rev 2010;39(1):228–40.
[5] Wang G, Yang J, Park J, et al. Facile synthesis and characterization of graphene nanosheets. J Phys Chem C 2008;112(22):8192–5.
[6] Wang G, Shen X, Wang B, Yao J, Park J. Synthesis and characterisation of hydrophilic and organophilic graphene nanosheets. Carbon 2009;47(5):1359–64.
[7] Li X, Wang X, Zhang L, Lee S, Dai H. Chemically derived, ultrasmooth graphene nanoribbon semiconductors. Science 2008;319(5867):1229–32.
[8] Blake P, Brimicombe PD, Nair RR, et al. Graphene-based liquid crystal device. Nano Lett 2008;8(6):1704–8.
[9] Li D, Müller MB, Gilje S, Kaner RB, Wallace GG. Processable aqueous dispersions of graphene nanosheets. Nat Nanotechnol 2008;3(2):101–5.
[10] Si Y, Samulski ET. Synthesis of water soluble graphene. Nano Lett 2008;8(6):1679–82.
[11] Quintana M, Vazquez E, Prato M. Organic functionalization of graphene in dispersions. Acc Chem Res 2013;46(1):138–48.
[12] Yan L, Zheng YB, Zhao F, et al. Chemistry and physics of a single atomic layer: strategies and challenges for functionalization of graphene and graphene-based materials. Chem Soc Rev 2011;41(1):97–114.
[13] Quintana M, Tapia JI, Prato M. Liquid-phase exfoliated graphene: functionalization, characterization, and applications. Beilstein J Nanotechnol 2014;5:2328–38.
[14] Chua CK, Pumera M. Covalent chemistry on graphene. Chem Soc Rev 2013;42(8):3222–33.
[15] Hernandez Y, Nicolosi V, Lotya M, et al. High-yield production of graphene by liquid-phase exfoliation of graphite. Nat Nanotechnol 2008;3(9):563–8.
[16] Boukhvalov DW, Katsnelson MI. Chemical functionalization of graphene. J Phys Condens Matter 2009;21(34):344205.

[17] Georgakilas V, Otyepka M, Bourlinos AB, et al. Functionalization of graphene: covalent and non-covalent approaches, derivatives and applications. Chem Rev 2012;112(11):6156–214.
[18] Rodríguez-Pérez L, Herranz MÁ, Martín N. The chemistry of pristine graphene. Chem Commun 2013;49(36):3721–35.
[19] Banhart F, Kotakoski J, Krasheninnikov AV. Structural defects in graphene. ACS Nano 2011;5(1):26–41.
[20] Johns JE, Hersam MC. Atomic covalent functionalization of graphene. Acc Chem Res 2013;46(1):77–86.
[21] Ma X, Li F, Wang Y, Hu A. Functionalization of pristine graphene with conjugated polymers through diradical addition and propagation. Chem – Asian J 2012;7(11):2547–50.
[22] Sarkar S, Bekyarova E, Haddon RC. Reversible grafting of α-naphthylmethyl radicals to epitaxial graphene. Angew Chem 2012;124(20):4985–8.
[23] Lomeda JR, Doyle CD, Kosynkin DV, Hwang W-F, Tour JM. Diazonium functionalization of surfactant-wrapped chemically converted graphene sheets. J Am Chem Soc 2008;130(48):16201–6.
[24] Sharma R, Baik JH, Perera CJ, Strano MS. Anomalously large reactivity of single graphene layers and edges toward electron transfer chemistries. Nano Lett 2010;10(2):398–405.
[25] Sinitskii A, Dimiev A, Corley DA, Fursina AA, Kosynkin DV, Tour JM. Kinetics of diazonium functionalization of chemically converted graphene nanoribbons. ACS Nano 2010;4(4): 1949–54.
[26] Georgakilas V. Covalent attachment of organic functional groups on pristine graphene. In: Functionalization of graphene, ed. V. Georgakilas, Wiley-VCH Verlag GmbH & Co. KGaA, Weinheim, Germany 2014:21–56.
[27] Mondal T, Bhowmick AK, Krishnamoorti R. Chlorophenyl pendant decorated graphene sheet as a potential antimicrobial agent: synthesis and characterization. J Mater Chem 2012;22(42):22481–7.
[28] Xu X, Chen J, Luo X, et al. Poly(9,9′-diheylfluorene carbazole) functionalized with reduced graphene oxide: convenient synthesis using nitrogen-based nucleophiles and potential applications in optical limiting. Chem – Eur J 2012;18(45):14384–91.
[29] Li P-P, Chen Y, Zhu J, et al. Charm-bracelet-type poly(N-vinylcarbazole) functionalized with reduced graphene oxide for broadband optical limiting. Chem – Eur J 2011;17(3):780–5.
[30] Quintana M, Spyrou K, Grzelczak M, Browne WR, Rudolf P, Prato M. Functionalization of graphene via 1,3-dipolar cycloaddition. ACS Nano 2010;4(6):3527–33.
[31] Chua CK, Ambrosi A, Pumera M. Introducing dichlorocarbene in graphene. Chem Commun 2012;48(43):5376–8.
[32] Choi J, Kim K, Kim B, Lee H, Kim S. Covalent functionalization of epitaxial graphene by azidotrimethylsilane. J Phys Chem C 2009;113(22):9433–5.
[33] Zhong X, Jin J, Li S, et al. Aryne cycloaddition: highly efficient chemical modification of graphene. Chem Commun 2010;46(39):7340–2.
[34] Sarkar S, Bekyarova E, Niyogi S, Haddon RC. Diels-Alder chemistry of graphite and graphene: graphene as diene and dienophile. J Am Chem Soc 2011;133(10):3324–7.
[35] Sarkar S, Bekyarova E, Haddon RC. Chemistry at the Dirac point: Diels–Alder reactivity of graphene. Acc Chem Res 2012;45(4):673–82.
[36] Park J, Yan M. Covalent functionalization of graphene with reactive intermediates. Acc Chem Res 2013;46(1):181–9.
[37] Georgakilas V, Bourlinos AB, Zboril R, et al. Organic functionalisation of graphenes. Chem Commun 2010;46(10):1766–8.
[38] Zhang X, Hou L, Cnossen A, et al. One-pot functionalization of graphene with porphyrin through cycloaddition reactions. Chem – Eur J 2011;17(32):8957–64.

[39] Economopoulos SP, Rotas G, Miyata Y, Shinohara H, Tagmatarchis N. Exfoliation and chemical modification using microwave irradiation affording highly functionalized graphene. ACS Nano 2010;4(12):7499–507.
[40] Campuzano S, Escamilla-Gómez V, Herranz MÁ, Pedrero M, Pingarrón JM. Development of amperometric biosensors using thiolated tetrathiafulvalene-derivatised self-assembled monolayer modified electrodes. Sens Actuators B Chem 2008;134(2):974–80.
[41] Servant A, Bianco A, Prato M, Kostarelos K. Graphene for multi-functional synthetic biology: The last "zeitgeist" in nanomedicine. Bioorg Med Chem Lett 2014;24(7):1638–49.
[42] Wojtaszek M, Tombros N, Caretta A, van Loosdrecht PHM, van Wees BJ. A road to hydrogenating graphene by a reactive ion etching plasma. J Appl Phys 2011;110(6):063715.
[43] Medeiros PVC, Mascarenhas AJS, de Brito Mota F, de Castilho CM. A DFT study of halogen atoms adsorbed on graphene layers. Nanotechnology 2010;21(48):485701.
[44] Robinson JT, Burgess JS, Junkermeier CE, et al. Properties of fluorinated graphene films. Nano Lett 2010;10(8):3001–5.
[45] Baraket M, Walton SG, Lock EH, Robinson JT, Perkins FK. The functionalization of graphene using electron-beam generated plasmas. Appl Phys Lett 2010;96(23):231501.
[46] Zbořil R, Karlický F, Bourlinos AB, et al. Graphene fluoride: a stable stoichiometric graphene derivative and its chemical conversion to graphene. Small 2010;6(24):2885–91.
[47] Hummers WS, Offeman RE. Preparation of graphitic oxide. J Am Chem Soc 1958;80(6):1339–1339.
[48] Marcano DC, Kosynkin DV, Berlin JM, et al. Improved synthesis of graphene oxide. ACS Nano 2010;4(8):4806–14.
[49] Barinov A, Malcioğlu OB, Fabris S, et al. Initial stages of oxidation on graphitic surfaces: photoemission study and density functional theory calculations. J Phys Chem C 2009;113(21):9009–13.
[50] Hossain MZ, Johns JE, Bevan KH, et al. Chemically homogeneous and thermally reversible oxidation of epitaxial graphene. Nat Chem 2012;4(4):305–9.
[51] Chua CK, Pumera M. Friedel–Crafts acylation on graphene. Chem – Asian J 2012;7(5):1009–12.
[52] Yuan C, Chen W, Yan L. Amino-grafted graphene as a stable and metal-free solid basic catalyst. J Mater Chem 2012;22(15):7456–60.
[53] Ran S, Guo Z, Han L, Fang Z. Effect of a Lewis acid catalyst on the performance of HDPE/BFR/GNPs composites. Ind Eng Chem Res 2014;53(12):4711–17.
[54] Kuila T, Bose S, Mishra AK, Khanra P, Kim NH, Lee JH. Chemical functionalization of graphene and its applications. Prog Mater Sci 2012;57(7):1061–105.
[55] Tasis D, Tagmatarchis N, Bianco A, Prato M. Chemistry of carbon nanotubes. Chem Rev 2006;106(3):1105–36.
[56] Meyer EA, Castellano RK, Diederich F. Interactions with aromatic rings in chemical and biological recognition. Angew Chem Int Ed 2003;42(11):1210–50.
[57] Singh NJ, Lee HM, Suh SB, Kim KS. De novo design approach based on nanorecognition toward development of functional molecules/materials and nanosensors/nanodevices. Pure Appl Chem 2007;79(6):1057–75.
[58] Kim KS, Tarakeshwar P, Lee JY. Molecular clusters of π-systems: theoretical studies of structures, spectra, and origin of interaction energies. Chem Rev 2000;100(11):4145–86.
[59] Riley KE, Pitoňák M, Jurečka P, Hobza P. Stabilization and structure calculations for noncovalent interactions in extended molecular systems based on wave function and density functional theories. Chem Rev 2010;110(9):5023–63.
[60] Green AA, Hersam MC. Solution phase production of graphene with controlled thickness via density differentiation. Nano Lett 2009;9(12):4031–6.
[61] Lotya M, King PJ, Khan U, De S, Coleman JN. High-concentration, surfactant-stabilized graphene dispersions. ACS Nano 2010;4(6):3155–62.

[62] Vadukumpully S, Paul J, Valiyaveettil S. Cationic surfactant mediated exfoliation of graphite into graphene flakes. Carbon 2009;47(14):3288–94.
[63] Wajid AS, Das S, Irin F, et al. Polymer-stabilized graphene dispersions at high concentrations in organic solvents for composite production. Carbon 2012;50(2):526–34.
[64] Das S, Irin F, Tanvir Ahmed HS, et al. Non-covalent functionalization of pristine few-layer graphene using triphenylene derivatives for conductive poly (vinyl alcohol) composites. Polymer 2012;53(12):2485–94.
[65] Parviz D, Das S, Ahmed HST, Irin F, Bhattacharia S, Green MJ. Dispersions of non-covalently functionalized graphene with minimal stabilizer. ACS Nano 2012;6(10):8857–67.
[66] Ghosh A, Rao KV, George SJ, Rao CNR. Noncovalent functionalization, exfoliation, and solubilization of graphene in water by employing a fluorescent coronene carboxylate. Chem – Eur J 2010;16(9):2700–4.
[67] Englert JM, Röhrl J, Schmidt CD, et al. Soluble graphene: generation of aqueous graphene solutions aided by a perylenebisimide-based bolaamphiphile. Adv Mater 2009;21(42):4265–9.
[68] Eigler S, Hirsch A. Chemistry with graphene and graphene oxide – challenges for synthetic chemists. Angew Chem Int Ed 2014;53(30):7720–38.
[69] Bai H, Xu Y, Zhao L, Li C, Shi G. Non-covalent functionalization of graphene sheets by sulfonated polyaniline. Chem Commun 2009;(13):1667–9.
[70] Yang H-W, Lu Y-J, Lin K-J, et al. EGRF conjugated PEGylated nanographene oxide for targeted chemotherapy and photothermal therapy. Biomaterials 2013;34(29):7204–14.
[71] Lee DY, Khatun Z, Lee J-H, Lee Y, In I. Blood compatible graphene/heparin conjugate through noncovalent chemistry. Biomacromolecules 2011;12(2):336–41.
[72] Szabó T, Berkesi O, Forgó P, et al. Evolution of surface functional groups in a series of progressively oxidized graphite oxides. Chem Mater 2006;18(11):2740–9.
[73] Lerf A, He H, Forster M, Klinowski J. Structure of graphite oxide revisited. J Phys Chem B 1998;102(23):4477–82.
[74] Kuila T, Bose S, Khanra P, Mishra AK, Kim NH, Lee JH. Recent advances in graphene-based biosensors. Biosens Bioelectron 2011;26(12):4637–48.
[75] Mohanty N, Berry V. Graphene-based single-bacterium resolution biodevice and DNA transistor: interfacing graphene derivatives with nanoscale and microscale biocomponents. Nano Lett 2008;8(12):4469–76.
[76] Yang K, Zhang S, Zhang G, Sun X, Lee S-T, Liu Z. Graphene in mice: ultrahigh in vivo tumor uptake and efficient photothermal therapy. Nano Lett 2010;10(9):3318–23.
[77] Yang K, Wan J, Zhang S, Tian B, Zhang Y, Liu Z. The influence of surface chemistry and size of nanoscale graphene oxide on photothermal therapy of cancer using ultra-low laser power. Biomaterials 2012;33(7):2206–14.
[78] Akhavan O, Ghaderi E. Toxicity of graphene and graphene oxide nanowalls against bacteria. ACS Nano 2010;4(10):5731–6.
[79] Akhavan O, Choobtashani M, Ghaderi E. Protein degradation and RNA efflux of viruses photocatalyzed by graphene–tungsten oxide composite under visible light irradiation. J Phys Chem C 2012;116(17):9653–9.
[80] Rebuttini V, Fazio E, Santangelo S, et al. Chemical modification of graphene oxide through diazonium chemistry and its influence on the structure–property relationships of graphene oxide–iron oxide nanocomposites. Chem – Eur J 2015;21(35):12465–74.
[81] Huang Y, Yan W, Xu Y, Huang L, Chen Y. Functionalization of graphene oxide by two-step alkylation. Macromol Chem Phys 2012;213(10–11):1101–6.
[82] Xu J, Xu M, Wu J, Wu H, Zhang WH, Li YX. Graphene oxide immobilized with ionic liquids: facile preparation and efficient catalysis for solvent-free cycloaddition of CO_2 to propylene carbonate. RSC Adv 2015;5:72361–8.

[83] Yu R, Zhang S, Luo Y, Bai R, Zhou J, Song H. Synthetic possibility of polystyrene functionalization based on hydroxyl groups of graphene oxide as nucleophiles. New J Chem 2015;39(7):5096–9.
[84] Mejias Carpio IE, Mangadlao JD, Nguyen HN, Advincula RC, Rodrigues DF. Graphene oxide functionalized with ethylenediamine triacetic acid for heavy metal adsorption and anti-microbial applications. Carbon 2014;77:289–301.
[85] Sydlik SA, Swager TM. Functional graphenic materials via a Johnson–Claisen rearrangement. Adv Funct Mater 2013;23(15):1873–82.
[86] Xu Y, Liu Z, Zhang X, et al. A graphene hybrid material covalently functionalized with porphyrin: synthesis and optical limiting property. Adv Mater 2009;21(12):1275–9.
[87] Yongsheng Liu JZ. Synthesis, characterization and optical limiting property of covalently oligothiophene-functionalized graphene material. Carbon 2009;47(13):3113–21.
[88] Zhu J, Li Y, Chen Y, et al. Graphene oxide covalently functionalized with zinc phthalocyanine for broadband optical limiting. Carbon 2011;49(6):1900–5.
[89] Tang X-Z, Li W, Yu Z-Z, et al. Enhanced thermal stability in graphene oxide covalently functionalized with 2-amino-4,6-didodecylamino-1,3,5-triazine. Carbon 2011;49(4):1258–65.
[90] Liu Z, Robinson JT, Sun X, Dai H. PEGylated nanographene oxide for delivery of water-insoluble cancer drugs. J Am Chem Soc 2008;130(33):10876–7.
[91] Zhang S, Yang K, Feng L, Liu Z. In vitro and in vivo behaviors of dextran functionalized graphene. Carbon 2011;49(12):4040–9.
[92] Hu H, Wang X, Wang J, Liu F, Zhang M, Xu C. Microwave-assisted covalent modification of graphene nanosheets with chitosan and its electrorheological characteristics. Appl Surf Sci 2011;257(7):2637–42.
[93] Mei KC, Rubio N, Costa PM, et al. Synthesis of double-clickable functionalised graphene oxide for biological applications. Chem Commun 2015;51(81): 14981–4.
[94] Yu D, Yang Y, Durstock M, Baek J-B, Dai L. Soluble P3HT-grafted graphene for efficient bilayer–heterojunction photovoltaic devices. ACS Nano 2010;4(10):5633–40.
[95] Shan C, Yang H, Han D, Zhang Q, Ivaska A, Niu L. Water-soluble graphene covalently functionalized by biocompatible poly-L-lysine. Langmuir 2009;25(20):12030–3.
[96] Stankovich S, Piner RD, Nguyen ST, Ruoff RS. Synthesis and exfoliation of isocyanate-treated graphene oxide nanoplatelets. Carbon 2006;44(15):3342–7.
[97] Wang G, Wang B, Park J, Yang J, Shen X, Yao J. Synthesis of enhanced hydrophilic and hydrophobic graphene oxide nanosheets by a solvothermal method. Carbon 2009;47(1):68–72.
[98] Zhang D-D, Zu S-Z, Han B-H. Inorganic–organic hybrid porous materials based on graphite oxide sheets. Carbon 2009;47(13):2993–3000.
[99] Dougherty DA. The cation–π interaction. Acc Chem Res 2013;46(4):885–93.
[100] Lee EC, Kim D, Jurečka P, Tarakeshwar P, Hobza P, Kim KS. Understanding of assembly phenomena by aromatic–aromatic interactions: benzene dimer and the substituted systems. J Phys Chem A 2007;111(18):3446–57.
[101] Grimme S. On the importance of electron correlation effects for the π–π interactions in cyclophanes. Chem – Eur J 2004;10(14):3423–9.
[102] Tarakeshwar P, Choi HS, Kim KS. Olefinic vs aromatic π–H interaction: a theoretical investigation of the nature of interaction of first-row hydrides with ethene and benzene. J Am Chem Soc 2001;123(14):3323–31.
[103] Jana M, Saha S, Khanra P, et al. Non-covalent functionalization of reduced graphene oxide using sulfanilic acid azocromotrop and its application as a supercapacitor electrode material. J Mater Chem A 2015;3(14):7323–31.
[104] Eigler S, Dotzer C, Hirsch A. Visualization of defect densities in reduced graphene oxide. Carbon 2012;50(10):3666–73.

[105] Pei S, Cheng H-M. The reduction of graphene oxide. Carbon 2012;50(9):3210–28.
[106] Chua CK, Pumera M. Chemical reduction of graphene oxide: a synthetic chemistry viewpoint. Chem Soc Rev 2013;43(1):291–312.
[107] Kar T, Devivaraprasad R, Singh RK, Bera B, Neergat M. Reduction of graphene oxide – a comprehensive electrochemical investigation in alkaline and acidic electrolytes. RSC Adv 2014;4(101):57781–90.
[108] Gao X, Jang J, Nagase S. Hydrazine and thermal reduction of graphene oxide: reaction mechanisms, product structures, and reaction design. J Phys Chem C 2010;114(2):832–42.
[109] Ganguly A, Sharma S, Papakonstantinou P, Hamilton J. Probing the thermal deoxygenation of graphene oxide using high-resolution in situ X-ray-based spectroscopies. J Phys Chem C 2011;115(34):17009–19.
[110] Wang H-X, Zhou K-G, Xie Y-L, et al. Photoactive graphene sheets prepared by “click” chemistry. Chem Commun 2011;47(20):5747–9.
[111] Castelaín M, Martínez G, Merino P, et al. Graphene functionalisation with a conjugated poly(fluorene) by click coupling: striking electronic properties in solution. Chem – Eur J 2012;18(16):4965–73.
[112] Ismaili H, Geng D, Sun AX, Kantzas TT, Workentin MS. Light-activated covalent formation of gold nanoparticle–graphene and gold nanoparticle–glass composites. Langmuir 2011;27(21):13261–8.
[113] Hsiao M-C, Liao S-H, Yen M-Y, et al. Preparation of covalently functionalized graphene using residual oxygen-containing functional groups. ACS Appl Mater Interfaces 2010;2(11):3092–9.
[114] Dreyer DR, Park S, Bielawski CW, Ruoff RS. The chemistry of graphene oxide. Chem Soc Rev 2009;39(1):228–40.
[115] Shen J, Hu Y, Li C, Qin C, Ye M. Synthesis of amphiphilic graphene nanoplatelets. Small 2009;5(1):82–5.
[116] Choudhary S, Mungse HP, Khatri OP. Hydrothermal deoxygenation of graphene oxide: chemical and structural evolution. Chem – Asian J 2013;8(9):2070–8.
[117] Makharza S, Cirillo G, Bachmatiuk A, et al. Graphene oxide-based drug delivery vehicles: functionalization, characterization, and cytotoxicity evaluation. J Nanoparticle Res 2013;15(12):1–26.
[118] Tamás Szabó OB. DRIFT study of deuterium-exchanged graphite oxide. Carbon 2005;43(15):3186–9.
[119] Stobinski L, Lesiak B, Malolepszy A, et al. Graphene oxide and reduced graphene oxide studied by the XRD, TEM and electron spectroscopy methods. J Electron Spectrosc Relat Phenom 2014;195:145–54.

Matilde Eredia, Artur Ciesielski and Paolo Samorì

6 Graphene via Molecule-Assisted Ultrasound-Induced Liquid-Phase Exfoliation: A Supramolecular Approach

Abstract: Graphene is a two-dimensional (2D) material holding unique optical, mechanical, thermal and electrical properties. The combination of these exceptional characteristics makes graphene an ideal model system for fundamental physical and chemical studies as well as technologically ground breaking material for a large range of applications. Graphene can be produced either following a *bottom-up* or *top-down* method. The former is based on the formation of covalent networks suitably engineered molecular building blocks undergoing chemical reaction. The latter takes place through the exfoliation of bulk graphite into individual graphene sheets. Among them, ultrasound-induced liquid-phase exfoliation (UILPE) is an appealing method, being very versatile and applicable to different environments and on various substrate types. In this chapter, we describe the recently reported methods to produce graphene via molecule-assisted UILPE of graphite, aiming at the generation of high-quality graphene. In particular, we will focus on the supramolecular approach, which consists in the use of suitably designed organic molecules during the UILPE of graphite. These molecules act as graphene dispersion-stabilizing agents during the exfoliation. This method relying on the joint effect of a solvent and ad hoc molecules to foster the exfoliation of graphite into graphene in liquid environment represents a promising and modular method toward the improvement of the process of UILPE in terms of the concentration and quality of the exfoliated material. Furthermore, exfoliations in aqueous and organic solutions are presented and discussed separately.

6.1 Introduction

Graphene, a two-dimensional (2D) honeycomb lattice of carbon atoms, has emerged as a fantastic material possessing outstanding electrical, optical, mechanical and thermal properties [1]. In view of its superior characteristics, this *wonder* material holds potential to influence future emerging technologies, including solar cells [2, 3], light-emitting devices [4], photodetectors [5–8], touch screens [9], spin valves [10, 11], ultrafast lasers [12, 13] and flexible and wearable electronics [14], to name a few. Moreover, its surface area, quantified experimentally being as high as 2,700 m^2/g [15, 16], has made graphene an attractive system for gas [17–19], and energy [15, 20, 21] storage, (micro-) optoelectronics [22–25], catalysis [26, 27], energy conversion [15], as well as in biological labeling [28].

Graphene can be produced and isolated either following the *bottom-up* or the *top-down* strategy [29, 30]. Graphene can be obtained in very high-quality sheets by exploiting the *bottom-up* covalent association of small molecular building blocks, undergoing chemical reaction to form 2D networks [31–33]; however, the quantity of materials produced with this method is limited. The growth on (catalytically active) solid surfaces achieved by chemical vapor deposition (CVD) [34, 35], or via silicon evaporation from silicon carbide [36], represents alternative *bottom-up* paths. *Top-down* approaches, which are based on the separation of graphene sheets from graphite, can be carried out under various environmental conditions [37, 38]. In particular, defect-free sheets can be obtained by making use of the micromechanical cleavage [39], and microwaves [40]. Among the *top-down* methods, liquid-phase exfoliation (LPE), which can be further divided into three submethods, that is ultrasound-induced LPE (UILPE) [41–47], electrochemical exfoliation (EE) [48–52], high-shear mixing [53, 54], is an attractive approach, being extremely versatile and applicable to various experimental conditions. Whereas *bottom-up* approaches, and in particular CVD, can yield large-sized graphene, LPE gives limited sheet sizes [43, 55–57]. Nonetheless, LPE has numerous advantages. It is a viable inexpensive process that can be easily upscaled to mass-produced dispersions processable by well-established methods, like spin coating, drop casting, screen printing and ink-jet printing [58–60]. High-yield exfoliation and dispersion of graphene in high quantities into the liquid phase is key for fundamental studies and for practical applications [14]. Hence, increasing research efforts are being dedicated to the production of graphene via LPE, and especially via UILPE and to improve the material's physicochemical and electrical properties. Remarkably, the versatility of the UILPE enables the exploration of various 2D-layered systems [41, 42, 61, 62], like transition metal dichalcogenides exhibiting different compositions, for example MoS_2, WS_2, $NbSe_2$, TaS_2, as well as graphene-like (hexagonal) structures like *h*-BN [63], thereby making it possible to modulate various physicochemical properties of 2D materials.

In this chapter, we highlight the recent advances in the field of production of high-quality graphene by means of UILPE of graphite. In particular, we discuss the mechanisms of exfoliation and methods that are used to characterize graphene. We then discuss numerous UILPE examples by subdividing them into two major classes, that is molecule-free and molecule-assisted UILPE. Moreover, exfoliation in aqueous and organic solutions is discussed separately.

6.2 Ultrasound-Induced Liquid-Phase Exfoliation

The effective exfoliation of graphite toward graphene requires the overcoming of the van der Waals attractions between the adjacent sheets. An efficient and straightforward method to lower the strength of the van der Waals interactions is the liquid immersion. Although the van der Waals interactions between adjacent sheets are

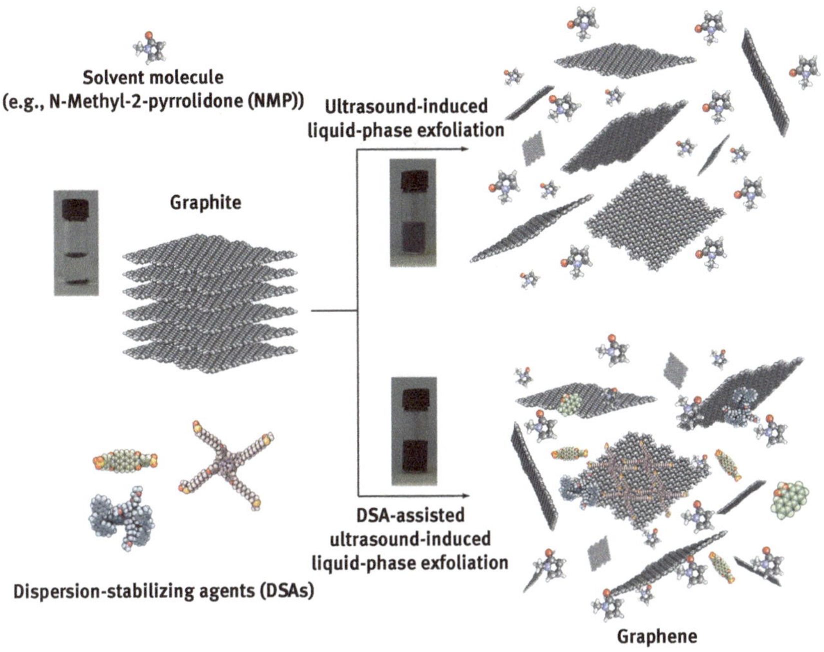

Figure 6.1: Schematic representation of the ultrasound-induced liquid-phase exfoliation (UILPE) process of graphite in the absence and presence of dispersion-stabilizing agents (DSAs). Reproduced from Ref. [43] with permission from the Royal Society of Chemistry.

weak enough to let them slide on each other in the direction perpendicular to the *c*-axis, the interactions are strong enough to inhibit the exfoliation of graphite into individual graphene sheets. This issue can be overcome by making use of external physical forces to the graphite immersed in the solvent. In particular, graphite can be exfoliated in liquid environments by using ultrasounds to extract individual layers [12, 43, 45, 55, 56, 64–67] (Figure 6.1).

After the exfoliation, the solvent–graphene interaction needs to balance the intersheet attractive forces. Solvents with surface tension (y) ~40 mJ/m^2 [45] have been found being ideal for dispersing graphene and graphitic flakes, since they minimize the interfacial tension between the graphene and the solvent, for example, *N*-methyl-2-pyrrolidone (NMP; y –40 mJ/m^2), *N*,*N*-dimethylformamide (DMF; y –37.1 mJ/m^2), *ortho*-dichlorobenzene (*o*-DCB; y – 37 mJ/m^2) [45]. Despite extensive efforts in this field, the UILPE of pristine graphite can be achieved in a limited number of solvents. The use of these solvents has some downsides that cannot be overlooked, for example, NMP and *o*-DCB can cause irritation of the eyes and respiratory tract. Furthermore, NMP and DMF are toxic for multiple organs [68, 69]. Therefore, the search of alternative solvents for graphene exfoliation has gathered considerable attention in the past few years.

In 2009, Bourlinos and collaborators [70] studied the efficiency of peculiar class of fluorinated solvents. In particular, perfluorinated analogous of hydrocarbon solvents, that is, benzene, toluene, nitrobenzene and pyridine, have been used. The performance of each solvent was reported as follows: octafluorotoluene ~ pentafluoropyridine < hexafluorobenzene < pentafluorobenzonitrile. Depending on the solvent, the concentrations of the graphene dispersions, mostly composed of few-layered graphene (FLG), ranged between 0.05 and 0.1 mg/mL.

Recently, Sun and collaborators [71] have shown that graphene can be efficiently dispersed in amine-based solvents, namely 3,3′-iminobis(*N*,*N*-dimethylpropylamine), *N*-[3-(dimethylamino)propyl]methacrylamide, 2-(*tert*-butylamino)ethyl methacrylate and 2-(dimethylamino)ethyl methacrylate. Albeit the concentration of graphene dispersions was found as high as 15 mg/mL (estimated via analysis of absorption coefficient), no information on the sheet(s) thickness was provided.

The comparison of values of graphene concentration obtained in different labs and using different methods is one of the hottest matters in the field. The concentration of graphene as well as the lateral flake size varies considerably from one article to another. This can be due to the fact that (slightly) dissimilar experimental conditions, like initial graphite concentration, sonication power, solvent volume and temperature employed by numerous groups, are (with a few exceptions) commonly not indicated or not discussed. Thus, it is of fundamental importance to define a reproducible procedure relying on the best experimental conditions for LPE and in particular for UILPE, as well as the exploits defined standards for describing the graphene dispersions. In particular, the yield of LPE is defined as the ratio between the weight of dispersed graphitic materials and that of the starting graphite flakes (Y_W [%]) [29]. The yield by single-layered graphene (SLG) (Y_S [%]) is defined as the ratio between the number of SLG and the total number of graphitic flakes in the dispersion. The yield by SLG weight (Y_{WS} [%]) is expressed as the ratio between the total mass of dispersed SLG and the total mass of all dispersed flakes. The Y_W does not provide information on the quantity of SLG, but only on the total amount of graphitic material. Yields by Y_S and Y_{WS} are more appropriate to quantify the amount of dispersed SLG sheets.

Various characterization techniques must be employed in parallel if one wants to perform a thorough qualitative and quantitative analysis of the exfoliated material. In particular, quantitative information can be evaluated by providing the yield of exfoliation expressed in terms of Y_W, the qualitative analysis provides more relevant details such as Y_S or Y_{SW}, the lateral size of the flakes and the presence/absence of defects. Presently, the only reliable method to identify the number of graphene layers (*N*) in material produced by UILPE is high-resolution transmission electron microscopy (HR-TEM) [56]. Together with the information coming from electron diffraction patterns, in HR-TEM the number of layers can be directly counted by analyzing the sheet edges [72]. The *N* can be also quantified by exploiting atomic force microscopy (AFM) imaging. However, it is worth noting that SLG height via AFM depends on the substrate and on the environmental conditions, for example, relative humidity. Actually, on mica SLG thickness amounts to ~0.4 nm [73], while on SiO_2 SLG appears to have a height of ~1 nm [39]. Raman spectroscopy allows identification of structural damages, electronic perturbations as well as non-covalent functionalization and chemical modifications (possibly) occurring during the UILPE, processing or deposition of graphene on various substrates [74, 75]. The analysis of Raman spectra can give insight into the number and position of broken conjugation areas in graphene, known as graphene atomic defects or point defects, which can influence the electronic properties of graphene. Over the past years, there has been a major step toward the understanding of Raman spectroscopy of graphene, powered by new results on doping [76–78], edge defects [79–81], electrical mobility [82, 83] and oxidation [84].

6.3 Molecule-Assisted UILPE

The use of properly selected organic molecules can enhance the exfoliation of bulk graphite into graphene, in particular when the molecules have a high energy of adsorption on the basal plane of graphene. These molecules mainly act as graphene dispersion-stabilizing agents (DSAs) interacting non-covalently with graphene, that is, through the physisorption of their hydrophobic tails on the graphene surface during the process of exfoliation of graphite via UILPE.

6.3.1 Dispersions in Aqueous Solutions

Water, the "natural" solvent, has y ~72 mJ/m^2 [85], which is too high for dispersing graphene and graphite [86], because of the hydrophobic nature of graphene sheets. Nevertheless, the use of water as a liquid medium in UILPE of graphite is of interest for the use of graphene as component in biocompatible materials [87, 88]. Remarkably, the low performance in UILPE when water is used as media can be overcome by employing DSA molecules, to help the exfoliated graphene sheets to remain dispersed and hinder their aggregation [89–97].

Among DSAs, polycyclic aromatic hydrocarbons (PAHs) [98–101] substituted with various side groups are the most studied compounds [102, 103]. Adsorption of PAHs onto the graphene surface takes place via π–π interactions between the planar π-conjugated surfaces. In these non-covalent interactions, both PAHs and graphene aromatic planar surfaces share the electrons of π-orbitals, which ultimately results in the reduction of the surface free energy of the dispersion.

In the past decades, pyrene derivatives have been successfully employed to stabilize carbon nanotube (CNT) dispersions [104], and as in the case of NMP, they have been adopted for UILPE of graphite [105–113]. Noteworthy, DSAs suitable for dispersing CNTs with a curved surface may not always be adapt for dispersing graphene,

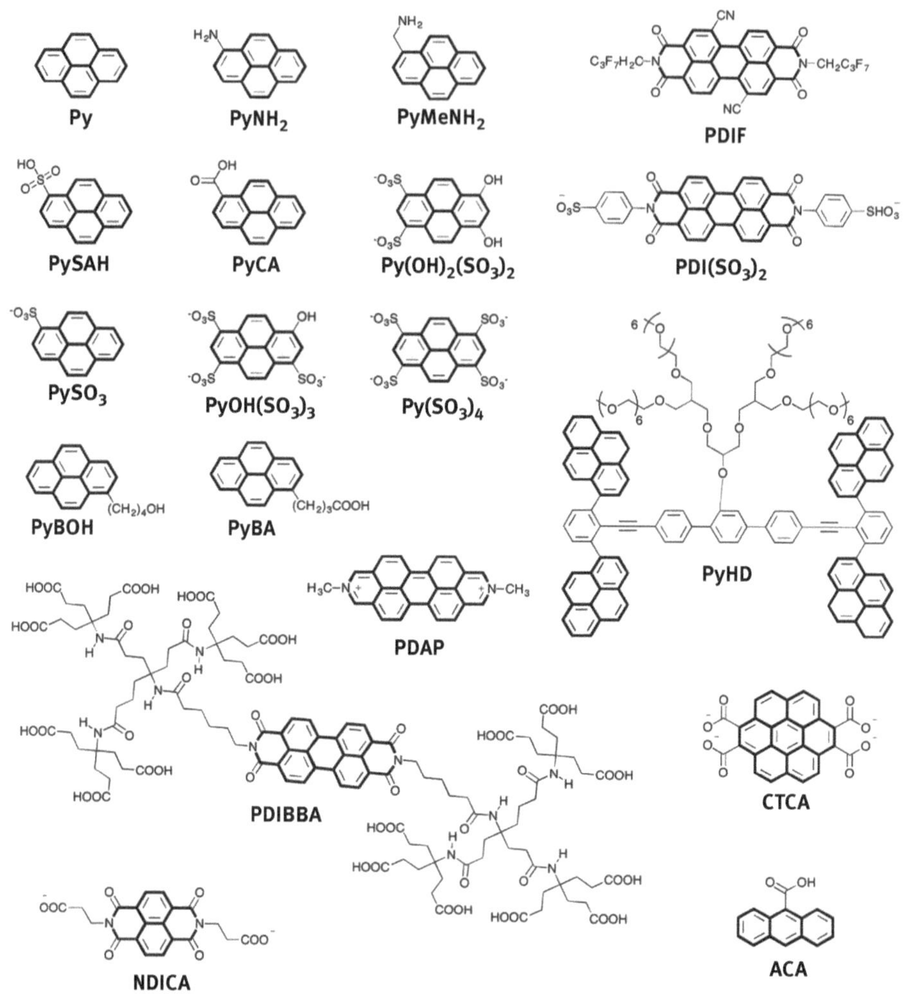

Figure 6.2: Chemical structure of functionalized polycyclic aromatic hydrocarbons (PAHs) used as dispersion-stabilizing agents (DSAs) in the ultrasound-induced liquid-phase exfoliation (UILPE) process, with their acronyms as used in the text.

which features a flat surface. Figure 6.2 depicts the chemical formulae of different PAH derivatives employed as DSAs.

In particular, in 2010, He and coworkers [113] dispersed SLG into an aqueous dispersion by using 1,3,6,8-pyrenetetrasulfonic acid tetrasodium salt ($Py(SO_3)_4$) and aminomethylpyrene ($PyMeNH_2$) as DSAs, and fabricated transparent conductive films. Yet, neither the yield nor the effectiveness of the protocol was discussed. In 2012, a number of pyrenes were utilized by Green and coworkers [110] to test their performance as DSAs. Among all investigated pyrenes, that is, pyrene (Py), 1-pyrenecarboxylic acid (PyCA), 1-pyrenebutyric acid (PyBA), 1-pyrenesulfonic acid hydrate

(PySAH), 1-aminopyrene ($PyNH_2$), 1-aminomethyl pyrene ($PyMeNH_2$), 1-pyrenebutanol (PyBOH), 1-pyrenesulfonic acid sodium salt ($PySO_3$) and 1,3,6,8-pyrenetetrasulfonic tetra acid tetra sodium salt ($Py(SO_3)_4$), the $PySO_3$ was found being the most efficient, yielding graphene dispersion concentration as high as 1 mg/mL. To quantify the amount of SLG and FLG in dispersions, the $PySO_3$-stabilized graphene samples were characterized by HR-TEM, which revealed the presence of two to four layers thick sheets, as commonly observed in UILPE samples.

Recently, Palermo and collaborators [108] went one step further and explored the thermodynamics of molecule-assisted UILPE of graphite. The authors investigated the mechanism of physisorption of different pyrenes on the surface of graphene, and successive UILPE in water. An in-depth analysis was carried out on pyrenes functionalized with sulfonic groups. In particular, $PySO_3$, 6,8-dihydroxy-1,3-pyrenedisulfonic acid disodium salt ($Py(OH)_2(SO_3)_2$), 8-hydroxy-1,3,6-pyrenetrisulfonic acid trisodium salt ($PyOH(SO_3)_3$) and 1,3,6,8-pyrenetetrasulfonic acid tetrasodium salt ($Py(SO_3)_4$) were tested. Experimental results supported by molecular dynamics simulations showed the correlation between molecule–graphene adsorption energy and the amount of dispersed graphene sheets. Remarkably, the results obtained imply that the performance of pyrene-assisted UILPE is indirectly driven by the molecular dipoles, which are not important per se, but since they facilitate the adsorption of pyrenes on graphene sheets by promoting the lateral displacement of the solvent molecules intercalating between the graphene sheets and pyrene cores.

Significantly, many other examples of increasing the yields UILPE of graphite in water by exploiting pyrene–graphene π–π interactions have been reported. In 2011, Lee and coworkers [107] revealed that an aromatic amphiphile based on a conformationally flexible aromatic segment including four pyrene moieties (PyHD) stabilizes graphene dispersions in water with the concentration of 1.5 mg/mL. In other works such as Shi et al. [114], Müllen et al. [96] and Honma et al. [106] used pyrenebutyrate (PyBA) and/or PySAH to stabilize graphene in water for use in electrochemical, solar cell and composite applications.

Recently, Liu and coworkers [115] designed naphthalene diimide (NDI) with ionic groups attached to the NDI unit through flexible alkyl spacers (see NDICA in Figure 6.2) and successfully used it as DSAs. It was demonstrated that NDICA exhibits an excellent capability to exfoliate graphite and disperse graphene in an aqueous solution, as revealed by the FLG concentration as high as 5 mg/mL after centrifuging at 1,000 rpm or 1.2 mg/mL after centrifuging at 5,000 rpm. The superior performance of the NDICA is ascribed to their chemical structures, which guarantee strong π–π interactions between the molecules and graphene and electrostatic (ionic) interactions between carboxylic groups of NDICA and water molecules.

Several perylene-based DSAs have been used to leverage the exfoliation of graphite in aqueous solutions, including sophisticated perylene diimide (PDI)-based bolaamphiphiles [94] (PDIBBA; Figure 6.4) and PDI-sulfonic acid ($PDI(SO_3)_2$) [96]. An efficient method for the preparation of graphene by UILPE in aqueous dispersions was reported by Stupp and coworkers [116], where *N*,*N′*-dimethyl-2,9-diazaperopyrenium dichloride (PDAP; see Figure 6.2) molecules were employed to stabilize dispersed graphene sheets in water. Nevertheless, the AFM study of the exfoliated material showed that the exfoliated graphene sheets are primarily two to four layers thick.

Because of the good performance of pyrenes, NDIs and PDIs as DSAs, other PAHs are also expected to stabilize graphene produced through UILPE of graphite. Some recent examples revealed that both anthracene [93] and coronene [95] can be used as DSAs. In particular, Lee and coworkers [93] proved that the exfoliation of graphite can also be achieved by non-covalent functionalization using 9-anthracene carboxylic acid (ACA). Amazingly, ACA–graphene-based composites exhibit unique electronic properties, that is, they hold high specific capacitance value of 148 F/g, as demonstrated by the performance of ACA–graphene-based ultracapacitor.

Over the past years, it has been shown that graphene can be dispersed in water by using various organic polymers as DSAs [12, 89, 91, 92, 110–112, 117–126], like cellulose acetate [127], ethyl cellulose (EC) [123], polyvinylpyrrolidone [118], lignin [124], gum arabic [119, 120], gelatin derived from animal skin and bones [121] and even more complex systems such as bovine serum albumin [117], hyaluronan (PyHA) [111] and DNA [92] functionalized with pyrene units.

While majority of DSAs employed in UILPE of graphite have ionic nature, in their pioneering work Guardia and coworkers [122] studied numerous nonionic DSAs and compared them with their ionic analogues. It was concluded that the nonionic DSAs significantly outperform their ionic counterparts. The best result, a graphene dispersion of ~1 mg/mL, was attained when the triblock copolymer Pluronic® P-123 (see Figure 6.3) was employed. AFM analysis showed that graphene sheets produced in the presence of P-123 had lateral sizes in the range of hundreds of nanometers, and almost all the sheets were thinner than five layers thick (ca. 15 % SLG), in accordance with other investigations of DSA-assisted UILPE.

More recently, Farris and coworkers [126] performed UILPE of graphite in water with the assistance of three polysaccharides, namely nonionic pullulan, cationic chitosan and anionic alginate. The effects of polymer type, initial concentration of graphite and ultrasonication time on the graphene concentration and quality were benchmarked. Upon ultrasonication treatment for 30 min, graphene dispersions with concentrations of up to 2.3 mg/mL in pullulan-stabilized dispersions and 5.5 mg/mL in the case of chitosan were produced. The obtained graphene sheets were characterized as low-defect SLG and FLG (<5). Findings arising from these studies suggest that pullulan and chitosan are outperforming alginate as DSAs, because of the different surface free energy and thermodynamic affinity.

The use of polymers as DSAs in the UILPE process is unquestionably more beneficial (from graphene dispersion concentration point of view) than the use of organic molecules; however, because of the strong polymer/graphene interactions the majority of graphene produced by making use of this approach cannot be fully separated from polymer/graphene composites.

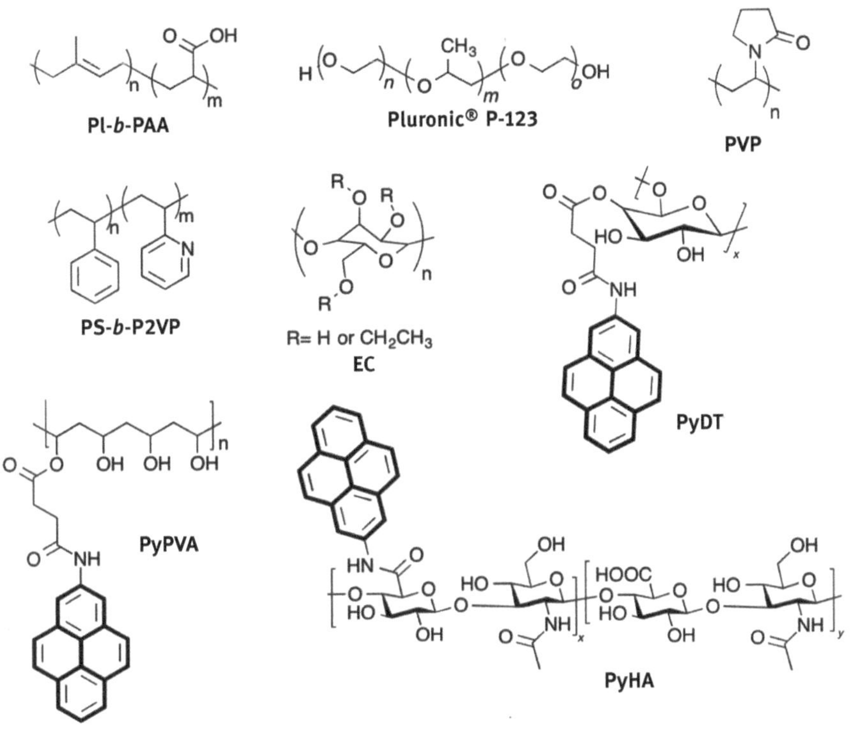

Figure 6.3: Chemical structure of polymeric DSAs used in the UILPE process, with their acronyms as used in the text.

Interestingly, in some cases the elimination of the polymeric DSAs is not compulsory, and in fact their presence can have some benefits. In particular, Shim et al. revealed that graphene dispersions can be stabilized by four different polymers based on either poly(vinyl alcohol) (PVA) or dextran functionalized with conjugated moieties, like phenyl or pyrene (see PyDT and PyPVA in Figure 6.3) [112]. The ability of these polymers to stabilize graphene dispersions was systematically explored. Moreover, graphene hydrogels and aerogels were prepared from the aqueous dispersion of the graphene/polymer. The cross-linking of the dispersed polymers in the solution gave hydrogels with embedded graphene flakes inside the polymeric networks, and the subsequent freeze-drying of the hydrogel resulted in an aerogel. Compared with a control experiment on a PVA gel electrolyte (84.2 ± 5.2 F/g), the use of graphene/polymer gel electrolyte (107.5 ± 3.1 F/g) allowed higher specific capacitances and long-term cycling stability, which was attributed to the fact that graphene embedded into the gel enhances the ionic conductivity of the gel [128]. These results clearly evidence the variety of possible applications for graphene/DSA composites.

Despite the few aforementioned examples, residuals of the PAHs and polymeric DSAs can utterly affect the electrical performance of graphene-based devices. Therefore, the search of inexpensive DSAs that have high stabilization efficiency, which

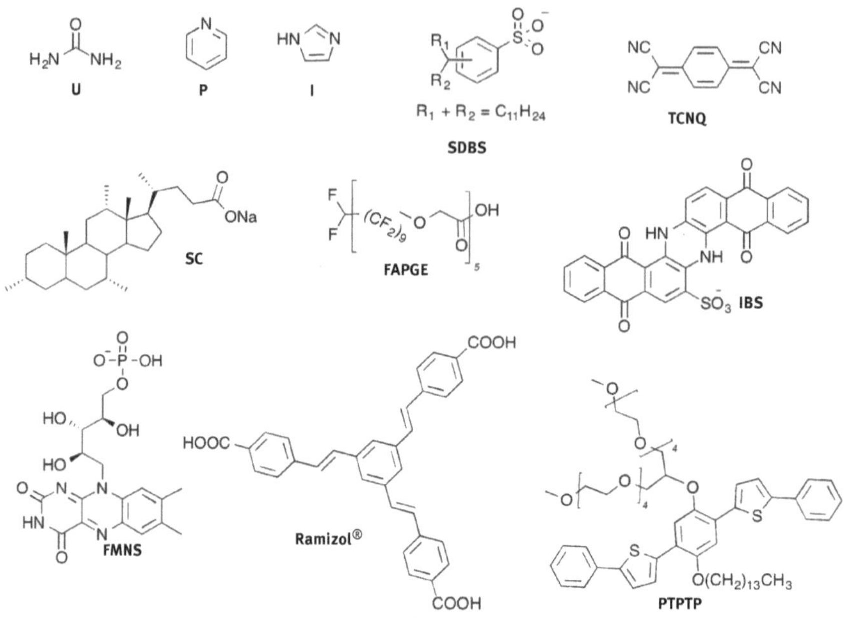

Figure 6.4: Chemical structure of small organic DSAs used in the UILPE process in water, with their acronyms as used in the text.

can be easily removed, has gathered a great attention in the field [90, 129–138], as it will largely simplify the industrialization and application of graphene produced in aqueous dispersions.

Various examples of small DSAs can be found in literature (see Figure 6.4), including urea (U) [134], sodium salt of flavin mononucleotide (FMNS) [129], 1*H*,1*H*,11*H*-eycosofluor-1-decanol polyglycidyl ether (FAPGE) [136], amphiphilic compound consisting of alternating phenylene and thienylene subunits (PTPTP) [135], and (1,3,5-tris[(1*E*)-2′-(4′-benzoic acid)vinyl]benzene) (Ramizol®) [138]. Lately, Palermo and coworkers [137] described the UILPE, processing and inclusion in polymer composites of FLG by using indanthrone blue sulfonic acid sodium salt (IBS), a very common industrial dye, and showed that IBS can be used to stabilize FLG dispersions in water. To establish that their method is suitable for applications in composites, graphene/IBS hybrids were processed into PVA, increasing its electrical conductivity by several orders of magnitude.

In 2014, Chen and coworkers [131] described the UILPE of graphite in water into SLG and FLG sheets via the direct exfoliation of highly oriented pyrolytic graphite using pyridine (P) as DSA. Electrical conductivity >5,100 S/cm was observed for filtered graphene paper, and the exfoliated graphene exhibited superior performance as a hole transport layer (HTL) compared to the commercial HTL, namely *N*,*N*-di(naphthalene-1-yl)-*N*,*N*-diphenylbenzidine (NBP), in a basic organic light-emitting diode (OLED) using tris-8-hydroxyquinolinealuminum (Alq_3) as the emissive layer.

The same group has also reported on another DSA, namely imidazole [130], which interacts with the exfoliated graphene sheets, dramatically improving the concentration of graphene dispersion (1 mg/mL) in water. Graphene film prepared from the exfoliated material revealed an electrical conductivity of 131.7 S/cm. Furthermore, an all-solid-state supercapacitor with a new design fabricated using the exfoliated graphene sheets delivered an ultrahigh area capacitance (~72 mF/cm^2).

Recently, Francis and coworkers [139] demonstrated fine patterning of graphene by screen printing using a silicon stencil and a high conductivity ink based on graphene dispersions prepared by EC-assisted UILPE in ethanol. The screen-printed graphene patterns on polyimide films showed high electrical conductivity of 1.86×10^4 S/m and remarkable mechanical flexibility, suitable for electronic applications.

Regardless of the exfoliation yields and the stability of graphene aqueous dispersions, the use of water as an exfoliation media is not recommended for the exploitation of graphene in electronic devices such as field-effect transistors. In particular, the presence of water remaining at the interface with dielectrics can augment the occurrence of charge-trapping phenomena [140]. Therefore, the use of DSA-assisted UILPE in organic solvents has to be explored.

6.3.2 Graphene Dispersions in Organic Solvents

Despite the increasing interest in the field, the knowledge gathered about the DSA-assisted UILPE of graphite in organic solvents is still relatively poor [123, 127, 141–147]. The first reported example dealing with this approach involves the use of 1,2-distearoyl-*sn*-glycero-3-phosphoethanolamine-*N*-[methoxy(polyethyleneglycol)-5000] (DSPE-mPEG; Figure 6.5) molecules in combination with tetrabutylammonium hydroxide (TBA)-inserted oleum-intercalated graphite [144]. Such mixture was sonicated in DMF to give homogeneous graphene dispersion, which was further deposited on various transparent substrates, including glass and quartz, by exploiting Langmuir–Blodgett (LB). The one-, two- and three-layer LB films on quartz afforded sheet resistances of 150, 20 and 8 kΩ/sq and transparencies of 93 %, 88 % and 83 %, respectively.

Porphyrins are known to interact with various carbon-based materials, such as graphite, CNTs and fullerenes through π-stacking that takes place between porphyrinic electron-abundant aromatic cores and conjugated surfaces [148–151]. Therefore, similar interactions are expected to take place between porphyrins and graphene [152, 153]. To explore this idea, Jung and coworkers dispersed graphene in the presence of 5,10,15,20-tetraphenyl-(4,11-acetyltioundecyl-oxyphenyl)-21*H*,23*H*-porphin (TATPP; Figure 6.5) in NMP-containing organic ammonium ions. It was found that the TATPP-assisted UILPE of graphite could be employed to produce SLG sheets with high quality.

Successful UILPE of graphite can also be attained in ethanol by exploiting a PEG terminated with a quinquethiophene moiety (5TN-PEG) as a DSA. The graphene films have been prepared via vacuum filtration, followed by removal of the 5TN-PEG molecules with tetrahydrofuran (THF) and by chemical treatment with HNO_3 and $SOCl_2$. The graphene film displayed interesting optoelectronic performance (a transmittance of 74 % at 550 nm, a sheet resistance of 0.3 kΩ/sq and $\sigma_{dc}/\sigma_{ac} = 3.65$).

Graphene directly exfoliated from graphite using UILPE and cetyltrimethylammonium bromide (CTAB) (see Figure 6.5) as a DSA has been demonstrated by Valiyaveettil and coworkers [146]. The sheets could be dispersed in common organic solvents like DMF. Characterization of the flakes by various techniques like TEM, AFM and scanning electron microscopy (SEM) revealed the exfoliation into graphene flakes of average ~1.2 nm thicknesses. Field emission measurements exhibited a turn on voltage of 7.5 V/μm and emission current densities of 0.15 mA/cm^2.

The use of small organic molecules such as DSAs is expected to promote the UILPE of graphite when the DSA molecules have a strong affinity for graphene, especially being stronger than that of the solvent/graphene interactions. A good starting point can be the use of alkanes, which are known to possess a high affinity for the basal plane of graphite/graphene [154]. In this framework, we have recently demonstrated that arachidic acid (C19CA; Figure 6.5) and *n*-octylbenzene (NOTB) can be efficaciously used to harness the exfoliation of graphene in NMP [141]. Notably, the addition of the C19CA and/or NOTB does not affect the quality and structure of graphene, when compared to the use of NMP alone, providing evidence for the non-invasive nature of the process. Furthermore, the use of alkyl chain-based DSAs lead to an enhancement of the percentage of SLG and bilayer graphene flakes. In particular, by using NOTB as DSA the amount of SLG increases by ca. 10 % and graphene concentration increases to 25 %, with respect to graphene exfoliated in pure NMP. Conversely, the use of C19CA resulted in slightly lower increase of percentage of SLG and 50 % increase of the concentration.

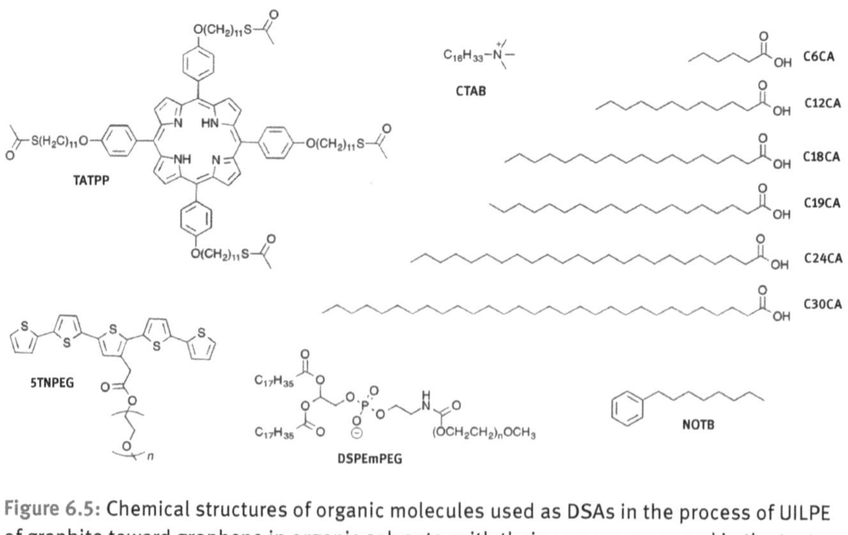

Figure 6.5: Chemical structures of organic molecules used as DSAs in the process of UILPE of graphite toward graphene in organic solvents, with their acronyms as used in the text.

We have also demonstrated that the performance of linear alkanes exposing a carboxylic acid head group as DSAs directly depends on the length of the alkane chain [143]. To this end, we explored five linear modules, that is, hexanoic acid (C6CA), lauric acid (C12CA), stearic acid (C18CA), lignoceric acid (C24CA) and melissic acid (C30CA) (Figure 6.5), whose different adsorption energies on graphene and marked tendency to form tightly packed self-assembled monolayers on such a surface affected their performances as DSAs. Analysis of the carboxylic acid-assisted UILPE showed that the concentration of graphene dispersions prepared in NMP, *o*-DCB and TCB increases linearly with the length of the alkyl tail.

The observed dependence of the UILPE yield with the length of the aliphatic chain has been interpreted by means of a thermodynamic model of molecular self-assembly on graphene. Our analysis shows that the shorter the aliphatic chain, the larger the (rotational and translational) entropic cost of forming a 2D structure will be. These results suggest that a model based on molecular mechanics for the energetics and a statistical mechanic treatment of entropy could be used to predict the efficiency of supramolecular building blocks as DSAs and guide the chemical design of the next generation of DSAs. Nevertheless, a contribution played by kinetics cannot be fully ruled out.

Besides the above examples on the use of simple DSAs in organic solvents, only a limited number of polymers have been exploited in the past years. In particular, through a modeling study, the group of Coleman have predicted that maximal graphene concentration can be attained when the polymer and solvent exhibit similar Hildebrand solubility parameters as the graphene sheets [127]. Albeit being effective, the graphene concentration in the dispersions obtained therein was unfortunately often very low; for example, 0.141 mg/mL in cyclohexanone or 0.02 mg/mL in THF. The search for a suitable polymers acting as DSAs in conventional low-polarity, low-boiling-point organic solvents is thus important to render highly concentrated graphene dispersions of high quality.

It has been demonstrated recently that the exfoliated graphene in NMP can be stabilized with an acidic solution of the poly(isoprene-*b*-acrylic acid) (PI-*b*-PAA) or poly[styrene-*b*-(2-vinylpyridine)] (PS-*b*-P2VP) block copolymers [91] (Figure 6.3). Unfortunately, the thickness of graphitic flakes amounts to 44 and 2.5 nm, respectively, implying that the PI-*b*-PAA- and/or PS-*b*-P2VP-assisted UILPE protocol requires further optimization.

6.4 Conclusion

UILPE of bulk graphite is a particularly mild, versatile and potentially upscalable approach to generate high-quality graphene inks using cheap tools available in all the labs. When exploited in the liquid phase, the supramolecular approach can be exploited to leverage the performance of UILPE of bulk graphite. UILPE in the presence of a given solvent molecule with the aid of an additional molecule acting as a DSA is a route not only to circumvent graphene reaggregation due to van der Waals attraction but also to enhance the efficiency of exfoliation, thereby compromising the effort made during exfoliation. Moreover, the presence of the DSAs interacting with the graphene through non-covalent forces can be exploited to modulate the properties of the graphene by conferring novel functions to the 2D material. UILPE approach is extremely interesting from technological point of view as many applications rely on large-scale mass production using low-cost methods such as ink-jet and screen printing or R2R. UILPE is attractive for the preparation of stable graphene inks that can be processed in thin conductive films and composites. A great deal of effort has been devoted to enhance the degree and the yield of UILPE of graphene. Yet, the yield of SLG sheets is still moderately low and requires long treatment with ultrasound. Additionally, the exfoliated material has quite an amount of graphitic waste, which adds another (purification) step into the UILPE process. In order to harness the yield and reduce the by-products, various alternative methods are being intensively investigated. Among the LPE approaches, electrochemical exfoliation (EE) emerged in the last years as the most promising tactic to disperse graphene in liquid media [48]. Differently from UILPE, EE may cause the edge oxidation of graphene flakes; nonetheless, the level of oxidation is still markedly lower than that of reduced graphene oxide (rGO), guaranteeing interesting opto-electronic properties to the materials. Moreover, the concentration of dispersion produced by ELPE can be as high as 50 mg/mL, being two orders of magnitude greater than the average concentration of UILPE dispersions.

Acknowledgments

The authors are grateful to Sébastien Haar, Thomas Mosciatti, Mirella El Gemayel, Markus Döbbelin, Jeffrey Mativetsky, Emanuele Orgiu, Emanuele Treossi, Andrea Liscio and Vincenzo Palermo for the joint research activity on graphene chemistry, which was an essential source of inspiration for this review. We acknowledge the financial support from the European Commission through the Graphene Flagship (GA-604391), the Agence Nationale de la Recherche through the LabEx project Chemistry of Complex Systems (ANR-10-LABX-0026_CSC) and the International Center for Frontier Research in Chemistry (icFRC).

References

[1] Geim AK, Novoselov KS. The rise of graphene. Nat Mater 2007;6:183–91.
[2] Wang X, Zhi LJ, Müllen K. Transparent, conductive graphene electrodes for dye-sensitized solar cells. Nano Lett 2008;8:323–7.
[3] Roy-Mayhew JD, Aksay IA. Graphene materials and their use in dye-sensitized solar cells. Chem Rev 2014;114:6323–48.
[4] Wu JB, Agrawal M, Becerril HA, et al. Organic light-emitting diodes on solution-processed graphene transparent electrodes. ACS Nano 2010;4:43–8.

[5] Xia FN, Mueller T, Lin YM, Valdes-Garcia A, Avouris P. Ultrafast graphene photodetector. Nat Nanotechnol 2009;4:839–43.
[6] Echtermeyer TJ, Britnell L, Jasnos PK, et al. Strong plasmonic enhancement of photovoltage in graphene. Nat Commun 2011;2:458–62.
[7] Konstantatos G, Badioli M, Gaudreau L, et al. Hybrid graphene-quantum dot phototransistors with ultrahigh gain. Nat Nanotechnol 2012;7:363–8.
[8] Mueller T, Xia FN, Avouris P. Graphene photodetectors for high-speed optical communications. Nat Photonics 2010;4:297–301.
[9] Bae S, Kim H, Lee Y, et al. Roll-to-roll production of 30-inch graphene films for transparent electrodes. Nat Nanotechnol 2010;5:574–8.
[10] Hill EW, Geim AK, Novoselov K, Schedin F, Blake P. Graphene spin valve devices. IEEE Trans Magn 2006;42:2694–6.
[11] Tombros N, Jozsa C, Popinciuc M, Jonkman HT, van Wees BJ. Electronic spin transport and spin precession in single graphene layers at room temperature. Nature 2007;448:571–4.
[12] Hasan T, Torrisi F, Sun Z, et al. Solution-phase exfoliation of graphite for ultrafast photonics. Phys Status Solidi B 2010;247:2953–7.
[13] Sun ZP, Hasan T, Torrisi F, et al. Graphene mode-locked ultrafast laser. ACS Nano 2010;4: 803–10.
[14] Ferrari AC, Bonaccorso F, Fal'ko V, et al. Science and technology roadmap for graphene, related two-dimensional crystals, and hybrid systems. Nanoscale 2015;7:4598–810.
[15] Stoller MD, Park SJ, Zhu YW, An JH, Ruoff RS. Graphene-based ultracapacitors. Nano Lett 2008;8:3498–502.
[16] Rao CNR, Sood AK, Subrahmanyam KS, Govindaraj A. Graphene: the new two-dimensional nanomaterial. Angew Chem Int Ed 2009;48:7752–77.
[17] Patchkovskii S, Tse JS, Yurchenko SN, Zhechkov L, Heine T, Seifert G. Graphene nanostructures as tunable storage media for molecular hydrogen. Proc Natl Acad Sci U S A 2005;102: 10439–44.
[18] Dua V, Surwade SP, Ammu S, et al. All-organic vapor sensor using inkjet-printed reduced graphene oxide. Angew Chem Int Ed 2010;49:2154–7.
[19] Schedin F, Geim AK, Morozov SV, et al. Detection of individual gas molecules adsorbed on graphene. Nat Mater 2007;6:652–5.
[20] Bonaccorso F, Colombo L, Yu GH, et al. Graphene, related two-dimensional crystals, and hybrid systems for energy conversion and storage. Science 2015;347:1246501–9.
[21] Chen KF, Song SY, Liu F, Xue DF. Structural design of graphene for use in electrochemical energy storage devices. Chem Soc Rev 2015;44:6230–57.
[22] Avouris P, Chen ZH, Perebeinos V. Carbon-based electronics. Nat Nanotechnol 2007;2:605–15.
[23] Wu JS, Pisula W, Müllen K. Graphenes as potential material for electronics. Chem Rev 2007;107:718–47.
[24] Müllen K, Rabe JP. Nanographenes as active components of single-molecule electronics and how a scanning tunneling microscope puts them to work. Acc Chem Res 2008;41:511–20.
[25] Eda G, Chhowalla M. Graphene-based composite thin films for electronics. Nano Lett 2009;9:814–8.
[26] Scheuermann GM, Rumi L, Steurer P, Bannwarth W, Mülhaupt R. Palladium nanoparticles on graphite oxide and its functionalized graphene derivatives as highly active catalysts for the Suzuki-Miyaura coupling reaction. J Am Chem Soc 2009;131:8262–70.
[27] Albero J, Garcia H. Graphenes in heterogeneous catalysis. In: Tiwari A, Titinchi S, editors. Advanced catalytic materials. 1st ed. New Jersey, and Scrivener Publishing LLC, Salem, MA: John Wiley & Sons 2015:69–120.
[28] Yang WR, Ratinac KR, Ringer SP, Thordarson P, Gooding JJ, Braet F. Carbon nanomaterials in biosensors: should you use nanotubes or graphene? Angew Chem Int Ed 2010;49:2114–38.

[29] Bonaccorso F, Lombardo A, Hasan T, Sun ZP, Colombo L, Ferrari AC. Production and processing of graphene and 2d crystals. Mater Today 2012;15:564–89.
[30] Papageorgiou DG, Kinloch IA, Young RJ. Graphene/elastomer nanocomposites. Carbon 2015;95:460–84.
[31] Chen L, Hernandez Y, Feng XL, Müllen K. From nanographene and graphene nanoribbons to graphene sheets: chemical synthesis. Angew Chem Int Ed 2012;51:7640–54.
[32] Narita A, Feng XL, Hernandez Y, et al. Synthesis of structurally well-defined and liquid-phase-processable graphene nanoribbons. Nat Chem 2014;6:126–32.
[33] Palma C-A, Samorì P. Blueprinting macromolecular electronics. Nat Chem 2011;3:431–6.
[34] Kim KS, Zhao Y, Jang H, et al. Large-scale pattern growth of graphene films for stretchable transparent electrodes. Nature 2009;457:706–10.
[35] Li XS, Cai WW, An JH, et al. Large-area synthesis of high-quality and uniform graphene films on copper foils. Science 2009;324:1312–4.
[36] Berger C, Song ZM, Li XB, et al. Electronic confinement and coherence in patterned epitaxial graphene. Science 2006;312:1191–6.
[37] Yi M, Shen ZG. A review on mechanical exfoliation for the scalable production of graphene. J Mat Chem A 2015;3:11700–15.
[38] Cai MZ, Thorpe D, Adamson DH, Schniepp HC. Methods of graphite exfoliation. J Mater Chem 2012;22:24992–5002.
[39] Novoselov KS, Jiang D, Schedin F, et al. Two-dimensional atomic crystals. Proc Natl Acad Sci USA 2005;102:10451–3.
[40] Matsumoto M, Saito Y, Park C, Fukushima T, Aida T. Ultrahigh-throughput exfoliation of graphite into pristine “single-layer” graphene using microwaves and molecularly engineered ionic liquids. Nat Chem 2015;7:730–6.
[41] Coleman JN, Lotya M, O’Neill A, et al. Two-dimensional nanosheets produced by liquid exfoliation of layered materials. Science 2011;331:568–71.
[42] Nicolosi V, Chhowalla M, Kanatzidis MG, Strano MS, Coleman JN. Liquid exfoliation of layered materials. Science 2013;340:1420–40.
[43] Ciesielski A, Samorì P. Graphene via sonication assisted liquid-phase exfoliation. Chem Soc Rev 2014;43:381–98.
[44] Parvez K, Yang S, Feng XL, Müllen K. Exfoliation of graphene via wet chemical routes. Synth Met 2015;210:123–132.
[45] Hernandez Y, Nicolosi V, Lotya M, et al. High-yield production of graphene by liquid-phase exfoliation of graphite. Nat Nanotechnol 2008;3:563–8.
[46] Zhong YL, Tian ZM, Simon GP, Li D. Scalable production of graphene via wet chemistry: progress and challenges. Mater Today 2015;18:73–8.
[47] Schlierf A, Samorì P, Palermo V. Graphene-organic composites for electronics: optical and electronic interactions in vacuum, liquids and thin solid films. J Mater Chem C 2014;2:3129–43.
[48] Abdelkader AM, Cooper AJ, Dryfe RAW, Kinloch IA. How to get between the sheets: a review of recent works on the electrochemical exfoliation of graphene materials from bulk graphite. Nanoscale 2015;7:6944–56.
[49] Parvez K, Wu ZS, Li RJ, et al. Exfoliation of graphite into graphene in aqueous solutions of inorganic salts. J Am Chem Soc 2014;136:6083–91.
[50] Wang GX, Wang B, Park J, Wang Y, Sun B, Yao J. Highly efficient and large-scale synthesis of graphene by electrolytic exfoliation. Carbon 2009;47:3242–6.
[51] Parvez K, Li RJ, Puniredd SR, et al. Electrochemically exfoliated graphene as solution-processable, highly conductive electrodes for organic electronics. ACS Nano 2013;7:3598–606.
[52] Munuera JM, Paredes JI, Villar-Rodil S, et al. High quality, low oxygen content and biocompatible graphene nanosheets obtained by anodic exfoliation of different graphite types. Carbon 2015;94:729–39.

[53] Paton KR, Varrla E, Backes C, et al. Scalable production of large quantities of defect-free few-layer graphene by shear exfoliation in liquids. Nat Mater 2014;13:624–30.
[54] Varrla E, Paton KR, Backes C, et al. Turbulence-assisted shear exfoliation of graphene using household detergent and a kitchen blender. Nanoscale 2014;6:11810–9.
[55] Coleman JN. Liquid-phase exfoliation of nanotubes and graphene. Adv Funct Mater 2009;19:3680–95.
[56] Coleman JN. Liquid exfoliation of defect-free graphene. Acc Chem Res 2013;46:14–22.
[57] Marago OM, Bonaccorso F, Saija R, et al. Brownian motion of graphene. ACS Nano 2010;4: 7515–23.
[58] Torrisi F, Hasan T, Wu WP, et al. Inkjet-printed graphene electronics. ACS Nano 2012;6: 2992–3006.
[59] Capasso A, Del Rio Castillo AE, Sun H, Ansaldo A, Pellegrini V, Bonaccorso F. Ink-jet printing of graphene for flexible electronics: an environmentally-friendly approach. Solid State Commun 2015;224:53–63.
[60] Torrisi F, Coleman JN. Electrifying inks with 2D materials. Nat Nanotechnol 2014;9:738–9.
[61] Cunningham G, Lotya M, Cucinotta CS, et al. Solvent exfoliation of transition metal dichalcogenides: dispersibility of exfoliated nanosheets varies only weakly between compounds. ACS Nano 2012;6:3468–80.
[62] Lee K, Kim HY, Lotya M, Coleman JN, Kim GT, Duesberg GS. Electrical characteristics of molybdenum disulfide flakes produced by liquid exfoliation. Adv Mater 2011;23:4178–+.
[63] Bari R, Parviz D, Khabaz F, et al. Liquid phase exfoliation and crumpling of inorganic nanosheets. Phys Chem Chem Phys 2015;17:9383–93.
[64] Khan U, O'Neill A, Lotya M, De S, Coleman JN. High-concentration solvent exfoliation of graphene. Small 2010;6:864–71.
[65] Khan U, O'Neill A, Porwal H, May P, Nawaz K, Coleman JN. Size selection of dispersed, exfoliated graphene flakes by controlled centrifugation. Carbon 2012;50:470–5.
[66] O'Neill A, Khan U, Nirmalraj PN, Boland J, Coleman JN. Graphene dispersion and exfoliation in low boiling point solvents. J Phys Chem C 2011;115:5422–8.
[67] Zhang XY, Coleman AC, Katsonis N, Browne WR, van Wees BJ, Feringa BL. Dispersion of graphene in ethanol using a simple solvent exchange method. Chem Commun 2010;46:7539–41.
[68] Solomon HM, Burgess BA, Kennedy GL, Staples RE. 1-Methyl-2-pyrrolidone (NMP): Reproductive and developmental toxicity study by inhalation in the rat. Drug Chem Toxicol 1995;18:271–93.
[69] Kennedy GL, Sherman H. Acute and subchronic toxicity of dimethylformamide and dimethylacetamide following various routes of administration. Drug Chem Toxicol 1986;9:147–70.
[70] Bourlinos AB, Georgakilas V, Zboril R, Steriotis TA, Stubos AK. Liquid-phase exfoliation of graphite towards solubilized graphenes. Small 2009;5:1841–5.
[71] Sun ZY, Huang X, Liu F, et al. Amine-based solvents for exfoliating graphite to graphene outperform the dispersing capacity of N-methyl-pyrrolidone and surfactants. Chem Commun 2014;50:10382–5.
[72] Ferrari AC, Meyer JC, Scardaci V, et al. Raman spectrum of graphene and graphene layers. Phys Rev Lett 2006;97:187401.
[73] Valles C, Drummond C, Saadaoui H, et al. Solutions of negatively charged graphene sheets and ribbons. J Am Chem Soc 2008;130:15802–4.
[74] Ferrari AC, Basko DM. Raman spectroscopy as a versatile tool for studying the properties of graphene. Nat Nanotechnol 2013;8:235–46.
[75] Casiraghi C. Raman spectroscopy of graphene. In: Yarwood J, Douthwaite R, Duckett S, editors. Spectroscopic properties of inorganic and organometallic compounds: techniques, materials and applications, Volume 43. Cambridge, UK: The Royal Society of Chemistry 2012:29–56.
[76] Chen CF, Park CH, Boudouris BW, et al. Controlling inelastic light scattering quantum pathways in graphene. Nature 2011;471:617–20.

[77] Das A, Chakraborty B, Piscanec S, Pisana S, Sood AK, Ferrari AC. Phonon renormalization in doped bilayer graphene. Phys Rev B 2009;79:155417–23.
[78] Das A, Pisana S, Chakraborty B, et al. Monitoring dopants by Raman scattering in an electrochemically top-gated graphene transistor. Nat Nanotechnol 2008;3:210–5.
[79] Basko DM. Boundary problems for Dirac electrons and edge-assisted Raman scattering in graphene. Phys Rev B 2009;79:205428.
[80] Casiraghi C, Hartschuh A, Qian H, et al. Raman spectroscopy of graphene edges. Nano Lett 2009;9:1433–41.
[81] You YM, Ni ZH, Yu T, Shen ZX. Edge chirality determination of graphene by Raman spectroscopy. Appl Phys Lett 2008;93:163112–5.
[82] Chen JH, Cullen WG, Jang C, Fuhrer MS, Williams ED. Defect scattering in graphene. Phys Rev Lett 2009;102:236805–8.
[83] Ni ZH, Ponomarenko LA, Nair RR, et al. On resonant scatterers as a factor limiting carrier mobility in graphene. Nano Lett 2010;10:3868–72.
[84] Gokus T, Nair RR, Bonetti A, et al. Making graphene luminescent by oxygen plasma treatment. ACS Nano 2009;3:3963–8.
[85] Israelachvili JN. Intermolecular and surface forces, 3rd ed. Waltham: Academic Press; 2011.
[86] Wang SR, Zhang Y, Abidi N, Cabrales L. Wettability and surface free energy of graphene films. Langmuir 2009;25:11078–81.
[87] Bianco A. Graphene: safe or toxic? The two faces of the medal. Angew Chem Int Ed 2013;52:4986–97.
[88] Kurapati R, Russier J, Squillaci MA, et al. Dispersibility-dependent biodegradation of graphene oxide by myeloperoxidase. Small 2015;11:3985–94.
[89] Lotya M, Hernandez Y, King PJ, et al. Liquid phase production of graphene by exfoliation of graphite in surfactant/water solutions. J Am Chem Soc 2009;131:3611–20.
[90] Lotya M, King PJ, Khan U, De S, Coleman JN. High-concentration, surfactant-stabilized graphene dispersions. ACS Nano 2010;4:3155–62.
[91] Skaltsas T, Karousis N, Yan H, Wang C-R, Pispas S, Tagmatarchis N. Graphene exfoliation in organic solvents and switching solubility in aqueous media with the aid of amphiphilic block copolymers. J Mater Chem 2012;22:21507–12.
[92] Liu F, Choi JY, Seo TS. DNA mediated water-dispersible graphene fabrication and gold nanoparticle-graphene hybrid. Chem Commun 2010;46:2844–6.
[93] Bose S, Kuila T, Mishra AK, Kim NH, Lee JH. Preparation of non-covalently functionalized graphene using 9-anthracene carboxylic acid. Nanotechnology 2011;22:1–7.
[94] Englert JM, Rohrl J, Schmidt CD, et al. Soluble graphene: generation of aqueous graphene solutions aided by a perylenebisimide-based bolaamphiphile. Adv Mater 2009;21:4265–9.
[95] Ghosh A, Rao KV, George SJ, Rao CNR. Noncovalent functionalization, exfoliation, and solubilization of graphene in water by employing a fluorescent coronene carboxylate. Chem Eur J 2010;16:2700–4.
[96] Su Q, Pang SP, Alijani V, Li C, Feng XL, Müllen K. Composites of graphene with large aromatic molecules. Adv Mater 2009;21:3191–5.
[97] Zhang H, Wen JQ, Meng XP, et al. An Improved Method to Increase the Concentration of Graphene in Organic Solvent. Chem Lett 2012;41:747–9.
[98] Feng XL, Marcon V, Pisula W, et al. Towards high charge-carrier mobilities by rational design of the shape and periphery of discotics. Nat Mater 2009;8:421–6.
[99] Li C, Liu MY, Pschirer NG, Baumgarten M, Müllen K. Polyphenylene-based materials for organic photovoltaics. Chem Rev 2010;110:6817–55.
[100] Rieger R, Müllen K. Forever young: polycyclic aromatic hydrocarbons as model cases for structural and optical studies. J Phys Org Chem 2010;23:315–25.

[101] Schmaltz B, Weil T, Müllen K. Polyphenylene-based materials: control of the electronic function by molecular and supramolecular complexity. Adv Mater 2009;21:1067–78.
[102] Björk J, Hanke F, Palma C-A, Samorì P, Cecchini M, Persson M. Adsorption of aromatic and anti-aromatic systems on graphene through pi-pi stacking. J Phys Chem Lett 2010;1:3407–12.
[103] Chen ZJ, Lohr A, Saha-Moller CR, Wurthner F. Self-assembled pi-stacks of functional dyes in solution: structural and thermodynamic features. Chem Soc Rev 2009;38:564–84.
[104] Fujigaya T, Nakashima N. Methodology for homogeneous dispersion of single-walled carbon nanotubes by physical modification. Polym J 2008;40:577–89.
[105] Dong X, Shi Y, Zhao Y, et al. Symmetry breaking of graphene monolayers by molecular decoration. Phys Rev Lett 2009;102:135501.
[106] Jang J-H, Rangappa D, Kwon Y-U, Honma I. Direct preparation of 1-PSA modified graphene nanosheets by supercritical fluidic exfoliation and its electrochemical properties. J Mater Chem 2010;21:3462–6.
[107] Lee DW, Kim T, Lee M. An amphiphilic pyrene sheet for selective functionalization of graphene. Chem Commun 2011;47:8259–61.
[108] Schlierf A, Yang H, Gebremedhn E, et al. Nanoscale insight into the exfoliation mechanism of graphene with organic dyes: effect of charge, dipole and molecular structure. Nanoscale 2013;5:4205–16.
[109] Yang H, Hernandez Y, Schlierf A, et al. A simple method for graphene production based on exfoliation of graphite in water using 1-pyrenesulfonic acid sodium salt. Carbon 2013;53:357–65.
[110] Parviz D, Das S, Ahmed HS, Irin F, Bhattacharia S, Green MJ. Dispersions of non-covalently functionalized graphene with minimal stabilizer. ACS Nano 2012;6:8857–67.
[111] Zhang F, Chen XJ, Boulos RA, et al. Pyrene-conjugated hyaluronan facilitated exfoliation and stabilisation of low dimensional nanomaterials in water. Chem Commun 2013;49:4845–7.
[112] Shim HW, Ahn K-J, Im K, et al. Effect of hydrophobic moieties in water-soluble polymers on physical exfoliation of graphene. Macromolecules 2015;48:6628–37.
[113] Zhang M, Parajuli RR, Mastrogiovanni D, et al. Production of graphene sheets by direct dispersion with aromatic healing agents. Small 2010;6:1100–7.
[114] Xu YX, Bai H, Lu GW, Li C, Shi GQ. Flexible graphene films via the filtration of water-soluble noncovalent functionalized graphene sheets. J Am Chem Soc 2008;130:5856–7.
[115] Zhang L, Zhang ZJ, He CZ, Dai LM, Liu J, Wang LX. Rationally designed surfactants for few-layered graphene exfoliation: ionic groups attached to electron-deficient pi-conjugated unit through alkyl spacers. ACS Nano 2014;8:6663–70.
[116] Sampath S, Basuray AN, Hartlieb KJ, Aytun T, Stupp SI, Stoddart JF. Direct exfoliation of graphite to graphene in aqueous media with diazaperopyrenium dications. Adv Mater 2013;25:2740–5.
[117] Ahadian S, Estili M, Surya VJ, et al. Facile and green production of aqueous graphene dispersions for biomedical applications. Nanoscale 2015;7:6436–43.
[118] Bourlinos AB, Georgakilas V, Zboril R, Steriotis TA, Stubos AK, Trapalis C. Aqueous-phase exfoliation of graphite in the presence of polyvinylpyrrolidone for the production of water-soluble graphenes. Solid State Commun 2009;149:2172–6.
[119] Chabot V, Kim B, Sloper B, Tzoganakis C, Yu AP. High yield production and purification of few layer graphene by Gum Arabic assisted physical sonication. Sci Rep 2013;3:1378.
[120] Fan JC, Shi ZX, Ge Y, Wang JL, Wang Y, Yin J. Gum arabic assisted exfoliation and fabrication of Ag-graphene-based hybrids. J Mater Chem 2012;22:13764–72.
[121] Ge Y, Wang JL, Shi ZX, Yin J. Gelatin-assisted fabrication of water-dispersible graphene and its inorganic analogues. J Mater Chem 2012;22:17619–24.
[122] Guardia L, Fernandez-Merino MJ, Paredes JI, et al. High-throughput production of pristine graphene in an aqueous dispersion assisted by non-ionic surfactants. Carbon 2011;49:1653–62.
[123] Liang YT, Hersam MC. Highly concentrated graphene solutions via polymer enhanced solvent exfoliation and iterative solvent exchange. J Am Chem Soc 2010;132:17661–3.

[124] Liu WS, Zhou R, Zhou D, et al. Lignin-assisted direct exfoliation of graphite to graphene in aqueous media and its application in polymer composites. Carbon 2015;83:188–97.
[125] Popescu MT, Tasis D, Tsitsilianis C. Ionizable star copolymer-assisted graphene phase transfer between immiscible liquids: organic solvent/water/ionic liquid. ACS Macro Lett 2014;3:981–4.
[126] Unalan IU, Wan CY, Trabattoni S, Piergiovanni L, Farris S. Polysaccharide-assisted rapid exfoliation of graphite platelets into high quality water-dispersible graphene sheets. RSC Adv 2015;5:26482–90.
[127] May P, Khan U, Hughes JM, Coleman JN. Role of solubility parameters in understanding the steric stabilization of exfoliated two-dimensional nanosheets by adsorbed polymers. J Phys Chem C 2012;116:11393–400.
[128] Yang X, Zhang F, Zhang L, Zhang TF, Huang Y, Chen YS. A high-performance graphene oxide-doped ion gel as gel polymer electrolyte for all-solid-state supercapacitor applications. Adv Funct Mater 2013;23:3353–60.
[129] Ayan-Varela M, Paredes JI, Guardia L, et al. Achieving extremely concentrated aqueous dispersions of graphene flakes and catalytically efficient graphene-metal nanoparticle hybrids with flavin mononucleotide as a high performance stabilizer. ACS Appl Mater Interfaces 2015;7:10293–307.
[130] Chen IWP, Chen YS, Kao NJ, Wu CW, Zhang YW, Li HT. Scalable and high-yield production of exfoliated graphene sheets in water and its application to an all-solid-state supercapacitor. Carbon 2015;90:16–24.
[131] Chen IWP, Huang C-Y, Jhou S-HS, Zhang Y-W. Exfoliation and performance properties of non-oxidized graphene in water. Sci Rep 2014;4:3928.
[132] De S, King PJ, Lotya M, et al. Flexible, transparent, conducting films of randomly stacked graphene from surfactant-stabilized, oxide-free graphene dispersions. Small 2010;6:458–64.
[133] Hao R, Qian W, Zhang LH, Hou YL. Aqueous dispersions of TCNQ-anion-stabilized graphene sheets. Chem Commun 2008:6576–8.
[134] He P, Zhou C, Tian SY, et al. Urea-assisted aqueous exfoliation of graphite for obtaining high-quality graphene. Chem Commun 2015;51:4651–4.
[135] Lee W, Lee DW, Lee M, Hong JI. Direct exfoliation of carbon allotropes with structural analogues of self-assembled nanostructures and their photovoltaic applications. Chem Commun 2014;50:14851–4.
[136] Samoilov VM, Danilov EA, Nikolaeva AV, et al. Formation of graphene aqueous suspensions using fluorinated surfactant-assisted ultrasonication of pristine graphite. Carbon 2015;84:38–46.
[137] Schlierf A, Cha K, Schwab MG, Samorì P, Palermo V. Exfoliation of graphene with an industrial dye: teaching an old dog new tricks. 2D Materials 2014;1:035006.
[138] Wahid MH, Stroeher UH, Eroglu E, et al. Aqueous based synthesis of antimicrobial-decorated graphene. J Colloid Interface Sci 2015;443:88–96.
[139] Hyun WJ, Secor EB, Hersam MC, Frisbie CD, Francis LF. High-resolution patterning of graphene by screen printing with a silicon stencil for highly flexible printed electronics. Adv Mater 2015;27:109–15.
[140] Chua LL, Zaumseil J, Chang JF, et al. General observation of n-type field-effect behaviour in organic semiconductors. Nature 2005;434:194–9.
[141] Ciesielski A, Haar S, El Gemayel M, et al. Liquid-phase exfoliation of graphene using intercalating compounds: a supramolecular approach. Angew Chem Int Ed 2014;53:10355–61.
[142] Geng J, Kong BS, Yang SB, Jung HT. Preparation of graphene relying on porphyrin exfoliation of graphite. Chem Commun 2010;46:5091–3.
[143] Haar S, Ciesielski A, Clough J, et al. A supramolecular strategy to leverage the liquid-phase exfoliation of graphene in the presence of surfactants: unraveling the role of the length of fatty acids. Small 2015;11:1691–702.

[144] Li X, Zhang G, Bai X, et al. Highly conducting graphene sheets and Langmuir–Blodgett films. Nat Nanotechnol 2008;3:538–42.
[145] Pathipati SR, Pavlica E, Schlierf A, et al. Graphene-induced enhancement of n-type mobility in perylenediimide thin films. J Phys Chem C 2014;118:24819–26.
[146] Vadukumpully S, Paul J, Valiyaveettil S. Cationic surfactant mediated exfoliation of graphite into graphene flakes. Carbon 2009;47:3288–94.
[147] Xu L, McGraw J-W, Gao F, et al. Production of high-concentration graphene dispersions in low-boiling-point organic solvents by liquid-phase noncovalent exfoliation of graphite with a hyperbranched polyethylene and formation of graphene/ethylene copolymer composites. J Phys Chem C 2013;117:10730–42.
[148] Hasobe T, Imahori H, Kamat PV, et al. Photovoltaic cells using composite nanoclusters of porphyrins and fullerenes with gold nanoparticles. J Am Chem Soc 2005;127:1216–28.
[149] Otsuki J, Nagamine E, Kondo T, Iwasaki K, Asakawa M, Miyake K. Surface patterning with two-dimensional porphyrin supramolecular arrays. J Am Chem Soc 2005;127:10400–5.
[150] El Garah M, Ciesielski A, Marets N, Bulach V, Hosseini MW, Samorì P. Molecular tectonics based nanopatterning of interfaces with 2D metal-organic frameworks (MOFs). Chem Commun 2014;50:12250–3.
[151] El Garah M, Marets N, Mauro M, et al. Nanopatterning of surfaces with monometallic and heterobimetallic 1D coordination polymers: a molecular tectonics approach at the solid/liquid interface. J Am Chem Soc 2015;137:8450–9.
[152] Brinkhaus L, Katsukis G, Malig J, et al. Tuning the stability of graphene layers by phthalocyanine-based oPPV oligomers towards photo- and redoxactive materials. Small 2013;9:2348–57.
[153] Malig J, Stephenson AW, Wagner P, Wallace GG, Officer DL, Guldi DM. Direct exfoliation of graphite with a porphyrin – creating functionalizable nanographene hybrids. Chem Commun 2012;48:8745–7.
[154] Rabe JP, Buchholz S. Commensurability and mobility in 2-dimensional molecular-patterns on graphite. Science 1991;253:424–7.

Mikhail Shekhirev and Alexander Sinitskii

7 Solution Synthesis of Atomically Precise Graphene Nanoribbons

Abstract: Bottom-up fabrication of narrow strips of graphene, also known as graphene nanoribbons or GNRs, is an attractive way to open a bandgap in semimetallic graphene. In this chapter, we review recent progress in solution-based synthesis of GNRs with atomically precise structures. We discuss a variety of atomically precise GNRs and highlight theoretical and practical aspects of their structural design and solution synthesis. These GNRs are typically synthesized through a polymerization of rationally designed molecular precursors followed by a planarization through a cyclodehydrogenation reaction. We discuss various synthetic techniques for polymerization and planarization steps, possible approaches for chemical modification of GNRs, and compare the properties of GNRs that could be achieved by different synthetic methods. We also discuss the importance of the rational design of molecular precursors to avoid isomerization during the synthesis and achieve GNRs that have only one possible structure. Significant attention in this chapter is paid to the methods of material characterization of solution-synthesized GNRs. The chapter is concluded with the discussion of the most significant challenges in the field and the future outlook.

7.1 Introduction

Graphene has attracted enormous attention due to its extraordinary combination of electronic, mechanical, thermal and optical properties [1]. In recent years, graphene research has been unprecedentedly active, and the progress in fabrication and characterization of graphene has been very significant. As a result, a decade after the first experimental studies of graphene, there is an active discussion of how graphene's properties will be implemented in a broad spectrum of applications in the near future [1a, 2]. One of the most highly praised properties of graphene, its exceptionally high charge-carrier mobilities, makes it promising for electronics applications [1a, 2–3]. However, since graphene is a semimetal, it does not have an energy bandgap, which prevents its use in logic devices [1b, 4]. Numerous studies have focused on different ways to open a bandgap in graphene. One approach for tuning electronic properties of graphene relies on making few-nanometer-wide strips of graphene that are typically referred to as graphene nanoribbons (GNRs).

Interestingly, GNRs have been a subject of theoretical studies long before the first experiments on isolated graphene samples by Novoselov, Geim et al. in 2004 [5]. A comprehensive theoretical analysis of GNRs has been reported by Nakada et al. in 1996 [6]. In general, GNRs can have two principle edge structures – armchair (Figure 7.1(a)) and zigzag (Figure 7.1(b)) – in addition to an infinite number of mixed, chiral and disordered edges. Figure 7.1(a) and (b) also shows the convention on how the carbon atoms are counted across zigzag and armchair GNRs to distinguish ribbons with different widths. For example, for the armchair GNR shown in Figure 7.1(a), $N = 10$, so this GNR is typically referred to as 10-AGNR, whereas for the zigzag GNR shown in Figure 7.1(b), $N = 5$, so this is 5-ZGNR.

It was shown that the electronic properties of armchair and zigzag GNRs are very different. Narrow armchair GNRs can exhibit substantial energy bandgaps that were predicted to be very sensitive to the ribbon's width. Figure 7.1(c) shows the results of tight-binding calculations by Nakada et al. for armchair GNRs with $N = 4$, 5 and 6 [6]. According to the results of these calculations, 4-AGNR is a semiconductor. However, if the width of this GNR is increased by just one carbon atom, the calculated energy bandgap decreases to zero, and the resulting 5-AGNR becomes metallic. If one more carbon atom is added to this ribbon, the resulting 6-AGNR becomes a semiconductor again (Figure 7.1(c)). While nanoscience knows many examples of nanomaterials, such as semiconductor quantum dots [8] and plasmonic metallic nanoparticles [9], which have size-dependent physical properties, this particular example, where the change in one dimension of a nanostructure by just one atom changes physical properties in such a dramatic way, is among the most impressive ones.

Figure 7.1(d) further illustrates the dependence of the bandgap of *N*-AGNRs on their width according to the results of tight-binding calculations [7]. Depending on the *N* value, an *N*-AGNR may belong to one of three different families with $N = 3p$, $N = 3p + 1$ or $N = 3p + 2$, where p is a positive integer (accordingly, *N*-AGNRs in Figure 7.1(c) provide examples of nanoribbons from all these three families). For *N*-AGNRs with $N = 3p$ or $N = 3p + 1$ the bandgap is inversely proportional to the ribbon's width, while *N*-AGNRs with $N = 3p + 2$ are metallic (Figure 7.1(d)) [7]. The results may change if a different computational model is used [7, 10]. Figure 7.1(e) shows energy bandgaps for the same armchair GNRs that were obtained using first-principle calculations [7]. In this model, there are no metallic AGNRs, although *N*-AGNRs from the $N = 3p + 2$ family still have much lower bandgaps compared to the *N*-AGNRs from $N = 3p$ and $N = 3p + 1$ families (Figure 7.1(e)). These three families of AGNRs are now well separated, and within each family the bandgap is inversely proportional to the ribbon's width.

Depending on the calculation method used, zigzag GNRs are predicted to be either metallic or narrow bandgap semiconductors with much less-pronounced width dependence than for armchair GNRs [7] (Figure 7.1(f) and (g)). However, unlike armchair GNRs, they possess nonbonding edge states that were predicted to exhibit a peculiar type of magnetic ordering [6, 7, 11]. The magnetically ordered state of zigzag GNRs can be controlled by means of an external electric field [12], which may find use in spintronics, an emerging area of electronics that uses electron's spin instead of charge for data transmission and storage [13].

Computational study of GNRs is an active area of research. Nanoribbons with different widths, geometries and edge structures have been investigated using various

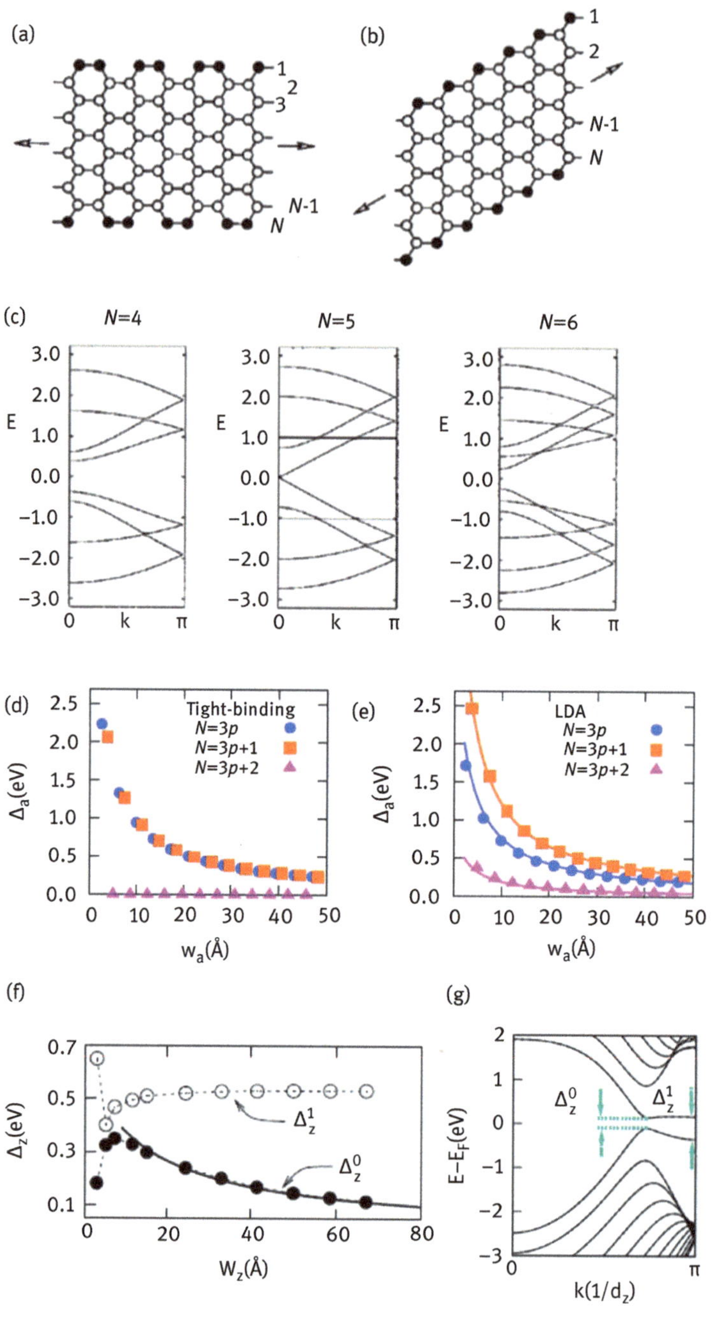

Figure 7.1. Electronic properties of atomically precise GNRs: (a, b) schemes of GNRs with (a) armchair and (b) zigzag edges. (c) Calculated band structures for armchair GNRs with various widths: $N = 4$, $N = 5$ and $N = 6$. (d, e) The variations of bandgaps of armchair GNRs as a function of width (w_a) obtained using (d) tight-binding and (e) first-principle calculations. (f, g) Electronic properties of zigzag GNRs: (f) the variations of Δ_z^0 and Δ_z^1 as a function of the width (w_z) of zigzag GNRs. Panel (g) explains Δ_z^0 and Δ_z^1 using the band structure of a 12-ZGNR as an example: Δ_z^0 and Δ_z^1 denote the direct bandgap and the energy splitting at $kd_z = \pi$, respectively. Panels (a)–(c) are reproduced from Ref. [6]; panels (d)–(g) are reproduced from Ref. [7].

calculation approaches; a discussion of advances in this area can be found in the recent review papers [14]. While we do not intend to provide an in-depth review of these studies in this chapter, there are several practically important points that have to be made. On the one hand, even a brief glimpse at the results of computational studies reveals a wide breadth of properties that could be potentially realized in properly designed GNRs, including a largely tunable energy bandgap, half metallicity and nanomagnetism. On the other hand, these computational results can be used to formulate requirements to experimental approaches for the fabrication of GNRs with desired properties. For example, the following requirements should be met to achieve semiconductor characteristics in armchair GNRs (Figure 7.1):

1. In order to achieve practical energy bandgaps of >0.5 eV, AGNRs should be less than 2 nm wide (Figure 7.1(d)).
2. Not only should the AGNRs be narrow, but their width should be the same along their length with single atom precision. As we have already discussed, addition or subtraction of a single carbon atom from the width of a GNR can make a dramatic difference to the ribbon's energy gap (Figure 1(c)).
3. The ribbons should have atomically precise edges. The discussed calculations show a clear difference between GNRs with armchair and zigzag edges. Other computational studies have specifically focused on the edge disorder in graphene ribbons and showed that even a weak edge roughness could have a strong effect on the charge carrier transport in GNRs [15].

In practice, these requirements are very difficult to meet. While with the state-of-the-art electron-beam lithography it is possible to achieve a ~10 nm resolution, this is not sufficient to carve a sub-2-nm wide GNR from a graphene sheet, not to mention to achieve an atomically precise width and edge structure of a ribbon. In addition to nanofabrication [16], other top-down approaches that rely on fabrication of GNRs from larger blocks of sp^2 carbon materials, such as graphite or carbon nanotubes, have been developed and tested over the last decade. These methods include sonochemical method [17], nanowire lithography [18], nanoscale cutting of graphene using nickel nanoparticles [19] or a diamond knife [20], meniscus-mask lithography [21], chemical etching of prefabricated graphene nanostructures [22] and unzipping of carbon nanotubes [23] among others. Top-down approaches typically yield GNRs with width >10 nm and have limited control over their edge structure. Although several groups demonstrated that such GNRs could exhibit an insulating state in electrical measurements, it was later argued that the observed transport bandgaps of up to ~200–400 meV [16b, 17–18] are likely to be caused by strong localization effects due to edge disorder, rather than a true gap between valence and conduction bands [15c, 24].

Despite the great progress in the development of top-down fabrication approaches for GNRs over the last decade, none of the currently existing methods meet the requirements (1–3). This is why the 2010 paper from Müllen and Fasel groups, which demonstrated the potential of the bottom-up chemical approaches for the synthesis of narrow atomically precise GNRs, was a game changer in the field [25]. In this paper, ribbons that are only a few benzene rings wide and have atomically smooth armchair edges were synthesized on a surface of either Au(111) or Ag(111) single crystal by coupling molecular precursors into linear polyphenylenes followed by cyclodehydrogenation. This paper demonstrated that bottom-up techniques could yield narrow atomically engineered GNRs that meet the requirements (1–3) and are currently unachievable by any existing top-down approach.

Since 2010, the surface-assisted synthesis has become a popular approach for the preparation of atomically precise GNRs [26]. While the surface-assisted synthesis produces very high-quality GNRs that can serve as ideal test objects for the fundamental studies of nanoscale graphene-based materials, this approach has several important practical limitations. First of all, the surface-assisted synthesis cannot be scaled up to produce bulk quantities of atomically precise GNRs for large-scale applications. Also, GNRs prepared on a conductive Au(111) or Ag(111) single crystal cannot be directly used for device fabrication and electrical testing and thus should be somehow transferred to a dielectric substrate. Because of these limitations, the original report on the on-surface synthesis of GNRs [25] stimulated interest in the development of alternative solution-based bottom-up approaches for atomically precise GNRs, which will be the topic of this chapter.

Chemistry-wise, synthesis of a GNR from molecular precursors is the synthesis of a conjugated polymer. Examples of successful fabrications of ladder-type polymers were well known for years [27] and Müllen's group in particular has been notably active in this area, attempting the bottom-up synthesis of “graphite ribbons” in 2000 and 2003 [28], several years before the start of the general interest in graphene in 2004 [5]. However, the ladder-type polymers are too narrow and do not show the desired properties, such as high carrier mobilities, while the early attempts to synthesize wider “graphite ribbons” showed a lack of structural uniformity. It should also be noted that carrier mobilities depend on the width of the ribbons [29]; therefore relatively wide and atomically precise GNRs are of considerable interest. Even though a number of different approaches, strategies and designs of GNR structures have been proposed and implemented since then, the field of bottom-up synthesized GNRs is still in its infancy and many fascinating discoveries and developments are yet to be made.

7.2 Structure of Solution-Synthesized GNRs

As previously mentioned, the edge type and the width determine the electronic properties of GNRs. Thus, one of the most attractive advantages of the bottom-up synthetic approaches is that they provide a complete control over the edge structure. Figure 7.2 shows some of the GNR core aromatic structures synthesized in solution. In addition to straight armchair GNRs (Figure 7.2(a)) [30], several types of nanoribbons, such as

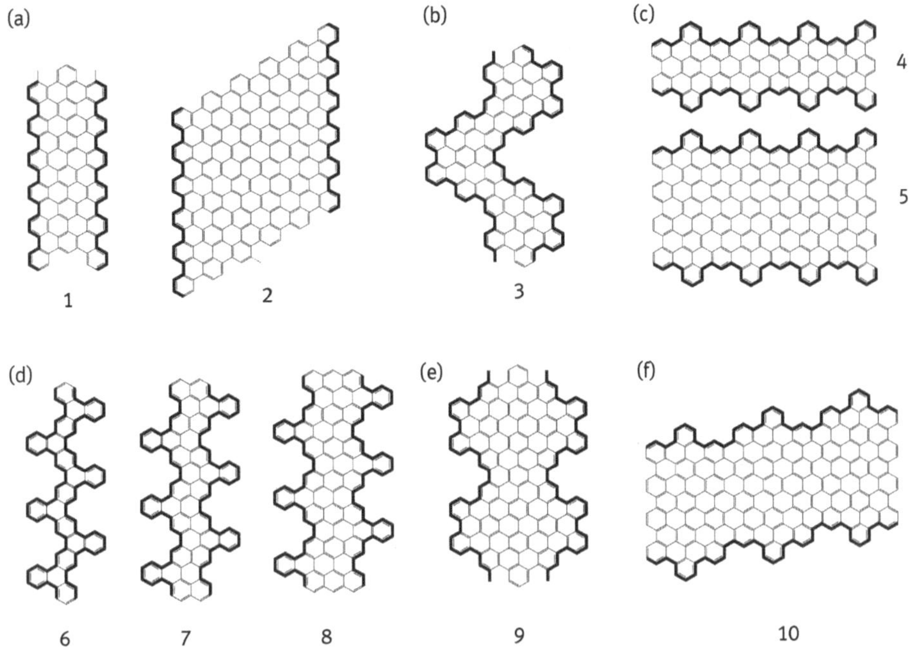

Figure 7.2: Variety of core aromatic structures of atomically precise GNRs that have been synthesized by solution-based approaches: (a) straight GNRs **1** and **2**, (b) chevron-type GNR **3**, (c) cove-type GNRs **4** and **5**, (d) "kinked" GNRs **6**, **7** and **8**, (e) necklace-like GNR **9**, (f) chiral GNR **10**. Possible solubilizing alkyl chains at the periphery of nanoribbons are omitted for the sake of clarity.

chevron-type GNRs (Figure 7.2(b)) [31], cove-type GNRs (Figure 7.2(c)) [32], "kinked" GNRs (Figure 7.2(d)) [33], necklace-like GNRs (Figure 7.2(e)) [34] and "chiral" or mixed GNRs (Figure 7.2(f)) [35], have been designed and synthesized. One particular type of bottom-up GNRs, poly-*peri*-naphthalene or 5-AGNR, which belongs to the $N = 3p + 2$ family and therefore, depending on the computational method, is predicted to either be metallic [6] or have a small bandgap [7], will not be discussed in this chapter, as the extensive studies of this and related GNRs have been recently reviewed in great detail [36]. Notably, the variety of possible structures is almost infinite, and tuning the precursor in order to achieve the optimum GNR structure for a specific application is another important advantage of the bottom-up approach. For example, chevron-type GNRs **3** were predicted to have larger exciton binding energies than straight armchair GNRs, which makes them promising for nano-optoelectronic applications [37].

Same GNR structure can often be synthesized using different routes and strategies with their own advantages and limitations. At this point, rational design of the synthesis and choice of the precursors play a very important role for achieving high-quality nanoribbons. For example, Figure 7.3 (a)–(e) shows five different routes to synthesize GNRs with the core structure **1**, where all methods have certain disadvantages [28b, 30b]. For example, because of the high steric hindrance in routes

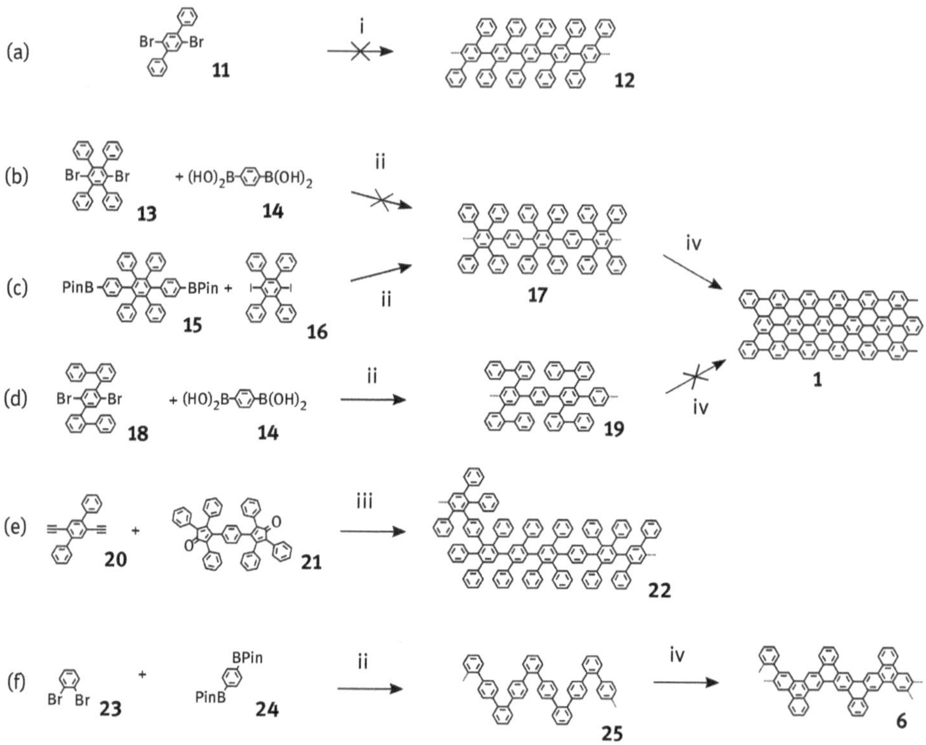

Figure 7.3: (a–e) Representative routes to straight armchair GNR with core aromatic structure **1**. (f) Scheme of synthesis of GNR **6**. Solubilizing alkyl chains at the edges of the nanoribbons are omitted for the sake of clarity. Conditions: (i) $Ni(COD)_2$, bipyridine, COD, toluene/DMF, 80°C, COD = cyclooctadiene; (ii) $Pd(PPh_3)_4$, K_2CO_3, 95 or 75°C; (iii) Ph_2O, reflux; (iv) $FeCl_3$, DCM/$MeNO_2$. See Refs [28b, 30b, 33a] for details.

(a) and (b), polymerizations of the monomer **11** or the monomers **13** and **14** do not provide the corresponding polymers **12** and **17**. In case of less-hindered monomers **18** and **14** (route (d)), only oligomers were found after the polymerization, which do not form the ribbon oligomer **1** upon the following cyclodehydrogenation step. When a successful polymerization is performed, such as the polymerization of monomers **20** and **21** in route (e), the formation of structural isomers during the Diels–Alder reaction results in a mixture of irregular GNRs, but not exclusively the GNR **1**. Finally, Suzuki coupling of precursors **15** and **16** to form polymer **17**, which was followed by the cyclodehydrogenation into GNR **1**, was successfully performed (route (c)) [30 b]; however, the resulting GNRs were only 12 nm long, which may be too short for device fabrication. It was argued that the synthesis of GNR **1** was not successful due to high steric hindrance, as well as due to rigid backbone of the polymers, resulting in limited solubility of the polymeric products. Alternatively, a new GNR **6** was designed and synthesized from less-hindered precursors **23** and **24** that produced a “kinked” and more soluble polymer **25**. The resulting GNRs **6** were 25 nm long in contrast to 12-nm-long GNRs **1** (Figure 7.3(f)) [30b, 33a]. The examples in Figure 7.3 show that the bottom-up solution synthesis of GNRs, despite the large number of powerful tools in organic chemistry, still requires development of new synthetic approaches and routes toward high-quality nanoribbons of appreciable length.

7.3 Synthetic Approaches Toward Atomically Precise GNRs

Several successful attempts to synthesize atomically precise GNRs in solution have been reported to date. In general, the synthesis consists of two steps: polymerization of rationally designed monomers and planarization of the resulting polymer structure to yield polyaromatic graphene molecules. The majority of the syntheses utilize Suzuki, Yamamoto or Diels–Alder polymerization reactions, and oxidative cyclodehydrogenation using $FeCl_3$ for the planarization step.

7.3.1 Suzuki Coupling

The synthesis of GNR **1** (Figure 7.3(c)), which was discussed earlier, was one of the first successful syntheses of atomically precise GNRs. It employed Suzuki cross-coupling as a polymerization reaction [30b]. Similarly, the synthesis of "kinked" GNRs **6** with improved solubility and polymerization efficiency was also performed through Suzuki polymerization (Figure 7.3(f)) [33a]. Later, the synthesis of "kinked" structures was extended to wider GNRs using naphthalene diboronic ester **26** or anthracene diboronic ester **28,** as shown in Figure 7.4(a) [33b]. Despite the successful cyclization step for the narrow version of GNRs **6**, the cyclodehydrogenation reaction for naphthalene and anthracene units to form GNRs **7** and **8** was shown to be less effective with degrees of cyclization of 78% and 75%, respectively. Notably, the "kinked" GNRs were used to fabricate thin-film field-effect transistors (FETs), and wider GNRs **8,** as expected, exhibited higher charge carrier mobilities.

Suzuki coupling was used to synthesize not only "kinked" GNRs but also necklace-like GNRs **9** (Figure 7.4(b)) [34]. Reactive sites in the precursor **30** are less hindered than in the precursor **16**, but the large size of the laterally extended structure **31** and the rigidity of the poly(*p*-phenylene) backbone are responsible for the relatively short length (13 nm) of GNRs **9** obtained in this synthesis [34].

7.3.2 Yamamoto Coupling

One of the problems with Suzuki cross-coupling polymerization reactions presented in Figure 7.4 is the need for two different monomers. This is why Yamamoto

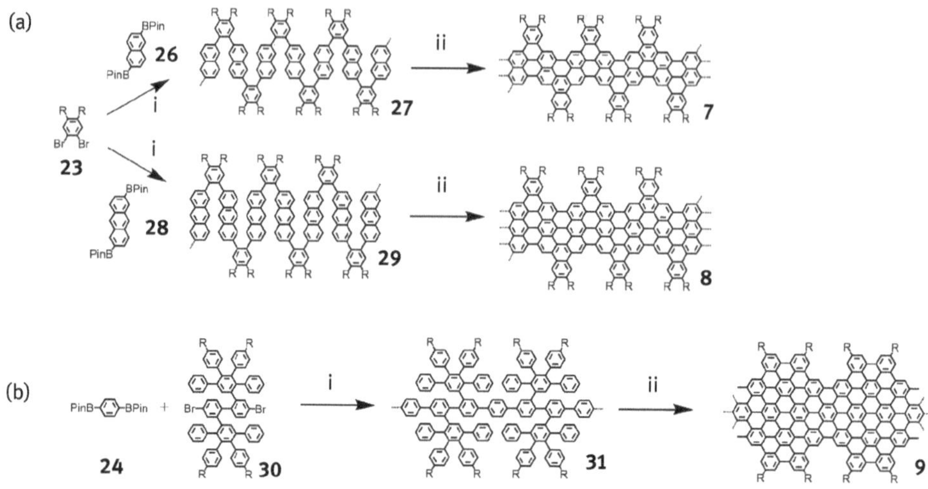

Figure 7.4: Synthesis of GNRs through Suzuki coupling: (a) scheme of synthesis of kinked GNRs **7** and **8**. (b) Scheme of synthesis of necklace-like GNR **9**. Conditions: (i) $Pd(PPh_3)_4$, K_2CO_3, toluene, Aliquat 336, reflux or microwave; (ii) $FeCl_3$, $DCM/MeNO_2$, rt. See Refs [33b, 34] for details.

homocoupling reaction, which requires only one dihalogenated monomer, became an attractive alternative to prepare long GNR precursors.

Several synthetic procedures for GNRs that involve Yamamoto coupling were recently reported (Figure 7.5). Historically, the first synthesis of GNRs employing Yamamoto coupling was reported in 2012 [35]. The monomer **32** was polymerized to form polymer **33** (Figure 7.5(a)) in a notably more efficient way in comparison to previously discussed Suzuki-coupled polymers, and the polymer **33** was then converted to the chiral GNR **10** [35]. Later, an even wider GNR **2** with an estimated width of ~2.1 nm was attempted, starting with the monomer **34** (Figure 7.5(b)), but the resulting polymers were rather short and could be considered as large nanographene molecules [30a]. The nanographenes were sublimed onto a substrate in order to fabricate FET devices, which showed better results in comparison to liquid-phase deposited molecules [38].

In 2014, our group reported a gram-scale synthesis procedure for chevron-type GNRs **3** and demonstrated their high structural quality [31]. Notably, the solubilizing alkyl groups that are usually employed to improve solubility of GNRs were not attached to the aromatic core of the GNRs **3**. This significantly facilitated characterization of solution-synthesized GNRs **3** using scanning tunneling microscopy (STM) to visualize their atomically precise aromatic structure. Even though the absence of the solubilizing groups limited the liquid-phase processability of the nanoribbons, the Yamamoto coupling as well as the following oxidative cyclodehydrogenation were shown to be very effective [31].

(a)

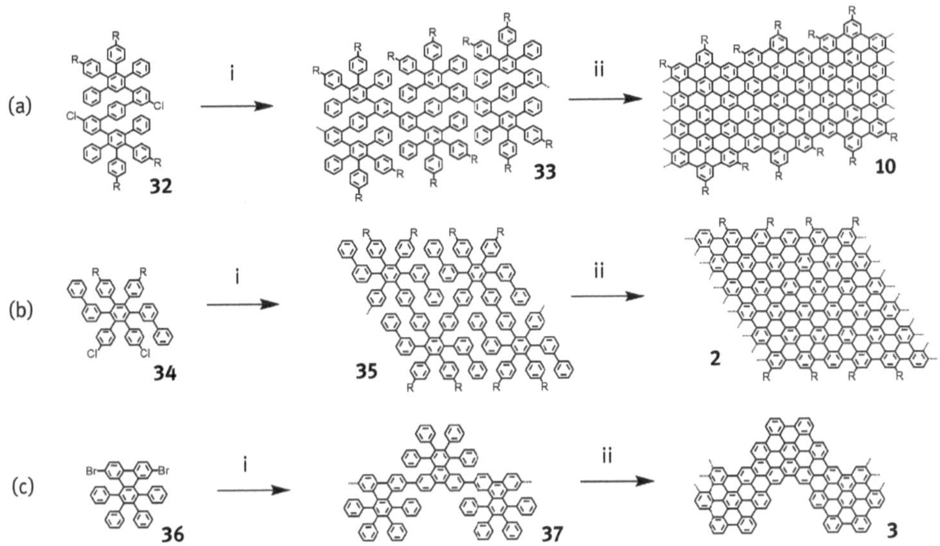

Figure 7.5: Synthesis of GNRs through Yamamoto coupling. Conditions: (i) $Ni(COD)_2$, bipyridine, COD, toluene/DMF, 80°C; (ii) $FeCl_3$, $DCM/MeNO_2$, rt. See Refs [30a, 31, 35] for details.

7.3.3 Diels–Alder Reaction

In attempts to achieve more efficient polymerization, excellent results were shown for GNRs synthesized via Diels–Alder polymerization reaction [32a]. Cove-type GNR precursor **39** was synthesized using a relatively simple procedure of refluxing monomer **38** in diphenyl ether or by heating it in a melt (Figure 7.6 (a)). This elegant approach, which does not require any reagents other than the monomer itself to perform the polymerization, not only yielded GNRs with novel "cove-type" edges, but also provided polymer precursors **39** and GNRs **4** with lengths of over 600 nm. These GNRs

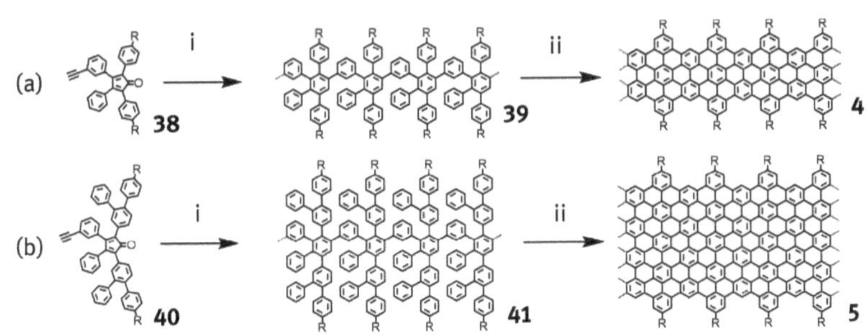

Figure 7.6: Synthesis of GNRs through Diels–Alder polymerization. Conditions: (i) Ph_2O, reflux, 20–28 h or melt, 260°C–270°C, 1.5–5 h; (ii) $FeCl_3$ (7 eq/H), DCM/$MeNO_2$, rt, 3 days. See Ref. [32] for details.

are not only significantly longer than nanoribbons produced using other coupling approaches, but also long enough to serve as channels of electronic devices prepared by the standard electron-beam lithography [39]. Another attractive feature of this GNR design is the possibility of the lateral extension of nanoribbons without changing the overall chemical approach. The wider cove-type GNR **5** was synthesized using the same Diels–Alder polymerization reaction from monomer **40,** see Figure 7.6 (b). Because of their larger width, GNRs **5** have a smaller optical bandgap of ~1.2 eV compared to ~1.9 eV for the original GNR **4** [32b]. Noncontact ultrafast terahertz photoconductivity measurements revealed high intrinsic mobility of GNRs **4** and **5** [32, 40], proving the effectiveness of the cyclodehydrogenation reaction and the high quality of both cove-type nanoribbons.

Recently, another type of Diels–Alder-like cycloaddition reaction, aza-Diels–Alder or Povarov reaction, was proposed to synthesize atomically precise GNRs in solution (Figure 7.7) [41]. It is based on the synthesis of polybenzoquinoline precursor **43** from alkyne- and aldimine-modified units **42**. However, the synthesis of GNR **44** is not reported yet.

7.3.4 Planarization of Synthesized Polymers

The last step in the solution synthesis of GNRs is the graphitization of a 3D polymer fabricated at the polymerization step. The most common procedure is the Scholl reaction, an oxidative cyclodehydrogenation using iron (III) chloride as an oxidant and a Lewis acid. This easy procedure has shown excellent to good yields for a number of polyaromatic systems [42]. However, in some cases, it results in a mixture of partially cyclized or chlorinated products preventing its use for the synthesis of atomically precise aromatic structures [43]. A number of other cyclodehydrogenation techniques and reagent systems have been developed for the synthesis of polyaromatic hydrocarbons (PAHs), which could be adapted for the synthesis of GNRs if for some reason the $FeCl_3$-based reaction does not provide a desired product at a good yield. Other possible reagents include $AlCl_3$/NaCl [44], $MoCl_5$ [45], dichlorodicyanobenzoquinone (DDQ)/CH_3SO_3H [46], $CuCl_2$ or $Cu(OTf)_2$/$AlCl_3$ [47], bis(trifluoroacetoxy)iodobenzene (PIFA)/$BF_3{\cdot}Et_2O$ [48] and others. Notably, the choice of the reagent is important not only to achieve a product of cyclodehydrogenation at a good yield (see the example

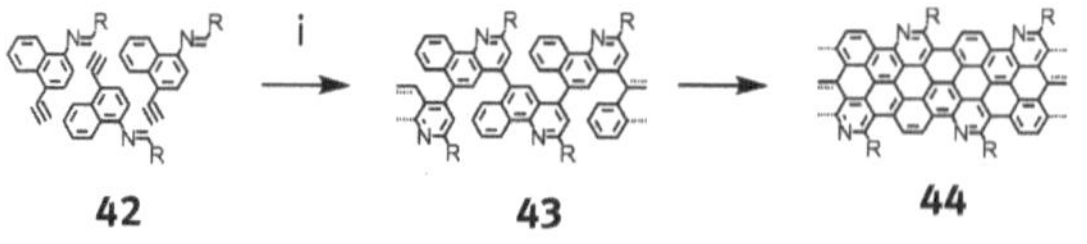

42 **43** **44**

Figure 7.7: Proposed scheme of synthesis of N-GNRs **44** using Povarov polymerization reaction. Conditions: (i) $BF_3{\cdot}OEt_2$, chloranil, Δ. See Ref. [41] for details.

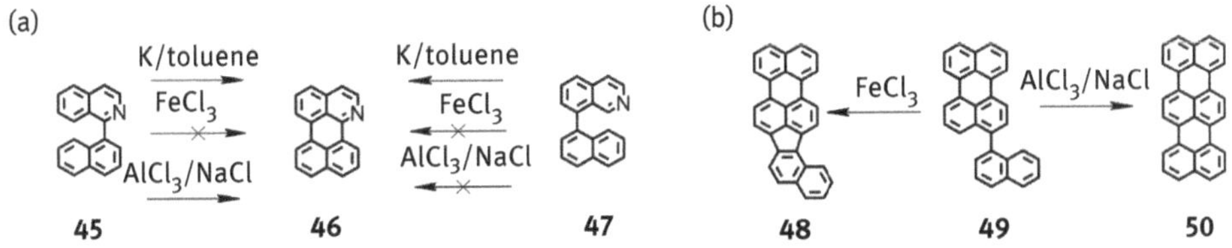

Figure 7.8: Importance of reagents for cyclodehydrogenation reaction: (a) reactivity of naphthylisoquinolines under different conditions, (b) cyclization of 3-(1-Naphthyl)perylene **49** under different conditions. See Refs [43b, 44, 49] for details.

in Figure 7.8(a) [43b, 49]), but also to control the direction of the reaction (see the example in Figure 7.8(b)) [44], which could potentially be used in syntheses of new types of polyaromatic structures. More details on the Scholl reaction can be found in the recent review paper by Butenschön, Gryko et al. [43b].

7.3.5 Alternative Approaches

As alternatives to the polymerization reactions described above, other polymerization techniques could also be employed, especially if the reactions fail or do not yield the desired product. For example, polymerization of monomer **51** using a Yamamoto coupling reaction was not successful, probably due to the high steric hindrance (Figure 7.9) [43a]. The oligomer products **52** were obtained via Ullmann coupling and purified by column chromatography. Furthermore, the following cyclodehydrogenation with $FeCl_3$ yielded mainly chlorinated products, and a DDQ/CF_3SO_3H mixture was employed to achieve cove-type GNR oligomers **53**.

In addition to the concept of alkyne cyclization from Swager and coworkers [50], interesting approaches have been recently proposed for the synthesis of polycyclic aromatic hydrocarbons (PAHs) and GNRs from polyynes (Figure 7.10). The first approach is based on the controllable radical cascade reactions to form central bonds of nanoribbons, while the bonds on the periphery could be formed via oxidative dehydrogenation (Figure 7.10(a)) [51]. Two reactions, namely "all-exo" and "all-endo", have been developed and tested for short oligomers (Figure 7.10(b) and (c)); however,

Figure 7.9: Synthesis of cove-type GNRs **53** using Ullmann coupling. Conditions: (i) Cu, $Pd(PPh_3)_4$, DMSO, 150°C, 48 h, (ii) DDQ/DCM, 0°C, methanesulfonic acid, 1–3 h. See Ref. [43a] for details.

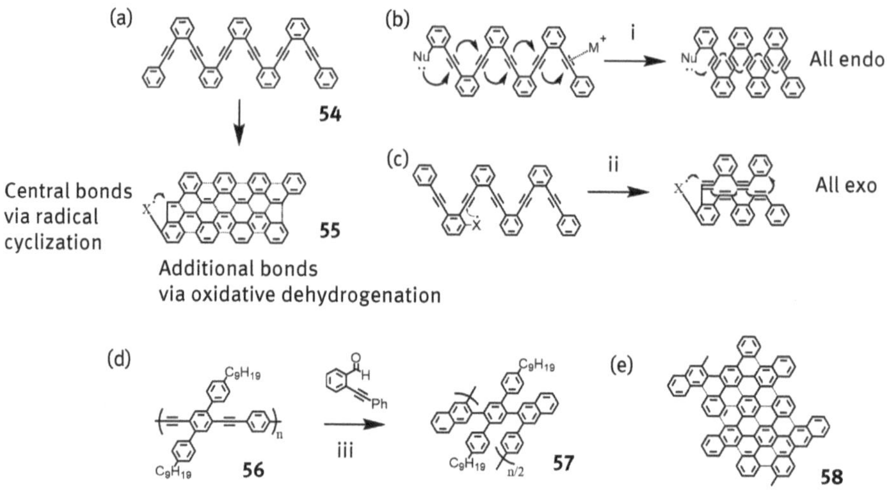

Figure 7.10: Concept of the GNR synthesis from polyynes: (a–c) Synthesis through radical cascade reactions. (a) Proposed GNR synthesis through selective radical attack at the central alkyne. Comparison of (b) "all-endo" and (c) "all-exo" strategies toward the preparation of graphene ribbons. Conditions: (i) $PdCl_2(PPh_3)_2$, $ClAuPPh_3$, AgOTf, o-xylene,150°C, (ii) $HSnBu_3$, AIBN. (d) Synthesis through benzannulation reaction. Conditions: (iii) $Cu(OTf)_2$, CF_3CO_2H, $CHCl_3$, 100°C. (e) Core aromatic structure of possible GNR. See Refs [51, 52] for details.

long GNRs have not been synthesized yet. The second approach utilizes $Cu(OTf)_2$-catalyzed benzannulation reaction of poly(phenylene ethynylene) precursor **56** with 2-(phenylethynyl)benzaldehyde (Figure 7.10(d)) to yield polymer **57** [52], which may eventually be converted to a GNR with the core structure **58** (Figure 7.10(e)).

7.4 Chemical Modification

As discussed in the previous sections of this chapter, bottom-up synthetic approaches allow a precise control over structural parameters of GNRs, such as their width and edge structure, which define GNRs' physical properties. Chemical modification is another important tool that can be used to alter GNRs' properties.

Two chemical modification strategies, edge functionalization and precise substitution of carbon atoms with heteroatoms, have been realized in solution-synthesized atomically precise GNRs. These approaches showed promise not only for the modification of electronic properties of GNRs, but also for improving solubility/processability and controlled self-assembly of nanoribbons.

7.4.1 Nitrogen Doping

For conventional semiconductors, such as silicon, substitutional doping is one of the key approaches to fine-tune electronic properties necessary for functional devices. Random substitution of carbons with heteroatoms, such as boron or nitrogen, has also been realized in graphene [53]. In contrast, the bottom-up synthetic approaches allow to not only dope a GNR with a certain amount of heteroatoms, but also incorporate them in specific positions in the ribbon's structure. Nitrogen-doped GNRs, which are also referred to as N-GNRs, have been considered in several theoretical studies [54], but only two types of solution-synthesized GNRs with precise nitrogen doping have been reported to date (Figure 7.11).

In 2013, Jo et al. reported modification of the GNRs **7**, employing 2,3-dibromo-5,6-dialkoxypyrazine molecule **59** during the Suzuki polymerization step as a nitrogen source [55]. In the resulting N-GNRs the level of doping could be controlled by changing the ratio of dibromopyrazine and dibromobenzene derivatives **59** and **23** during the synthesis (Figure 7.11(a)). It was shown that the nitrogen doping can tune the electronic transport behavior of the GNRs changing it from ambipolar for undoped GNR **7** to n-type upon nitrogen substitution. In the device measurements, N-doped GNRs with the highest level of nitrogen substitution (GNR **61**) showed highest electron mobility values [55].

In 2014 and 2015, our group reported nitrogen doping of previously synthesized chevron-type GNRs (Figure 7.11(b)) [56]. Unlike in the case of "kinked" GNRs (Figure 7.11(a)), here the level of doping was varied by changing the amount of nitrogen substitution in the monomer, not by changing the ratio of N-doped and undoped

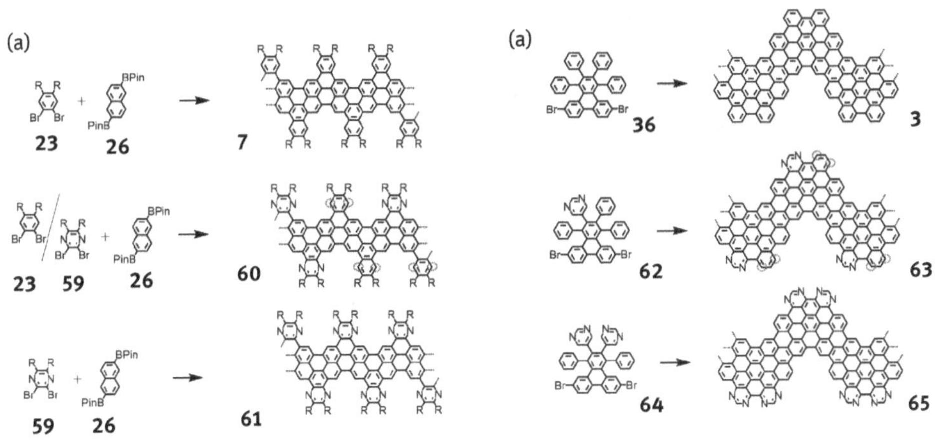

Figure 7.11: Bottom-up solution synthesis of N-doped GNRs: (a) "kinked" GNRs and (b) chevron-type GNRs. For N-GNRs **60** and **63** the positions of nitrogen atoms could vary and may also occupy positions highlighted with dotted circles.

precursors. In principle, copolymerization of monomers **36** and **62/64** should also result in random chevron-type N-GNRs, in which the average nitrogen concentration could be tuned by the ratio of N-doped and undoped precursors, but this possibility is not yet explored experimentally for solution reactions. The solution-synthesized chevron-type N-GNRs **63** and **65** were visualized by STM and characterized by a number of spectroscopic techniques. The optical properties N-GNRs **63** and **65** were similar to those of undoped chevron-type GNRs **3** [56].

Because of the solubilizing alkyl chains that effectively screen edge nitrogen atoms from the environment, incorporation of nitrogens in the "kinked" GNRs **60** and **61**, while changing their electronic properties, did not significantly affect interactions between the nanoribbons. However, in the case of chevron-type N-GNRs **63** and **65**, where no side alkyl chains are present, the situation is very different. In addition to slight changes in electrical properties of the nanoribbons, the nitrogen substitution triggers the hierarchical self-assembly of N-GNRs into ordered structures. The phenomenon was first observed for GNR **63** [56a] and then studied in detail using surface-assisted and solution approaches to compare the behaviors of undoped GNRs **3** and N-doped GNRs **65** (Figure 7.12) [56b]. On a surface of Au(111) substrate, after the thermal polymerization of the monomer **64** and cyclodehydrogenation of the resulting polymer, STM analysis reveals the formation of hydrogen-bonded assemblies of N-GNRs **65**, as shown in Figure 7.12(a–c). In addition to the same hydrogen bonding, solution-synthesized GNRs have van der Waals and π–π interactions between the

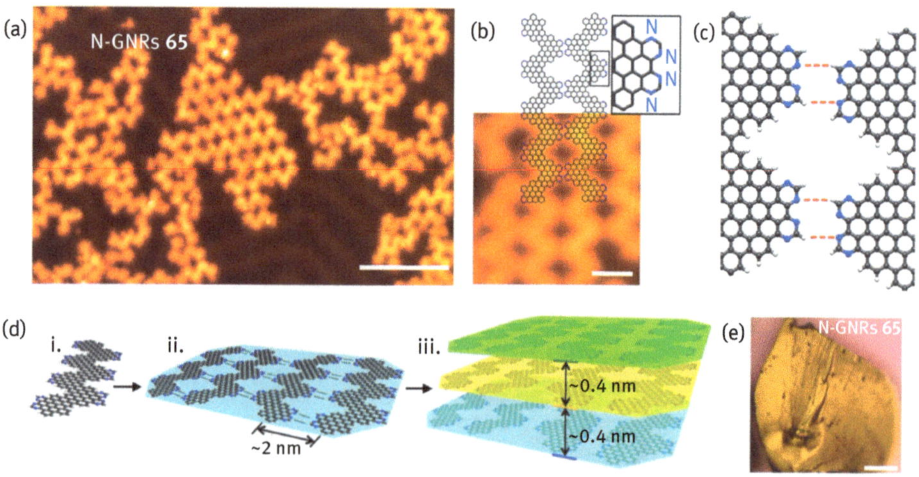

Figure 7.12: Hierarchical self-assembly of N-doped GNRs **65**: (a) STM image of the morphology of the N-GNRs on Au(111), scale bar: 10 nm, (b) close-up view of one of 2D assemblies of the N-GNRs, scale bar: 2 nm. (c) Schematic structure of laterally coordinated N-GNRs, which are offset slightly along their axes to enable hydrogen bonding between adjacent ribbons. (d) Scheme of the N-GNR self-assembly in a 3D solution environment. (e) Optical image of crystals of solution-synthesized N-GNRs, scale bar: 50 μm. Reproduced from Ref. [56b].

basal planes of GNRs (Figure 7.12(d)). These interactions ultimately result in the formation of macroscopic crystalline structures of N-GNRs, which is schematically shown in Figure 7.12(d). These results demonstrate that substitutional doping of GNRs could be a powerful tool to promote self-assembly of nanoribbons and eventually form new carbon nanostructures [56b].

7.4.2 Edge Chlorination

Edge chlorination is another method of precise chemical modification of GNRs. An important advantage of this method is that it allows a post-synthetic change of the GNR structure. Edge chlorination was originally demonstrated for PAH molecules but it also works well for bottom-up solution-synthesized GNRs [57]. A scheme of the edge chlorination of a GNR **4** is shown in Figure 7.13. The chlorination affects electronic properties of the GNRs decreasing the optical bandgap from 1.9 eV for the pristine GNR **4** to 1.7 eV for the chlorinated GNR **66.** Additionally, the edge chlorination promotes solubility of the GNRs as the edge Cl atoms induce deformation of the structure and force the GNRs to adopt non-planar geometry.

7.5 Characterization Techniques for Solution-Synthesized GNRs

Materials characterization is one of the most important and still one of the most challenging aspects of the studies of atomically precise GNRs. Low solubility of GNRs in many cases precludes their complete structural characterization and a combination of different techniques is required to gain insights into the structure of the nanoribbons.

Several techniques can be utilized to determine the average length of solution-synthesized GNRs. Matrix-assisted laser desorption/ionization time of flight mass spectrometry (MALDI-TOF MS), which is usually applied to determine masses of PAHs, has limitations for high molecular weight polymers with a broad molecular weight distribution [58]. As a result, the largest mass detected with MS does not necessarily correspond to the largest mass of a polymer. In order to estimate the molecular weight and the length of GNRs, size exclusion chromatography (SEC) of polymer

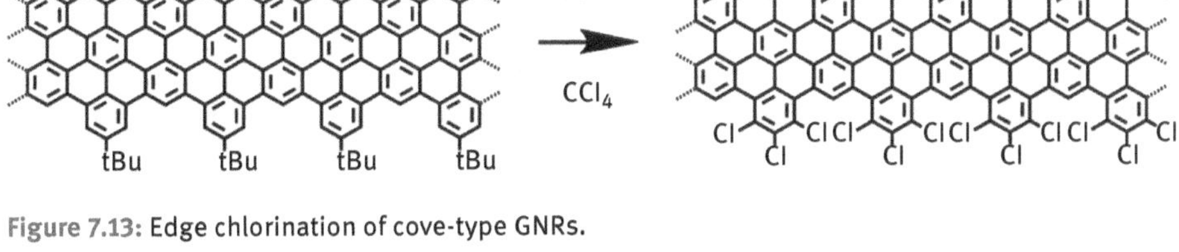

Figure 7.13: Edge chlorination of cove-type GNRs.

precursors can be used. Typically, polymers with non-rigid core structure are analyzed against polystyrene standards, but poly(p-phenylene) standard was also suggested as a second estimate of molecular weight [32]. Although not giving absolute molecular weights, SEC measurements provide approximate values for comparison with different polyphenylene precursors. Preparative SEC or re-precipitation may also help to remove small oligomers and improve polydispersity index [32]. In order to determine absolute molecular weights of polymers, dynamic light scattering measurements could be employed [59].

When possible, liquid-phase nuclear magnetic resonance (NMR) spectroscopy provides important structural information about polymeric precursors, GNRs and their transformations. For example, a comparison of NMR spectra of polymer **29** and GNR **8** was used to estimate the degree of the polymer cyclization (Figure 7.14(a)) [33b]. However, GNRs are often insoluble and solid-state NMR techniques could be employed, like in case of GNR **5** [32b]. Typically, solid-state NMR spectroscopy suffers from peak broadening and overall lower quality of spectra, providing less information than liquid-state NMR (Figure 7.14(b)). In many cases, synthesis of representative model PAH molecules is very helpful as they are usually suitable for conclusive characterization via conventional techniques, such as liquid-phase NMR or mass

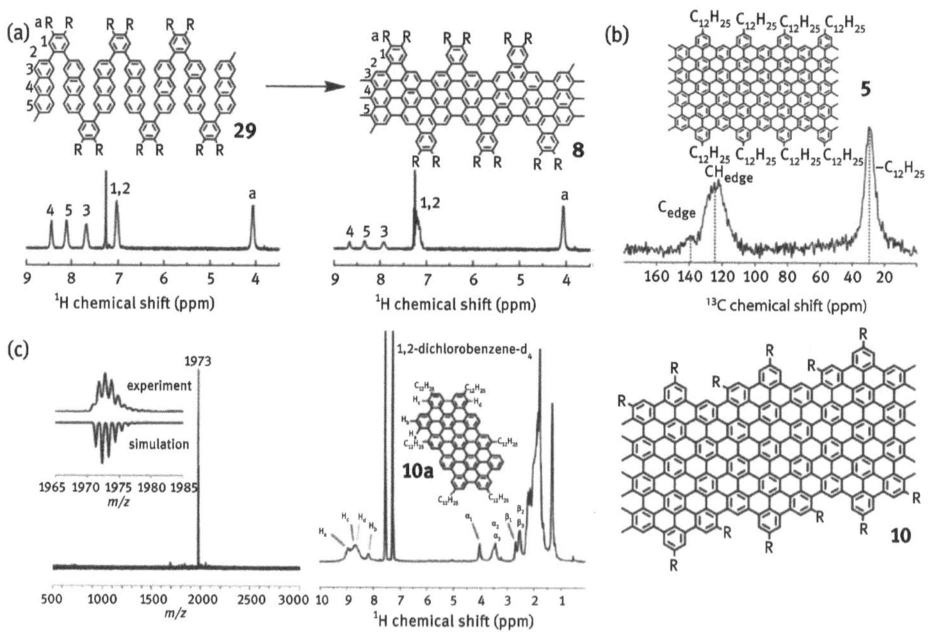

Figure 7.14: NMR studies of solution-synthesized GNRs: (a) ^{1}H NMR spectra of polymer **29** and GNR **8**, (b) ^{13}C NMR spectrum of GNR **5**, (c) MALDI-TOF MS spectrum (left), isotopic distribution (inset) and ^{1}H NMR spectrum (right) of PAH **10a** as a model compound for GNR **10**. Panel (a) is reproduced from Ref. [33b], panel (b) is reproduced from Ref. [32b], panel (c) is reproduced from Ref. [35].

spectroscopy. For example, polyaromatic compound **10a** was synthesized as a model for GNR **10** (Figure 7.14(c)) [35]. Since no partially fused or chlorinated products were detected after the last cyclodehydrogenation step, the same protocol was used for the synthesis of GNR **10** avoiding potential problems with the Scholl reaction, such as undesirable ring formation or chlorination.

STM becomes an increasingly important characterization method, as it provides a direct way to visualize the atomically precise structure of a GNR. STM is particularly helpful in case of insoluble GNRs, for which other options of characterization methods may be limited. Figure 7.15 shows the STM characterization of solution-synthesized chevron-type GNRs **3** [31]. Notably, STM images show a clear difference between the non-planar polymer structure and the flat GNR (Figure 7.15(a)). Although STM has enormous potential for the GNR research, its use for the visualization of solution-synthesized atomically precise nanoribbons is rather rare. A cost of a high-resolution STM instrument is one limiting factor, but even when the access to such piece of equipment is available, STM characterization of solution-synthesized GNRs is very challenging. This is due to the fact that a typical sample preparation involves drying a droplet of a GNR dispersion in an appropriate solvent on a conductive substrate, such as Au(111). If GNRs have poor solubility, the resulting sample will mostly contain various aggregates of nanoribbons, and finding a rare individual GNR on a substrate for STM imaging could be very difficult. Also, since the sample preparation is done in air, it inevitably results in the presence of atmospheric adsorbates and other forms of contamination on a substrate in addition to the residual solvent molecules, which further complicates imaging. Sample contamination can be seen in Figure 7.15(b). When GNRs are functionalized with solubilizing alkyl chains, this, on the one hand, improves dispersibility of nanoribbons, but on the other hand makes GNR structures more difficult to resolve.

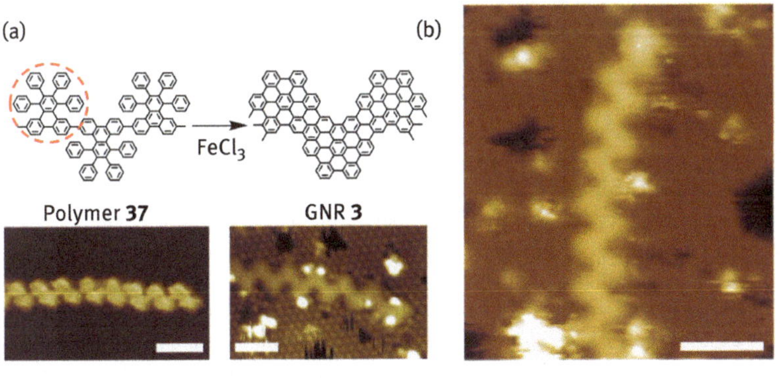

Figure 7.15: STM studies of chevron-type GNRs: (a) atomic structures and corresponding STM images of a polymer **37** and GNR **3**. (b) Another STM image of GNR **3**. Scale bars: 3 nm. Reproduced from Ref. [31].

Despite the difficulties with high-resolution imaging of nanoribbons, STM and other microscopy techniques, such as atomic force microscopy (AFM), scanning electron microscopy (SEM) and transmission electron microscopy (TEM), are used in the studies focused on the aggregation behavior of GNRs (Figure 7.16). Supramolecular assembly of large PAH molecules with solubilizing groups have been reported and investigated as an important premise for organic electronic devices [60]. Similarly, the assembly of extended PAHs – GNRs – also attracted a considerable attention. Depending on the structure of GNRs and alkyl chains attached to the aromatic core, GNRs show different self-assembly patterns on surfaces. For example, PAH **67** adopts columnar structure when deposited by drop-casting on a surface of highly oriented pyrolytic graphite (HOPG) (Figure 7.16(a)) [42d]. In the observed columnar arrangement the molecular planes are oriented perpendicular to the HOPG surface and parallel to each other, which facilitates electronic transport between PAH molecules **67** and makes these assemblies promising for electronic applications. In contrast, long cove-type GNRs **4** with bulky branched alkyl chains installed in the peripheral positions

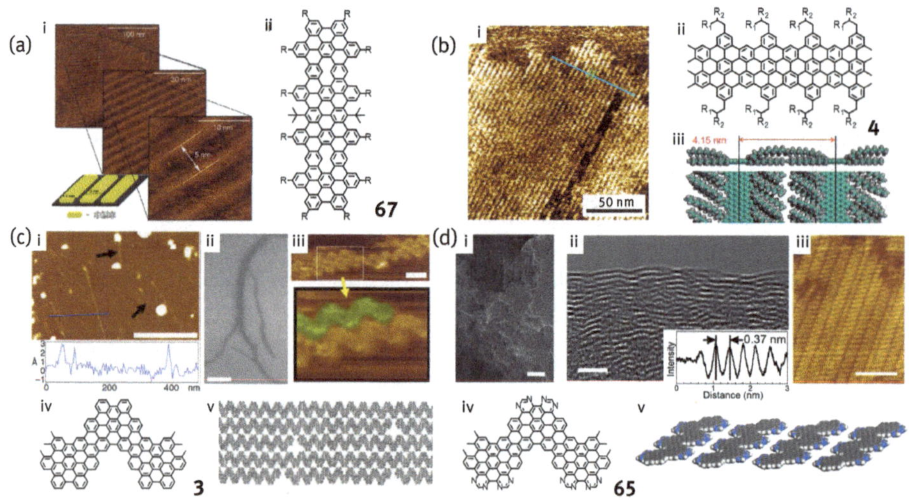

Figure 7.16: Microscopy studies of solution-synthesized GNRs: (a) STM image and corresponding model of molecular arrangement (i) of necklace-type GNRs **67** (ii) on HOPG. (b) AFM image (i) of cove-type GNRs **4** (ii) on HOPG with corresponding molecular model of the GNR assembly (iii). (c) Assembly of chevron-type GNRs **3**. AFM image with corresponding height profile (i), SEM (ii) and STM (iii) images of GNRs **3** (iv) deposited on mica, Si/SiO_2 and Au(111) substrates, respectively. Scale bars: 500 nm (i), 200 nm (ii), 3 nm (iii), the proposed structure of a GNR "nanobelt" (iv). (d) Assembly of N-doped chevron GNRs **65**. (i) TEM image of GNRs **65**, showing distinct flakes, scale bar: 20 nm; (ii) TEM image of GNRs **65** near the surface, showing layered structure with a period of ~0.37 nm, scale bar: 2 nm; (iii) STM image of side-by-side aggregated GNRs **65** deposited on Au(111), scale bar: 3 nm; (iv) chemical structure of GNRs **65** and (v) scheme of side-by-side arrangement. Panel (a) is reproduced from Ref. [42d], panel (b) is reproduced from Ref. [32a], panel (c) is reproduced from Ref. [31] and panel (d) is reproduced from Ref. [56b].

show highly ordered self-assembled monolayers on a HOPG surface (Figure 7.16(b)) [32a]. The branched alkyl chains prevent π-stacking between the co-adsorbed nanoribbons, which however was observed for the same GNRs with linear dodecyl chains.

In the case of chevron GNRs **3**, which were synthesized without solubilizing alkyl chains at the edges, microscopy studies suggested formation of "nanobelts" – side-by-side arrangements of GNRs on substrates (Figure 7.16(c)) [31]. The "nanobelts" are one graphene layer thick, according to AFM measurements (Figure 7.16(c-i)), but wider than one GNR, which is shown by SEM (Figure 7.16(c-ii)). STM images show side-by-side arrangement of chevron GNRs (Figure 16(c-iii)), where the protrusions of one ribbon perfectly fit into the grooves of another one. With this side-by-side assembly, the GNRs form wide and long nanostructures that can be visualized by AFM and SEM; a possible structure of these "nanobelts" of GNRs **3** is shown in Figure 7.16(c-v).

As discussed in Section 7.4.1, nitrogen substitution in GNRs triggers additional mechanism of aggregation, the hydrogen bonding. In this case, TEM images reveal flake-like structures of bulk material with interlayer distance similar to the interplanar distance in graphite, 0.34 nm (Figure 7.16(d-i,ii)) [56b]. The flakes consist of N-GNRs **65** that are arranged in a side-by-side fashion (see STM image in Figure 7.16(d-iii)), but in this case the convex elbows of adjacent GNRs project toward each other as shown in Figure 7.16(d-v).

Since solution methods allow synthesis of atomically precise nanoribbons in appreciable quantities, GNRs can often be characterized by various spectroscopic techniques, such as X-ray photoelectron spectroscopy (XPS), Fourier transform infrared spectroscopy (FTIR), UV-vis absorption spectroscopy and Raman spectroscopy. These methods are often helpful to confirm the efficiency of the synthetic procedure and assess the structural quality of GNRs.

XPS is widely used in the studies of graphene, graphene oxide and related materials [61]. For example, in case of graphene oxide, *i.e.* oxidized graphene sheets with hydroxyl and epoxy groups at the basal planes and carboxyl and carbonyl groups at the edges [61b, 62], XPS can be used to determine elemental composition and coordination of carbons and oxygens and thus estimate the abundance of different functional groups. Similarly, XPS can be helpful for the characterization of heteroatom-substituted GNRs. As an example, Figure 7.17(a) shows XPS spectra for pristine chevron-type GNRs **3** and nitrogen-doped chevron N-GNRs **65**. The GNR **3** exhibits only one component in the C1s spectrum, which corresponds to the sp^2 carbons, and, as expected, nitrogen doping introduces additional peaks in the N1s and C1s regions of the XPS spectrum of N-GNRs **65** (Figure 7.17(a)).

FTIR spectroscopy is also a very informative tool to investigate the efficiency of the last cyclodehydrogenation step of the synthesis. For example, the successful transformation of polymer **41** into GNR **5** was confirmed through the comparison of their FTIR spectra (Figure 7.17(b)) [32b]. The strong attenuation of the signal triad from aromatic C–H stretching vibrations, the disappearance of out-of-plane C–H deformation bands, as well as the band appearing at 863 cm^{-1} (the band from the aromatic

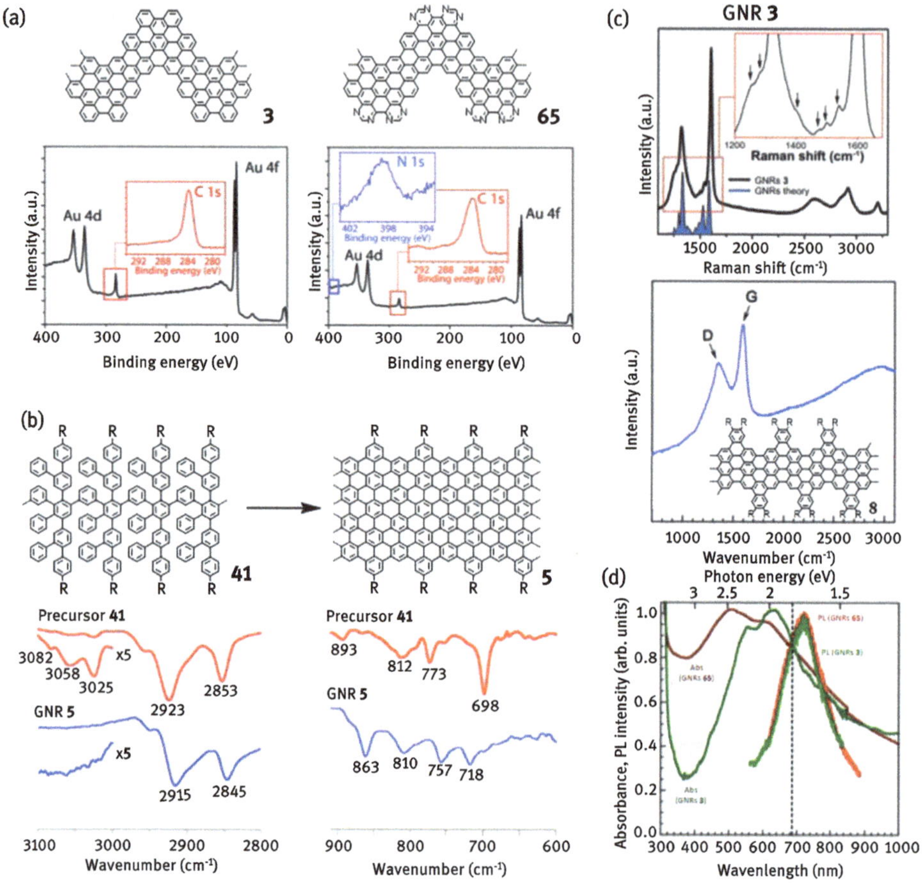

Figure 7.17: Spectroscopic characterization of solution-synthesized GNRs. (a) Comparison of XPS survey spectra of GNRs **3** (left) and N-GNRs **65** (right). Insets show the C1s and N1s regions. (b) Comparison of FTIR spectra of polymer **41** and cove-type GNR **5**. (c) Raman spectra of chevron-type GNR **3** (top) and "kinked" GNR **8** (bottom). The arrows in the inset in the top panel show small peaks around D and G bands. (d) UV-vis absorbance (Abs) and photoluminescence (PL) intensity plotted on the same scale for undoped GNRs **3** and nitrogen-doped GNRs **65** crystals. The vertical dashed line, corresponding to a photon energy of 1.82 eV, represents the estimated energy of the (0-0) transition in both GNRs. Panels (a) and (d) are reproduced from Ref. [56b], panel (b) is reproduced from Ref. [32b] and panel (c) is reproduced from Refs [66] (top) and [33b] (bottom).

C–H moiety at the cove) all support the successful conversion of the polymer **41** to atomically precise GNR **5**.

Raman spectroscopy is one of the most important characterization techniques in graphene research [63], as it could be used to study the number and orientation of layers, the edge type, disorder, doping and chemical functionalization. While the Raman spectra of graphene samples are typically dominated by D, G and 2D bands, the most intense lines around 1,300, 1,600 and 2,700 cm^{-1}, respectively, the spectra of low-dimensional graphene materials, PAH molecules and GNRs, are usually much more complex and exhibit fine structures of smaller peaks. The positions and intensities of these peaks can often be accurately predicted using density functional theory (DFT) simulations [64]. However, the fine structures could only be observed for high-quality GNRs, as with increasing concentration of defects, nanoribbons become qualitatively similar to other disordered carbon nanomaterials, such as carbon black, amorphous carbon and graphene oxide, which have only very broad D and G bands in their Raman spectra [61a, 65]. An example of a Raman spectrum of a solution-synthesized GNR **3** is shown in Figure 7.17(c); smaller bands around most intense D and G bands are shown by the arrows in the inset. Additional bands were also experimentally observed for other types of GNRs, such as cove-type GNR **4** [32a, 66]. The number and the positions of the side bands are characteristic for a particular nanoribbon structure, and the experimental observation of the fine structure in a Raman spectrum could be considered as an evidence for the high structural quality of synthesized GNRs. As shown in Figure 7.17(c), for GNR **3** the number and the positions of the side bands agree well with the spectrum obtained by DFT simulations. On the other hand, for a number of solution-synthesized GNRs Raman spectra exhibit only broad D and G bands; an example of such spectrum is shown in Figure 7.17(c) for a "kinked" GNR **8** [33b].

Spectroscopic techniques can be used to probe not only the structure of GNRs, but also their electronic properties. UV-vis absorption spectroscopy is often used to estimate optical bandgaps of solution-synthesized GNRs. However, the shape of an absorption spectrum may strongly depend on the measurement conditions, such as the solvent, concentration of GNRs and the temperature, as the aggregation of GNRs gives rise to excitonic effects [67]. Therefore, in order to assess electronic properties of GNRs it may be useful to combine UV-vis spectroscopy with other techniques, such as scanning tunneling spectroscopy (STS) or photoluminescence (PL) spectroscopy. For example, PL spectroscopy can also be used to estimate a bandgap of nanoribbons, as shown in Figure 7.17(d) for GNRs **3** and **65** [56b]. Very interesting results can also be obtained for atomically precise GNRs using ultrafast THz photoconductivity measurements. Despite the progress in the device fabrication and several reports on GNR-based FETs [33b, 55, 68], study of the intrinsic electronic properties of GNRs via transport measurements is still challenging. The THz spectroscopy provides a straightforward and noninvasive approach to evaluate the photoconductivity within individual GNRs on ultrashort timescales [32, 40].

7.6 Challenges in Solution-Based Synthesis of GNRs

The field of atomically precise GNRs is still in its infancy and there is still a long way before it will reach the point when a bottom-up synthesis of all-carbon electronic circuits will become a reality. However, there are a number of immediate problems to solve, including limited solubility of solution-synthesized GNRs and the development of more effective chemical routes toward high-quality GNRs. On the other hand, these synthetic challenges might be considered as an opportunity for new and fascinating developments.

7.6.1 Solubility

High lateral extension and length of nanoribbons are two very important structural parameters of solution-synthesized GNRs for practical applications. However, increasing aromatic core of GNRs results in stronger π–π interactions, which in practice leads to insolubility of a material. Reduced solubility of GNRs complicates their synthesis and nullifies one of the advantages of the solution-based methods, a liquid-phase processing of nanoribbons for fabrication of functional devices.

One of the most common ways to improve solubility and liquid-phase processability of GNRs is based on the attachment of alkyl chains to the periphery of molecules. This concept was successfully applied to solubilize large PAHs, such as hexabenzocoronene [60, 69], and was also used in many designs of solution-synthesized GNRs [70]. For example, in cove-type GNRs **4** the installed alkyl chains on the periphery significantly enhanced dispersability of nanoribbons, which made possible further deposition of the GNRs on different substrates, materials characterization and fabrication of FETs and chemical sensors [68, 71].

Despite the efficiency of alkyl chains, it was argued that the solubilizing groups may affect packing of the molecules, reduce electrical contacts between nanoribbons in bulk and result in inferior properties of GNR-based materials [67]. Thus, development of other approaches to improve solubility of GNRs is very important.

An interesting approach was developed to reduce the aggregation of the large PAH molecule **68**, which employs *tert*-butyl groups introduced in the indentations of the "cove-type" edge structure (Figure 7.18(a)) [72]. These groups introduce a distortion into the planar structure of the PAH, suppressing aggregation and improving solubility. The same concept was partially adopted for the synthesis of elongated homologues of PAH **68** [42d], as well as for the edge-chlorinated GNRs **65** (Figure 7.18(b)). With further development, synthesis of non-planar GNRs with improved solubility might be another advantage of solution bottom-up syntheses over the surface-assisted techniques.

A number of techniques to improve solubility of bottom-up synthesized GNRs could also be learned from graphene research. For example, some methods of dispersing graphenes in strong acids [73], appropriate organic solvents [74], surfactant-stabilized aqueous solutions [75] and other media [76] could be potentially adapted for processing and further investigation of solution-synthesized GNRs.

(a)

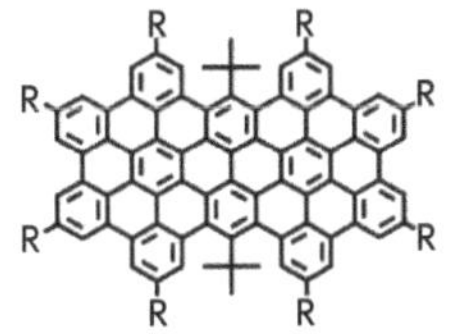

68

(b)

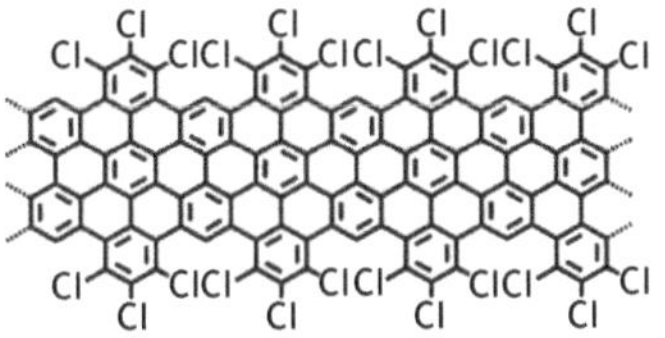

66

Figure 7.18: Non-planar polyaromatic structures with improved solubility.

Overall, the synthesis of GNRs with improved solubility, as well as the development of controlled self-assembly strategies for GNRs to form 1D, 2D and 3D functional materials, appear to be promising areas of research in polyaromatic materials in general and GNRs in particular [77].

7.6.2 Isomerization

Clearly, isomerization is highly undesirable during the GNR synthesis. As it leads to a number of different final structures, isomerization essentially breaks the idea of atomically precise GNRs. Thus, the rational design of monomers and the use of appropriate chemistries to avoid the formation of isomers are very important.

We have already discussed several examples of isomerization problems. In one example, isomerization during the synthesis of the GNR **1** by Diels–Alder polymerization reaction from precursors **20** and **21** resulted in a random mixture of three regioisomeric repeating units (Figure 7.3(e)). Even though the following cyclodehydrogenation provided a graphitized polymer, the atomically precise synthesis was not possible with this design of monomer structures. In another example, previously discussed "kinked" GNRs **7** and **8** are also subject to isomerization. Rotation along single carbon–carbon bonds could result in rather irregular aromatic structures **69** and **70** after the cyclodehydrogenation step in addition to the proposed GNRs **7** or **8** (Figure 7.19(b) and (c)). Notably, the free rotation in the case of phenyl-based GNR **6** is not a problem as the cyclization is possible only for one position (Figure 7.19(a)). This isomerization could explain the reduced degree of cyclodehydrogenation for wider GNRs in comparison to GNR **6** [33b].

The isomerization could be avoided with proper designs of the GNR precursors. One way to solve the problem is to use substituents in order to direct the desired bond formation and block the unwanted ones [78]. For example, alkyl chains on the periphery of the polymer precursor **39**, in addition to improved solubility, also serve as blocking groups to prevent isomerization during the cyclodehydrogenation step (Figure 7.19(d)).

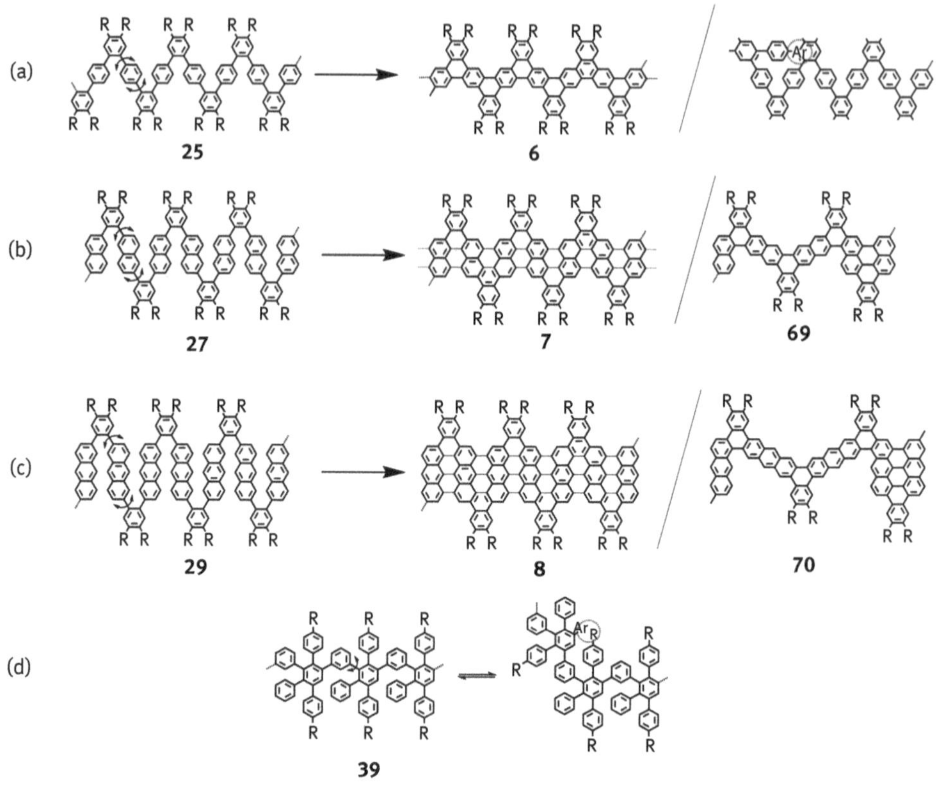

Figure 7.19: (a–c) Single-bond rotation as a reason for isomerization of bottom-up solution-synthesized GNRs, (d) use of blocking groups to prevent the formation of an undesired structure.

In addition to design problems, isomerization could also happen during the last cyclodehydrogenation step. Rearrangements in the Scholl reaction were reported for some polyphenylenes [79], making synthesis of some structures even more challenging. Related to the problem, complete cyclization and planarization of some structures are not possible, as the reaction yields a mixture of partially cyclized or chlorinated products [43b].

One of the ways to overcome the problems of the Scholl reaction could be the use of a surface-assisted cyclodehydrogenation step. Combination of solution-based and surface-assisted methods has a number of advantages, utilizing the best of both techniques. On the one hand, the solution synthesis is more versatile and flexible, and it is possible to synthesize structures that cannot be synthesized on surface. Also, some solution-based polymerization techniques are more effective than on-surface polymerizations, providing access to longer polymers. Yet, the use of a surface-assisted thermal cyclization may help to avoid some solution-based cyclization problems. After the cyclization, the resulting GNRs could be transferred to other substrates using transfer techniques well established in graphene research [80].

7.7 Summary and Future Outlook

Over the last few years there has been a great progress in the field of atomically precise GNRs. Several synthetic strategies that we have reviewed in this chapter made a considerable number of different types of GNRs available for materials studies. Because of their small size and proneness to aggregation, solution-synthesized GNRs are difficult to characterize. Yet, there has been a significant progress in the materials characterization of GNRs, which resulted in better understanding of their properties.

As the field of solution-synthesis of atomically precise GNRs will grow, future studies will likely focus on the following research topics. While we reviewed a number of examples of solution syntheses of atomically precise GNRs with various edge types (Figure 7.2), none of these examples showed a solution synthesis of GNRs with pure zigzag edges, which are predicted to show interesting electronic and magnetic properties [11d, 12]. Therefore, at the present, the solution synthesis of zigzag GNRs appears to be one of the most important challenges in the field. Despite the significant progress in the studies of acenes, which could be considered as the narrowest type of ZGNRs [81], and recent reports on the surface-assisted synthesis of certain ZGNRs [70], the development of a versatile solution-based approach for ZGNRs remains a formidable challenge.

In addition to ZGNRs, GNRs with non-six-membered rings, such as five- or seven-membered rings, integrated into the structure in atomically precise fashion, could also be very interesting. In recent years, a significant progress in the synthesis of PAHs with seven- and eight- membered rings has been shown [70], and some of these approaches may be potentially adapted for the synthesis of GNRs with distorted structures.

Doping of GNRs with heteroatoms has also shown promise for tuning electronic properties of GNRs, as well as triggering their self-assembly [56b]. As we discuss in this chapter, so far the reported examples of solution-synthesized nanoribbons are limited to nitrogen-doped GNRs. Therefore, the development of solution-based approaches for GNRs doped with other elements, such as boron, could be an interesting research direction. Further studies focused on varying concentration of heteroatoms and their positions in a GNR structure are very important for understanding the effect of substitutional doping on GNR properties.

Also of great interest are the optimization of the existing polymerization techniques and the development of new approaches to achieve longer polymers. So far, Diels–Alder polymerization has shown very promising results for the synthesis of longitudinally extended GNRs without any additional catalyst [32a]. From this prospective, other cycloaddition reactions and radical polymerization techniques could be found practical for facile and effective synthesis of longer GNRs with novel structures. Furthermore, post-synthetic treatment of GNRs and their edge-modification is a promising area of research for fine-tuning GNRs' optoelectronic properties and self-assembly behavior.

Finally, while GNRs are expected to be relevant for a number of applications, such as organic electronics and photovoltaics, there is still a very limited number of reports where solution-synthesized GNRs are used in device studies. As the number of available GNRs grows and their structural quality improves, more examples of GNR-based devices will be reported. A significant progress in the device fabrication will be impossible without deeper understanding of the aggregation behavior of solution-synthesized GNRs, as well as the development of effective approaches for GNR self-assembly, which means that these research topics are very important for this growing field.

References

[1] (a) Ferrari AC, Bonaccorso F, Fal'ko V, Novoselov KS, Roche S, Boggild P, et al. Science and technology roadmap for graphene, related two-dimensional crystals, and hybrid systems. Nanoscale 2015;7(11):4598–810; (b) Geim AK, Novoselov KS, The rise of graphene. Nat Mater 2007;6(3):183–91; (c) Castro Neto AH, Guinea F, Peres NM, Novoselov KS, Geim AK. The electronic properties of graphene. Rev Mod Phys 2009;81(1):109–62.
[2] Novoselov KS, Fal'ko VI, Colombo L, Gellert PR, Schwab MG, Kim K. A roadmap for graphene. Nature 2012;490(7419):192–200.
[3] Fiori G, Bonaccorso F, Iannaccone G, Palacios T, Neumaier D, Seabaugh A, et al. Electronics based on two-dimensional materials. Nat Nano 2014;9(10):768–79.
[4] Guisinger NP, Arnold MS. Beyond silicon: carbon-based nanotechnology. MRS Bull 2010;35(4):273–6.
[5] Novoselov KS, Geim AK, Morozov SV, Jiang D, Zhang Y, Dubonos SV, et al. Electric field effect in atomically thin carbon films. Science 2004;306(5696):666–9.
[6] Nakada K, Fujita M, Dresselhaus G, Dresselhaus MS. Edge state in graphene ribbons: Nanometer size effect and edge shape dependence. Phys Rev B 1996;54(24):17954–61.
[7] Son YW, Cohen ML, Louie SG. Energy gaps in graphene nanoribbons. Phys Rev Lett 2006;97(21):216803.
[8] Kim JY, Voznyy O, Zhitomirsky D, Sargent EH. 25th anniversary article: colloidal quantum dot materials and devices: a quarter-century of advances. Adv Mater 2013;25(36):4986–5010.
[9] Lu X, Rycenga M, Skrabalak SE, Wiley B, Xia Y. Chemical synthesis of novel plasmonic nanoparticles. Ann Rev Phys Chem 2009;60(1):167–92.
[10] Barone V, Hod O, Scuseria GE. Electronic structure and stability of semiconducting graphene nanoribbons. Nano Lett 2006;6(12):2748–54.
[11] (a) Fujita M, Wakabayashi K, Nakada K, Kusakabe K, Peculiar localized state at zigzag graphite edge. J Phys Soc Jpn 1996;65(7):1920–3; (b) Ezawa M, Peculiar width dependence of the electronic properties of carbon nanoribbons. Phys Rev B 2006;73(4):045432; (c) Yazyev OV, Katsnelson MI, Magnetic correlations at graphene edges: basis for novel spintronics devices. Phys Rev Lett 2008;100(4):047209; (d) Yazyev, OV, Emergence of magnetism in graphene materials and nanostructures. Rep Prog Phys 2010;73(5):056501.
[12] Son YW, Cohen ML, Louie SG. Half-metallic graphene nanoribbons. Nature 2006;444(7117): 347–9.
[13] Žutić I, Fabian J, Das Sarma S. Spintronics: fundamentals and applications. Rev Mod Phys 2004;76(2):323–410.
[14] (a) Palacios JJ, Fernández-Rossier J, Brey L, Fertig HA, Electronic and magnetic structure of graphene nanoribbons. Semicon Sci Technol 2010;25(3):033003; (b) Yazyev OV, A guide to the design of electronic properties of graphene nanoribbons. Acc Chem Res 2013;46(10):2319–28.

[15] (a) Areshkin DA, Gunlycke D, White CT. Ballistic transport in graphene nanostrips in the presence of disorder: importance of edge effects. Nano Lett 2006;7(1):204–10; (b) Gunlycke D, Areshkin DA, White CT, Semiconducting graphene nanostrips with edge disorder. Appl Phys Lett 2007;90(14):142104; (c) Mucciolo ER, Castro Neto AH, Lewenkopf CH, Conductance quantization and transport gaps in disordered graphene nanoribbons. Phys Rev B 2009;79(7):075407.

[16] (a) Chen ZH, Lin YM, Rooks MJ, Avouris P, Graphene nano-ribbon electronics. Phys E 2007;40(2):228–32; (b) Han MY, Özyilmaz B, Zhang YB, Kim P, Energy band-gap engineering of graphene nanoribbons. Phys Rev Lett 2007;98(20):206805.

[17] Li XL, Wang XR, Zhang L, Lee SW, Dai HJ. Chemically derived, ultrasmooth graphene nanoribbon semiconductors. Science 2008;319(5867):1229–32.

[18] (a) Bai JW, Duan XF, Huang Y, Rational fabrication of graphene nanoribbons using a nanowire etch mask. Nano Lett 2009;9(5):2083–7; (b) Sinitskii A, Tour JM, Patterning graphene nanoribbons using copper oxide nanowires. Appl Phys Lett 2012;100(10):103106.

[19] (a) Ci LJ, Xu ZP, Wang LL, Gao W, Ding F, Kelly KF, et al. Controlled nanocutting of graphene. Nano Res 2008;1(2):116–22; (b) Campos LC, Manfrinato VR, Sanchez-Yamagishi JD, Kong J, Jarillo-Herrero P, Anisotropic etching and nanoribbon formation in single-layer graphene. Nano Lett 2009;9(7): 2600–4.

[20] Mohanty N, Moore D, Xu ZP, Sreeprasad TS, Nagaraja A, Rodriguez AA, et al. Nanotomy-based production of transferable and dispersible graphene nanostructures of controlled shape and size. Nat. Commun 2012;3:844.

[21] Abramova V, Slesarev AS, Tour JM. Meniscus-mask lithography for narrow graphene nanoribbons. ACS Nano 2013;7(8):6894–8.

[22] (a) Wang X, Dai H, Etching and narrowing of graphene from the edges. Nat Chem 2010;2(8):661–5; (b) Yang R, Shi Z, Zhang L, Shi D, Zhang G, Observation of Raman g-peak split for graphene nanoribbons with hydrogen-terminated zigzag edges. Nano Lett 2011;11(10):4083–8.

[23] (a) Kosynkin DV, Higginbotham AL, Sinitskii A, Lomeda JR, Dimiev A, Price BK, et al. Longitudinal unzipping of carbon nanotubes to form graphene nanoribbons. Nature 2009;458(7240):872–6; (b) Jiao LY, Zhang L, Wang XR, Diankov G, Dai HJ, Narrow graphene nanoribbons from carbon nanotubes. Nature 2009;458(7240):877–80; (c) Cano-Marquez AG, Rodriguez-Macias FJ, Campos-Delgado J, Espinosa-Gonzalez CG, Tristan-Lopez F, Ramirez-Gonzalez D, et al. Ex-MWNTs: graphene sheets and ribbons produced by lithium intercalation and exfoliation of carbon nanotubes. Nano Lett 2009;9(4):1527–33; (d) Sinitskii A, Fursina AA, Kosynkin DV, Higginbotham AL, Natelson D, Tour JM, Electronic transport in monolayer graphene nanoribbons produced by chemical unzipping of carbon nanotubes. Appl Phys Lett 2009;95(25):253108; (e) Jiao LY, Wang XR, Diankov G, Wang HL, Dai HJ, Facile synthesis of high-quality graphene nanoribbons. Nat Nanotechnol 2010;5(5):321–5; (f) Kosynkin DV, Lu W, Sinitskii A, Pera G, Sun ZZ, Tour JM. Highly conductive graphene nanoribbons by longitudinal splitting of carbon nanotubes using potassium vapor. ACS Nano 2011;5(2):968–74; (g) Sinitskii A, Dimiev A, Kosynkin DV, Tour JM, Graphene nanoribbon devices produced by oxidative unzipping of carbon nanotubes. ACS Nano 2010;4(9):5405–13.

[24] (a) Han MY, Brant JC, Kim P. Electron transport in disordered graphene nanoribbons. Phys Rev Lett 2010;104(5):056801; (b) Oostinga JB, Sacepe B, Craciun MF, Morpurgo AF. Magnetotransport through graphene nanoribbons. Phys Rev B 2010;81(19):193408.

[25] Cai J, Ruffieux P, Jaafar R, Bieri M, Braun T, Blankenburg S, et al. Atomically precise bottom-up fabrication of graphene nanoribbons. Nature 2010;466(7305):470–3.

[26] Talirz L, Ruffieux P, Fasel R. On-surface synthesis of atomically precise graphene nanoribbons. Adv Mater 2016, 28, 6222–6231.

[27] Schlüter A-D. Ladder Polymers: The new generation. Advanced Materials 1991;3(6):282–91.

[28] (a) Shifrina ZB, Averina MS, Rusanov AL, Wagner M, Müllen K. Branched polyphenylenes by repetitive Diels–Alder cycloaddition. Macromolecules 2000;33(10):3525–9; (b) Wu JS, Gherghel L, Watson MD, Li JX, Wang ZH, Simpson CD. et al. From branched polyphenylenes to graphite ribbons. Macromolecules 2003;36(19):7082–7089.
[29] (a) Obradovic B, Kotlyar R, Heinz F, Matagne P, Rakshit T, Giles MD, et al. Analysis of graphene nanoribbons as a channel material for field-effect transistors. Appl Phys Lett 2006;88(14):142102; (b) Wang J, Zhao R, Yang M, Liu Z, Liu Z. Inverse relationship between carrier mobility and bandgap in graphene. J Chem Phys 2013;138(8):084701.
[30] (a) El Gemayel M, Narita A, Dössel LF, Sundaram RS, Kiersnowski A, Pisula W, et al. Graphene nanoribbon blends with P3HT for organic electronics. Nanoscale 2014;6(12):6301–14; (b) Yang X, Dou X, Rouhanipour A, Zhi L, Räder HJ, Müllen K. Two-dimensional graphene nanoribbons. J Am Chem Soc 2008;130(13):4216–7.
[31] Vo TH, Shekhirev M, Kunkel DA, Morton MD, Berglund E, Kong L, et al. Large-scale solution synthesis of narrow graphene nanoribbons. Nat Commun 2014;5:3189.
[32] (a) Narita A, Feng X, Hernandez Y, Jensen SA, Bonn M, Yang H. et al. Synthesis of structurally well-defined and liquid-phase-processable graphene nanoribbons. Nat Chem 2014;6(2):126–32; (b) Narita A, Verzhbitskiy IA, Frederickx W, Mali KS, Jensen SA, Hansen MR, Bonn M, De Feyter S, Casiraghi C, Feng X, Müllen K. Bottom-up synthesis of liquid-phase-processable graphene nanoribbons with near-infrared absorption. ACS Nano 2014;8(11):11622–30.
[33] (a) Dössel L, Gherghel L, Feng X, Müllen K. Graphene nanoribbons by chemists: nanometer-sized, soluble, and defect-free. Angew Chem Int Ed Engl 2011;50(11):2540–3; (b) Kim KT, Jung JW, Jo WH. Synthesis of graphene nanoribbons with various widths and its application to thin-film transistor. Carbon 2013;63:202–209.
[34] Schwab MG, Narita A, Osella S, Hu Y, Maghsoumi A, Mavrinsky A, et al. Bottom-up synthesis of necklace-like graphene nanoribbons. Chem Asian J 2015;10(10):2134–8.
[35] Schwab MG, Narita A, Hernandez Y, Balandina T, Mali KS, De Feyter S, et al. Structurally defined graphene nanoribbons with high lateral extension. J Am Chem Soc 2012;134(44):18169–72.
[36] Markiewicz JT, Wudl F. Perylene, oligorylenes, and aza-analogs. ACS Appl Mater Interfaces 2015;7(51):28063–85.
[37] Wang SD, Wang JL. Quasiparticle energies and optical excitations in chevron-type graphene nanoribbon. J Phys Chem C 2012;116(18):10193–7.
[38] Abbas AN, Liu B, Narita A, Dössel LF, Yang B, Zhang W, et al. Vapor-phase transport deposition, characterization, and applications of large nanographenes. J Am Chem Soc 2015;137(13):4453–9.
[39] Chen Y, Pepin A. Nanofabrication: conventional and nonconventional methods. Electrophoresis 2001;22(2):187–207.
[40] Jensen SA, Ulbricht R, Narita A, Feng X, Müllen K, Hertel T, et al. Ultrafast photoconductivity of graphene nanoribbons and carbon nanotubes. Nano Lett 2013;13(12):5925–30.
[41] Dibble DJ, Park YS, Mazaheripour A, Umerani MJ, Ziller JW, Gorodetsky AA. Synthesis of polybenzoquinolines as precursors for nitrogen-doped graphene nanoribbons. Angew Chem Int Edit 2015;54(20):5883–7.
[42] (a) Bohme T, Simpson CD, Müllen K, Rabe JP. Current-voltage characteristics of a homologous series of polycyclic aromatic hydrocarbons. Chemistry 2007;13(26):7349–57; (b) Ito S, Wehmeier M, Brand JD, Kubel C, Epsch R, Rabe JP et al. Synthesis and self-assembly of functionalized hexa-peri-hexabenzocoronenes. Chemistry 2000;6(23):4327–42; (c) Wu J, Pisula W, Müllen K. Graphenes as potential material for electronics. Chem Rev 2007;107(3):718–47; (d) Fogel Y, Zhi LJ, Rouhanipour A, Andrienko D, Räder HJ, Müllen K. Graphitic nanoribbons with dibenzo[e,l]pyrene repeat units: synthesis and self-assembly. Macromolecules 2009;42(18):6878–84.

[43] (a) Liu J, Li BW, Tan YZ, Giannakopoulos A, Sanchez-Sanchez C, Beljonne D, et al. Toward cove-edged low bandgap graphene nanoribbons. J Am Chem Soc 2015;137(18):6097–103; (b) Grzybowski M, Skonieczny K, Butenschön H, Gryko DT. Comparison of oxidative aromatic coupling and the Scholl reaction. Angew Chem Int Ed Engl 2013;52(38):9900–30; (c) Waldvogel SR, Trosien S. Oxidative transformation of aryls using molybdenum pentachloride. Chem Commun (Camb) 2012;48(73):9109–19.

[44] Avlasevich Y, Kohl C, Müllen K. Facile synthesis of terrylene and its isomer benzoindenoperylene. Journal of Materials Chemistry 2006;16(11):1053–7.

[45] Kramer B, Frohlich R, Waldvogel SR. Oxidative coupling reactions mediated by MoCl5 leading to 2,2′-cyclolignans: The specific role of HCl. Eur J Org Chem 2003;2003(18):3549–54.

[46] Zhai L, Shukla R, Rathore R. Oxidative C–C bond formation (Scholl reaction) with DDQ as an efficient and easily recyclable oxidant. Org Lett 2009;11(15):3474–7.

[47] Morgenroth F, Kübel C, Müller M, Wiesler UM, Berresheim AJ, Wagner M, et al. From three-dimensional polyphenylene dendrimers to large graphite subunits. Carbon 1998;36(5–6):833–7.

[48] Takada T, Arisawa M, Gyoten M, Hamada R, Tohma H, Kita Y. Oxidative biaryl coupling reaction of phenol ether derivatives using a hypervalent iodine(III) reagent. J Org Chem 1998;63(22):7698–706.

[49] Gryko DT, Piechowska J, Galezowski M. Strongly emitting fluorophores based on 1-azaperylene scaffold. J Org Chem 2010;75(4):1297–300.

[50] Goldfinger MB, Swager TM. Fused polycyclic aromatics via electrophile-induced cyclization reactions – application to the synthesis of graphite ribbons. J Am Chem Soc 1994;116(17):7895–6.

[51] (a) Byers PM, Alabugin IV. Polyaromatic ribbons from oligo-alkynes via selective radical cascade: stitching aromatic rings with polyacetylene bridges. J Am Chem Soc 2012;134(23):9609–14; (b) Byers PM, Rashid JI, Mohamed RK, Alabugin IV. Polyaromatic ribbon/benzofuran fusion via consecutive endo cyclizations of enediynes. Org Lett 2012;14(23):6032–5; (c) Mohamed RK, Mondal S, Gold B, Evoniuk CJ, Banerjee T, Hanson K. Alkenes as alkyne equivalents in radical cascades terminated by fragmentations: overcoming stereoelectronic restrictions on ring expansions for the preparation of expanded polyaromatics. J Am Chem Soc 2015;137(19):6335–49.

[52] Arslan H, Saathoff JD, Bunck DN, Clancy P, Dichtel WR. Highly efficient benzannulation of poly(phenylene ethynylene)s. Angew Chem Int Ed Engl 2012;51(48):12051–4.

[53] (a) Jin Z, Yao J, Kittrell C, Tour JM. Large-scale growth and characterizations of nitrogen-doped monolayer graphene sheets. ACS Nano 2011;5(5):4112–17; (b) Lv R, Li Q, Botello-Mendez AR, Hayashi T, Wang B, Berkdemir A. Nitrogen-doped graphene: beyond single substitution and enhanced molecular sensing. Sci Rep 2012;2:586; (c) Wang HB, Maiyalagan T, Wang X. Review on recent progress in nitrogen-doped graphene: synthesis, characterization, and its potential applications. ACS Catal 2012;2(5):781–94.

[54] (a) Yu SS, Zheng WT, Wen QB, Jiang Q. First principle calculations of the electronic properties of nitrogen-doped carbon nanoribbons with zigzag edges. Carbon 2008;46(3):537–43; (b) Li Y, Zhou Z, Shen P, Chen Z. Spin gapless semiconductor–metal–half-metal properties in nitrogen-doped zigzag graphene nanoribbons. ACS Nano 2009;3(7):1952–8; (c) Wei X-L, Fang H, Wang R-Z, Chen Y-P, Zhong J-X. Energy gaps in nitrogen delta-doping graphene: A first-principle study. Appl Phys Lett 2011;99(1):012107; (d) Kim H, Lee K, Woo SI, Jung Y. On the mechanism of enhanced oxygen reduction reaction in nitrogen-doped graphene nanoribbons. Phys Chem Chem Phys 2011;13(39):17505–10; (e) Cruz-Silva E, Barnett ZM, Sumpter BG, Meunier V. Structural, magnetic, and transport properties of substitutionally doped graphene nanoribbons from first principles. Phys Rev B 2011;83(15):155445.

[55] Kim KT, Lee JW, Jo WH. Charge-transport tuning of solution-processable graphene nanoribbons by substitutional nitrogen doping. Macromol Chem Phys 2013;214(23):2768–73.
[56] (a) Vo TH, Shekhirev M, Kunkel DA, Orange F, Guinel MJ, Enders A, Sinitskii A. Bottom-up solution synthesis of narrow nitrogen-doped graphene nanoribbons. Chem Commun (Camb) 2014;50(32):4172–4; (b) Vo TH, Perera UG, Shekhirev M, Mehdi Pour M, Kunkel, DA, Lu H, et al. Nitrogen-doping induced self-assembly of graphene nanoribbon-based two-dimensional and three-dimensional metamaterials. Nano Lett 2015;15(9):5770–7.
[57] Tan YZ, Yang B, Parvez K, Narita A, Osella S, Beljonne D, et al. Atomically precise edge chlorination of nanographenes and its application in graphene nanoribbons. Nat Commun 2013;4:2646.
[58] Martin K, Spickermann J, Räder HJ, Müllen K. Why does matrix-assisted laser desorption/ionization time-of-flight mass spectrometry give incorrect results for broad polymer distributions? Rapid Commun Mass Spect 1996;10(12):1471–4.
[59] Kroeger A, Deimede V, Belack J, Lieberwirth I, Fytas G, Wegner G. Equilibrium length and shape of rod-like polyelectrolyte micelles in dilute aqueous solutions. Macromolecules 2007;40(1):105–15.
[60] Feng XL, Pisula W, Müllen K. Large polycyclic aromatic hydrocarbons: synthesis and discotic organization. Pure Appl Chem 2009;81(12):2203–24.
[61] (a) Stankovich S, Dikin DA, Piner RD, Kohlhaas KA, Kleinhammes A, Jia Y, et al. Synthesis of graphene-based nanosheets via chemical reduction of exfoliated graphite oxide. Carbon 2007;45(7):1558–65; (b) Sinitskii A, Tour JM. Chemical approaches to produce graphene oxide and related materials. In: Murali R, editors. *Graphene Nanoelectronics*. Springer US: 2012:205–234.
[62] Dreyer DR, Park S, Bielawski CW, Ruoff RS. The chemistry of graphene oxide. Chem. Soc. Rev 2010;39(1):228–40.
[63] Ferrari AC, Basko DM. Raman spectroscopy as a versatile tool for studying the properties of graphene. Nat. Nanotechnol 2013;8(4):235–46.
[64] (a) Negri F, Castiglioni C, Tommasini M, Zerbi G. A computational study of the Raman spectra of large polycyclic aromatic hydrocarbons: toward molecularly defined subunits of graphite. J Phys Chem A 2002;106(14):3306–17; (b) Castiglioni C, Mapelli C, Negri F, Zerbi G. Origin of the D line in the Raman spectrum of graphite: a study based on Raman frequencies and intensities of polycyclic aromatic hydrocarbon molecules. J Chem Phys 2001;114(2):963–74.
[65] (a) Ferrari AC, Robertson J, Interpretation of Raman spectra of disordered and amorphous carbon. Phys Rev B 2000;61(20):14095–107; (b) Wilson PM, Orange F, Guinel MJ, Shekhirev M, Gao Y, Colon Santana JA, et al. Oxidative peeling of carbon black nanoparticles. RSC Adv 2015;5(112):92539–44.
[66] Vo TH, Shekhirev M, Lipatov A, Korlacki RA, Sinitskii A. Bulk properties of solution-synthesized chevron-like graphene nanoribbons. Faraday Discuss 2014;173(0):105–13.
[67] Hughes JM, Hernandez Y, Aherne D, Dössel L, Müllen K, Moreton B, et al. High quality dispersions of hexabenzocoronene in organic solvents. J Am Chem Soc 2012;134(29):12168–79.
[68] (a) Abbas AN, Liu G, Narita A, Orosco M, Feng X, Müllen K, et al. Deposition, characterization, and thin-film-based chemical sensing of ultra-long chemically synthesized graphene nanoribbons. J Am Chem Soc 2014;136(21):7555–8; (b) Zschieschang U, Klauk H, Mueller IB, Strudwick AJ, Hintermann T, Schwab MG, Narita A, Feng XL, Müllen K, Weitz RT, Electrical characteristics of field-effect transistors based on chemically synthesized graphene nanoribbons. Adv Electron Mater 2015;1(3):1400010.
[69] Kastler M, Pisula W, Wasserfallen D, Pakula T, Müllen K. Influence of alkyl substituents on the solution- and surface-organization of hexa-peri-hexabenzocoronenes. J Am Chem Soc 2005;127(12):4286–96.

[70] Narita A, Wang XY, Feng X, Müllen K. New advances in nanographene chemistry. Chem Soc Rev 2015;44(18):6616–43.

[71] Konnerth R, Cervetti C, Narita A, Feng X, Müllen K, Hoyer A, et al. Tuning the deposition of molecular graphene nanoribbons by surface functionalization. Nanoscale 2015;7(30):12807–11.

[72] Wasserfallen D, Kastler M, Pisula W, Hofer WA, Fogel Y, Wang Z, et al. Suppressing aggregation in a large polycyclic aromatic hydrocarbon. J Am Chem Soc 2006;128(4):1334–9.

[73] Behabtu N, Lomeda JR, Green MJ, Higginbotham AL, Sinitskii A, Kosynkin DV, et al. Spontaneous high-concentration dispersions and liquid crystals of graphene. Nat Nanotechnol 2010;5(6):406–11.

[74] (a) Khan U, O'Neill A, Lotya M, De S, Coleman JN, High-concentration solvent exfoliation of graphene. Small 2010;6(7):864–71; (b) Hamilton CE, Lomeda JR, Sun Z, Tour JM, Barron AR, High-yield organic dispersions of unfunctionalized graphene. Nano Lett 2009;9(10):3460–2.

[75] (a) Lotya M, Hernandez Y, King PJ, Smith RJ, Nicolosi V, Karlsson LS, et al. Liquid phase production of graphene by exfoliation of graphite in surfactant/water solutions. J Am Chem Soc 2009;131(10):3611–20; (b) Lotya M, King PJ, Khan U, De S, Coleman JN, High-concentration, surfactant-stabilized graphene dispersions. ACS Nano 2010;4(6):3155–62.

[76] Cai MZ, Thorpe D, Adamson DH, Schniepp HC. Methods of graphite exfoliation. J Mater Chem 2012;22(48):24992–5002.

[77] Müllen K. Chemistry in a materials world. Angew Chem Int Ed Engl 2015;54(35):10040–2.

[78] King BT, Kroulik J, Robertson CR, Rempala P, Hilton CL, Korinek JD, et al. Controlling the Scholl reaction. J Org Chem 2007;72(7):2279–88.

[79] Ormsby JL, Black TD, Hilton CL, Kumar B, King BT. Rearrangements in the Scholl oxidation: implications for molecular architectures. Tetrahedron 2008;64(50):11370–8.

[80] Kang J, Shin D, Bae S, Hong BH. Graphene transfer: key for applications. Nanoscale 2012;4(18):5527–37.

[81] (a) Tönshoff C, Bettinger HF, Beyond pentacenes: synthesis and properties of higher acenes. In: Siegel SJ, Wu Y-T, editors. Polyarenes I. Springer, Berlin, Heidelberg: Berlin, Heidelberg, 2014: 1–30; (b) Anthony JE, The larger acenes: versatile organic semiconductors. Angew Chem Int Ed Engl 2008;47(3):452–83.

Yuzhou Wu and Tanja Weil

8 Nanodiamonds for Biological Applications

8.1 Introduction

Nanodiamonds (NDs) represent a new class of carbon nanoparticles with unique physical and chemical properties. They are nanometer-sized carbon crystals with highly stable sp^3 carbon in the core and reactive sp^2 and amorphous carbons on the surface. The production of NDs is mainly from two methods, namely high-temperature/high-pressure (HTHP) synthesis and detonation method. The detonation method allows producing large quantities of NDs with low cost [1]. However, the detonation NDs normally provide single-digit nanometer sizes and form aggregation clusters in solution. The HTHP-synthesized NDs could be produced with different sizes ranging from a few nanometers to micrometers with highly crystalline structure [2]. A perfect diamond crystal is transparent to visible light. However, on rare occasions, natural diamond is found to display vivid color due to substitutional and vacancy defects in the lattice structure that are capable of absorbing and emitting visible light [3]. These defects could also be produced with synthetic NDs to obtain fluorescent NDs (FNDs) with desired optical properties. The most common and well-studied color center is the nitrogen–vacancy (NV) center [4] that offers spin-dependent photoluminescence. The NV center is a point defect in the diamond lattice that consists of a substitutional nitrogen atom that is directly adjacent to a lattice vacancy (Figure 8.1). The fluorescence from the NV center is independent of the size of the host diamond particles (Figure 8.1), and it is extremely stable without bleaching or blinking, which even allows their long-term observation at the single particle level in live cells [5]. Therefore, FNDs are particularly attractive for bioimaging applications. In addition, both the ground and excited states of the NV center display unique spin states with remarkably long coherence (T_2) and relaxation times (T_1), which are particularly attractive for developing sensitive magnetic and electron spin sensing in biological samples [6]. Recent studies suggest that NDs offer promising biocompatibility particularly in comparison with other fluorescent nanoparticles consisting of metals [7]. Taking together all these attractive features, NDs have emerged as promising nanoparticles for biological applications ranging from drug delivery to biosensing and theranostics [8]. The recent developments in this field will be discussed in this chapter.

8.2 Biocompatibility of NDs

Diamonds are generally considered stable, bioinert and biocompatible [7] due to their highly stable chemical structure, which makes them attractive for biological

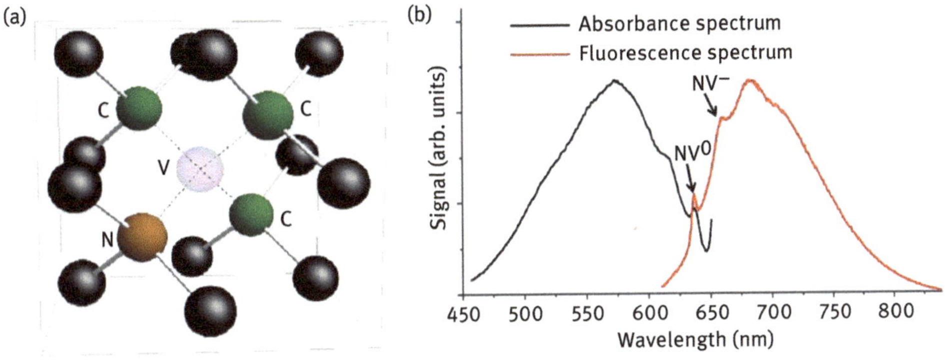

Figure 8.1: (a) The NV center in diamond – a substitutional nitrogen atom adjacent to a vacancy formed by a missing carbon atom. (b) The absorbance and fluorescence spectrum of an NV center.

applications. Unlike other fluorescent nanoparticles containing heavy metal elements, for example, quantum dots and gold nanoparticles, NDs are composed of carbon, which is compatible in live systems. In comparison to other carbon-based nanomaterials, for example, fullerenes and carbon nanotubes, NDs are considered one of the most biocompatible forms since they consist of sp^3 carbon, which is relatively unreactive. The biocompatibility of NDs has been tested both in vitro in cell lines and in vivo in animal models. Most of these experiments gave positive results [7, 9], while some reports also indicated potential risks [10]. Notably, their biocompatibility seemed to depend on the properties of the ND material, such as size, surface functionalities and concentrations. Therefore, the data from a particular experiment can only indicate the biocompatibility of a certain type of NDs. Optimization of the production and purification methods as well as surface functionalization could further improve the biocompatibility of ND materials.

8.2.1 In Vitro Biocompatibility Evaluation of NDs

Many in vitro cytotoxicity studies suggest that in comparison to other nanomaterials such as carbon nanotubes, semiconductor or metal particles, NDs represent the most biocompatible materials with no cytotoxicity to different cell lines [7, 9c, d]. Schrand et al. [9d] compared the biocompatibility of different carbon nanomaterials in aqueous suspensions, such as detonation NDs, single- and multi-walled carbon nanotubes (SWNTs and MWNTs) and carbon black (CB), in both neuronal and lung cell lines at concentrations ranging from 25 to 100 μg/mL for 24 h. The biocompatibility was evaluated with regard to the morphological and subcellular effects of these nanomaterials on mitochondrial membrane permeability and reactive oxygen species generation [9d]. The greatest biocompatibility was found after incubation with NDs and both cell types followed the trend: ND > CB > MWNT > SWNT [9d]. In another study [7], they have also shown that NDs with different surface functionalities, such as –COOH, $-COO^-Na^+$, $-SO_3^-Na^+$, are equally nontoxic in different cell lines, including even neuroblastoma and macrophages, which are normally much more sensitive to chemical reagents. Notably, in these experiments detonation NDs with sizes ranging from 2 to 10 nm were applied that are known to form aggregates in blood circulation and they do not exist as individual particles [11]. In another study, Liu et al. [9c] tested carboxylated NDs (cNDs) with 5 and 100 nm sizes and also compared their toxicity with carbon nanotubes [12]. They found that nanotubes induced cytotoxicity in human lung cells. However, treatment with NDs of both sizes did not induce cell death and apoptosis, although cND particles were retained in lung cells. Furthermore, cNDs did not alter the protein expression profile in these cells [12]. The 100 nm cNDs were taken into cells by macropinocytosis and clathrin-mediated endocytosis pathways. Even after long-term cell culture for 10 days in both A549 lung cancer cells and 3T3-L1 embryonic fibroblasts, no toxicity was observed. Cell division into two daughter cells was unaffected by the presence of cNDs. Finally, the cells retained NDs in the cytoplasm after cultivation for several generations without interfering with gene or protein expression on the regulation of cell cycle progression and adipogenic differentiation [12]. However, genotoxicity studies of detonation NDs with embryonic stem cells indicated that NDs induce slightly increased expression of DNA repair proteins (p53 and MOGG-1) and the oxidized NDs might cause more DNA damage than the pristine NDs. Though compared to other carbon nanomaterials such as nanotubes, the DNA damage caused by NDs is considered less severe [10a].

8.2.2 In Vivo Biocompatibility Evaluation of NDs

A few studies have also investigated NDs in living animal models, such as in rats and *Caenorhabditis elegans* (*C. elegans*) [9a, b]. The long-term stability and biocompatibility of 100 nm FNDs in rats through intraperitoneal injection was investigated over 5 months. Histopathological analysis of various tissues and organs indicated that FNDs are nontoxic even at very high dosing of up to 75 mg/kg body weight. The measurements of water and fodder consumption, body weight, and organ index also revealed no significant differences between the control and FND-treated groups [9a]. Another study reported long-term in vivo imaging of FNDs (0–10 nm) in *C. elegans* by either feeding them with a colloidal FND solution or microinjecting an FND suspension into the gonads of the worms [9b]. On feeding, bare FNDs remained in the intestinal lumen, while FNDs conjugated with biomolecules (such as dextran and bovine serum albumin) were absorbed into the intestinal cells. On microinjection, FNDs were dispersed in the gonads and delivered to the embryos and eventually into the hatched larvae of the next generation. The toxicity assessments, performed by employing longevity and reproductive potential as physiological indicators and measuring stress responses with use of reporter genes, suggested that FNDs are stable and nontoxic and do not cause any detectable stress to the worms [9b]. Therefore, the combination of many attractive features such as biocompatibility and high chemical and photophysical stability points to a promising applicability of NDs as contrast agents for long-term in vivo imaging.

Similar as other nanomaterials, one potential risk of NDs could be the long-time accumulation in liver, spleen and lung due to slow excretion rates [10b]. However, based on the thus far observed low toxicity, administration of NDs in vivo with small dosage is still promising. In addition, NDs of different sizes, surface modifications and administration routes will most likely also have an impact on their biodistribution. Therefore, by tuning the size distribution, enhancing the colloidal stability and optimizing the surface functionalities, the in vivo performance of NDs could be improved and adjusted to the respective in vivo requirements.

8.3 Surface Coating for Improving ND Stability and Biocompatibility in Biological Environments

The raw, as-synthesized NDs easily form aggregates especially in biological media due to their large surface to volume ratio, which could be one reason for long-term toxicity [11]. Although direct treatment of NDs surface, for example, oxidation, could improve the colloidal stability in aqueous solution [13], coating NDs with a biocompatible shell, such as a polymer shell, is still the most efficient strategy to avoid aggregation in complicated biological environments. With different properties of the coating materials, the biocompatibility and biodistribution of NDs could also be flexibly optimized. The coating of NDs could be achieved by covalent conjugation or non-covalent interaction. In addition, a silica shell could also be grown outside NDs to change their surface properties. A direct comparison of these approaches is given below.

8.3.1 Covalent Coating of NDs

The sp^2 carbon on the ND surface could be oxidized to impart functional groups such as hydroxyl groups and carboxylic acid groups. Thus, polymers could be directly attached to these functionalities to form a protecting outer shell. Polyethylene glycol (PEG) is one of the most widely used polymers for this purpose, including linear PEG, branched PEG and polyglycerol, which is known to provide a highly hydrated shell with low immunogenicity and low plasma protein adsorption [14]. The conjugation of PEG and its derivatives to NDs could be achieved by both "grafting to" and "grafting from" strategies [15] (Figure 8.2(a)). The "grafting to" method allows fast conjugation of different types

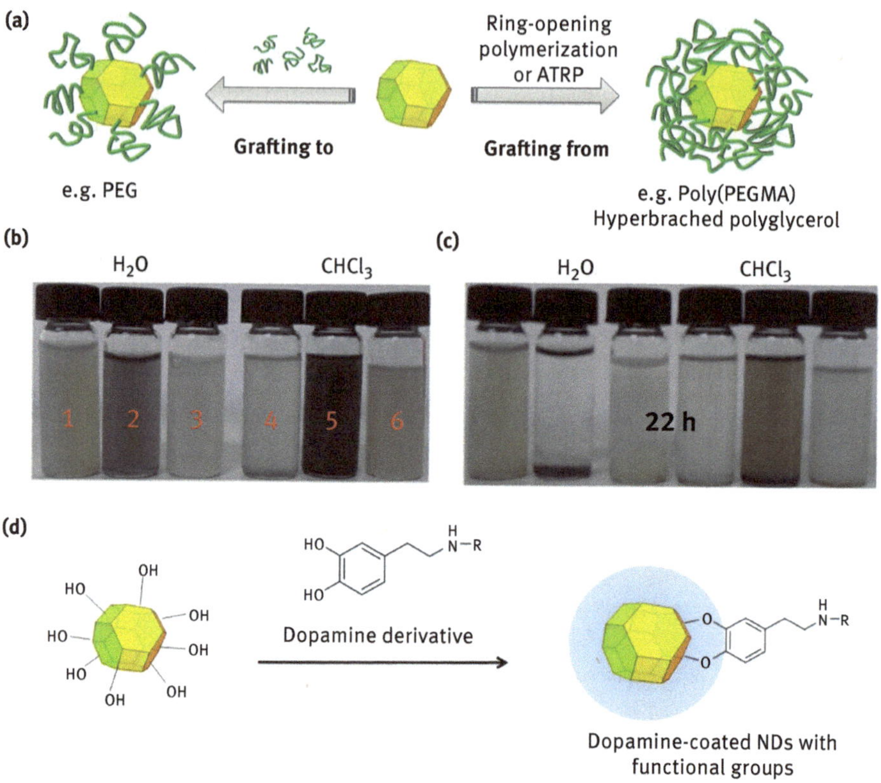

Figure 8.2: Methods for covalent coating of NDs with polymers. (a) Illustration of "grafting to" and "grafting from" methods for the preparation of PEG-like polymer coatings. (b) and (c) Photographs of raw NDs (1,4), PEG-coated NDs prepared by the "grafting to" strategy (2,5), and poly(PEGMA)-coated NDs prepared by the "grafting from" strategy (3,6) in water and chloroform at room temperature upon mixing and after 22 h. (d) Preparation of dopamine-coated NDs. (b) and (c) are reprinted from Ref. [15a] with permission.

of PEG-like polymers directly on the surface functional groups of NDs (e.g., carboxylic acid groups) and the polymer length could be rationally selected and well characterized before conjugating to NDs. The limitation of this method is a relatively low polymer density, which might result in insufficient protection of the inner ND core. For instance, PEG4000 was conjugated to NDs by the "grafting to" strategy, which could reduce the high tendency for aggregation of pristine ND and small clusters were obtained in phosphate buffered saline (PBS) and cell culture medium [15b]. However, the coated NDs still precipitated from aqueous solution after some time [15a] (Figure 8.2(b)). To achieve even more stable polymer coatings, the "grafting from" strategy turned out beneficial to obtain a denser shell. By atom transfer radical polymerization, a poly(PEG methyl ether methacrylate) layer could be grown on NDs. The resulting NDs were much more stable in organic and aqueous solution with negligible precipitation after 22 h [15a] (Figure 8.2(b)). Polyglycerol could also be grafted on NDs via ring opening polymerization [16]. The coated NDs are highly stable in aqueous condition, which allows size exclusion purification to achieve really narrowly distributed NDs. The purified NDs appeared as single particles on transmission electron microscope (TEM) without aggregations. Due to the high number of hydroxyl groups on the polyglycerol layer, these NDs could be easily functionalized with desired functional entities, such as anticancer drugs [15c]. Drug delivery applications will be discussed in later sections. Another special type of "grafting from" strategy is based on the biomimetic dopamine self-polymerization. Dopamine [17] (Figure 8.2(c)), chemically known as 4-(2-aminoethyl)benzene-1,2-diol, is one of the crucial catechol amine neurotransmitters that is widely distributed in mammalian brain tissues. In addition, it is also a biomimetic anchor for the functionalization of surfaces due to its self-polymerization nature under oxidative condition. The hydroxyl group on ND surface could be covalently reacted with dopamine during their self-polymerization process, thus resulting in a stable covalent coating of dopamine polymer on ND. By using dopamine derivatives with functional groups, this method could be a simple and a versatile strategy to functionalize ND surfaces with hydroxyl groups, azide groups and even poly-*N*-isopropylacrylamide [17].

8.3.2 Non-covalent Coating of NDs

NDs could also be coated by non-covalent interactions, which is the primary method to bring a protein coating onto NDs. It has been demonstrated that NDs, after oxidization in strong acid, could have very high affinity to various types of proteins, such as cytochrome c, myoglobin and albumin [18]. This is an integrated result from electrostatic forces, hydrogen bonding and hydrophobic interactions. In this way, NDs could absorb a layer of bovine serum albumin (BSA) or a-lactalbumin [18]. The stimulated emission depletion (STED) microscopy has shown that the albumin-coated NDs could be taken up into cells and homogeneously distributed without aggregation inside live cells [19]. However, the major limitation of non-covalent absorption is the stability of the coated layer, especially in complicated environments. Introducing multivalent interactions might be a possibility to overcome this limitation. For instance, the highly positively charged polyelectrolyte polyethyleneimine (PEI) could interact strongly with negatively charged NDs to form a stable complex for gene delivery [20]. However, the random electrostatic interaction between PEI and NDs creates inhomogeneous clusters instead of coating of single particles. Recently, an innovative concept based on a multivalent polymer [21a] as well as polypeptide biopolymer coating derived from the human blood plasma protein albumin was proposed [21b] (Figure 8.3(a)). The biopolymer was obtained after unfolding native serum albumin in urea and in situ attachment of PEG to the thiol groups reduced from disulfide bridges [22] (Figure 8.3(a)).

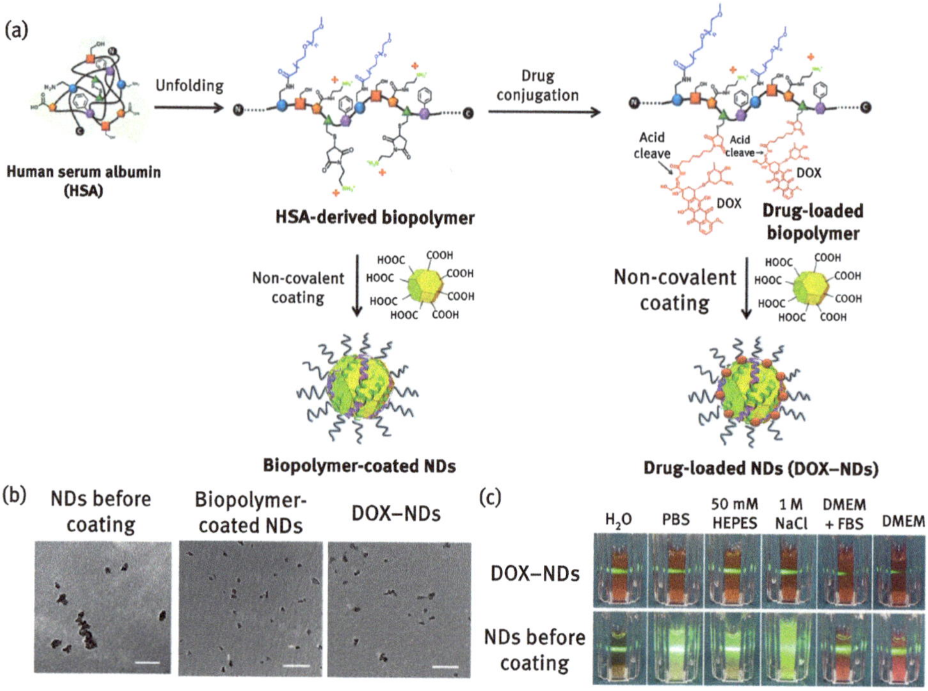

Figure 8.3: (a) Preparation of NDs coated with albumin-derived biopolymer and loading of anticancer drug doxorubicin (DOX). (b) Transmission electron microscope image of NDs before coating, after biopolymer coating and after DOX loading. Scale bar: 200 nm. (c) Photographs of 1 mg/mL DOX-NDs and raw NDs suspended in different physiologically relevant solutions, respectively. The green laser beam does not scatter in DOX-ND samples and in ND aqueous solution indicating no visible particle aggregation in solution. In contrast, the laser beam strongly scatters in raw ND samples suspended in PBS, 50 mM HEPES pH 7.5, 1 M NaCl and DMEM (Dulbecco's modified Eagle's medium) with/without 20 % fetal bovine serum (FBS) due to aggregate formation. Reprinted from Ref. [21b] with permission.

All the carboxylic acid groups from native albumin were converted to primary amine groups in this biopolymer, thus allowing efficient multivalent electrostatic interactions with the negatively charged ND surface. The PEG side chains decrease aggregations of the particles. Therefore, NDs coated by this biopolymer were found to be extraordinarily stable inside different kinds of aqueous buffers, including PBS, cell culture medium and even 1 M NaCl solution [21b] (Figure 8.3(c)). This biopolymer offers high numbers of orthogonal functional groups, which allow attachment of different active components onto coated NDs. This has been demonstrated by attaching drug molecules for anti-tumor drug delivery [21b]. Particularly, even after loading of high numbers of hydrophobic drug molecules (doxorubicin, DOX), NDs were still found to be extremely stable in all physiological media. Moreover, due to the biocompatibility of the biopolymer and the high colloidal stability, the coated NDs showed no cytotoxicity even with concentrations as high as 3 mg/mL [21b].

8.3.3 Silica Coating

Another strategy to functionalize ND surface focuses on growing an inorganic shell. For instance, NDs could be encapsulated with a silica shell by either direct silica shell growth from an ND core or by liposome-based encapsulation processes [23] (Figure 8.4). Cigler et al. demonstrated a multiple layer coating method to cover NDs with a first silica shell layer, a second layer of thin cross-linked aminopropyl-silica shell and a third layer of PEG with ethynyl end groups (Figure 8.4(b)). By this method, the obtained NDs are stable over a broad pH range (pH 2–10) and even in 1 M NaCl and cell culture medium [24]. This method also provides a high density of ethynyl functional groups on the surface that could allow efficient click reaction to achieve more than 2,000 dyes or peptides on an individual ND. Similarly, if the first layer of the silica shell contains methacrylate, the second layer could also be a polymeric layer by direct in situ radical polymerization [25]. In addition, a liposome-based method was reported by Neuman et al. [23] to decorate NDs with a silica shell (Figure 8.4(a)). The liposomes help to select for a desired size of particles, thus producing nearly monodispersed particles after coating that is independent of the surface properties of original NDs [23]. The silica shell imparts an anionic character and stable, monodispersed NDs were achieved over a relevant working pH range (pH = 5–8), particularly in the pH range above 3, where the strong negative zeta potential (–35 mV)

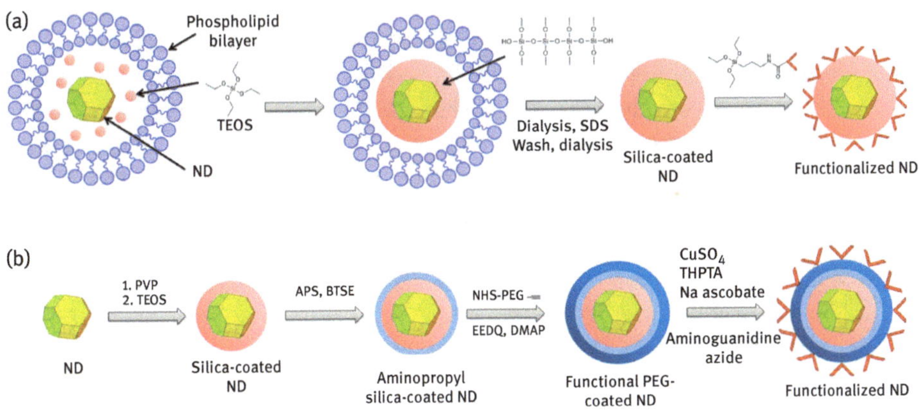

Figure 8.4: Preparation of silica-coated NDs via (a) liposome-based method and (b) multilayer coating. TEOS, tetraethoxysilane; BTSE, 1,2-bis(triethoxysilyl)ethane; APS, (3-aminopropyl) triethoxysilane; PVP, polyvinylpyrrolidone; EEDQ, 2-ethoxy-1-ethoxycarbonyl-1,2-dihydroquinoline; DMAP, dimethylaminopyridine.

allowed the particles to have strong electron repulsion, thus obtaining high colloidal stability [23].

8.4 Functionalization of NDs with Biomolecules

To advance the biological applications of NDs, another essential step is to attach biomolecules of interest onto NDs. Several methods to attach biomolecules, such as proteins and DNAs, to NDs for protein delivery, gene delivery and biosensing have been already reported, which will be discussed in this section.

8.4.1 Introduction of Proteins onto NDs

The most straightforward method to attach proteins onto NDs is physical adsorption. The large surface to volume ratio allows efficient adsorption of proteins. In addition, the functional groups at the ND surfaces, such as hydroxyl groups and carboxyl groups, could also provide electrostatic interactions and hydrogen binding with proteins. By integration of all these effects, the physical adsorption of proteins on NDs could be sufficiently stable for many desirable applications such as protein delivery, cell targeting and sensing. For instance, the therapeutic protein insulin was adsorbed to detonation NDs, thus forming insulin–ND complexes [26]. The adsorbed insulin could also be released under alkaline conditions [26]. Cell viability assays and real-time polymerase chain reaction quantifying the expression of Ins1 and Csf3/G-csf gene revealed that ND-bound insulin remained inactive while the desorbed insulin could recover their activity [26]. These results suggest great potential of NDs as facile platform for protein delivery. The HTHP NDs with average sizes of around 100 nm could also adsorb proteins through the interplay of electrostatic forces, hydrogen bonding and hydrophobic interactions [27]. By adsorbing different proteins, the modified NDs could be taken up selectively by a specific cell type for targeted cell imaging [27]. The physical adsorption approach has also facilitated sensing of proteins. For instance, ferritin was adsorbed on NDs with stable NV centers. Using the magnetosensing techniques with NV centers (see Section 8.5.2), the iron concentration inside the ferritin could be detected without damaging the protein cage [28].

However, to achieve more stable and specific binding of proteins to NDs, covalent conjugation is still a more reliable strategy in comparison to physical adsorption. By direct functionalization of NDs with different functional groups, proteins could be covalently conjugated to ND surfaces [8c, 29]. For instance, thiolated antibodies could be conjugated to thiol-reactive *N*-succinimidyl 3-(2-pyridyldithio) propionate (SPDP)-modified NDs [30]. However, it should be noted that direct chemical modification of NDs often leads to agglomeration. If well-dispersed individual ND particles are desired, it is necessary to conjugate the proteins on polymer-coated NDs.

For instance, NDs could be first coupled with functionalized PEG, which could on the one hand impart improved stability of the NDs and, on the other hand, allow easy conjugation of proteins. Streptavidin was conjugated to the functionalized PEG to further serve as an adaptor. Thus, commercially available biotinylated antibodies could be quickly labeled onto these NDs for immunostaining [27]. Notably, even if PEG linker is used, small amounts of agglomeration are still difficult to avoid. It was necessary to perform the labeling experiment with inversed cell culture to avoid deposition of such aggregates on the cell surface [27]. The coating strategies discussed above could provide novel opportunities to achieve more stable protein functionalized NDs, which still need to be further tested.

8.4.2 Modification of DNA on NDs

In comparison to the preparation of protein-modified NDs, it is even more challenge to prepare DNA-functionalized NDs with high colloidal stabilities. For some applications, such as gene delivery or drug delivery, controlled small agglomeration is not problematic. In these cases, the modification of NDs with DNA molecules could also be achieved by simply physical adsorption or direct covalent conjugation. For instance, highly positively charged polymer PEI was coated onto acid-treated NDs via electrostatic interaction to cover the NDs with high density of positive charges. Thus, negatively charged DNA plasmids formed dense complexes with PEI-coated NDs, which have shown high gene transfection efficiency with relatively low cytotoxicity [20]. Thiolated synthetic ssDNA was also covalently conjugated to SPDP-modified NDs similarly for protein conjugation [30]. However, both methods do not allow producing NDs with sufficient colloidal stability for advanced biosensing applications, where individual ND particles are preferred.

Recently, following the biopolymer coating strategy introduced above, highly stable NDs functionalized with DNA were achieved [31] (Figure 8.5). Biotinylated PEG was used as biopolymer side chains, which formed a biotin-functionalized shell on coated NDs. Thus, the streptavidin adaptor could be assembled on the NDs, which allowed easy conjugation of biotinylated ssDNA. Since the biopolymer coating is highly stable with excellent water solubility, single ND particles were found even after all these functionalization steps [31]. This achievement is particularly important for producing high-quality biofunctionalized NDs for NV center-based high-resolution sensing techniques.

8.4.3 Self-Assembled NDs on Bionanostructures

One unique application of protein- and DNA-conjugated NDs is to prepare defined self-assembled ND arrays on bionanostructures. Periodic arrangements of NDs with NV centers are an essential and highly challenging step toward efficient quantum

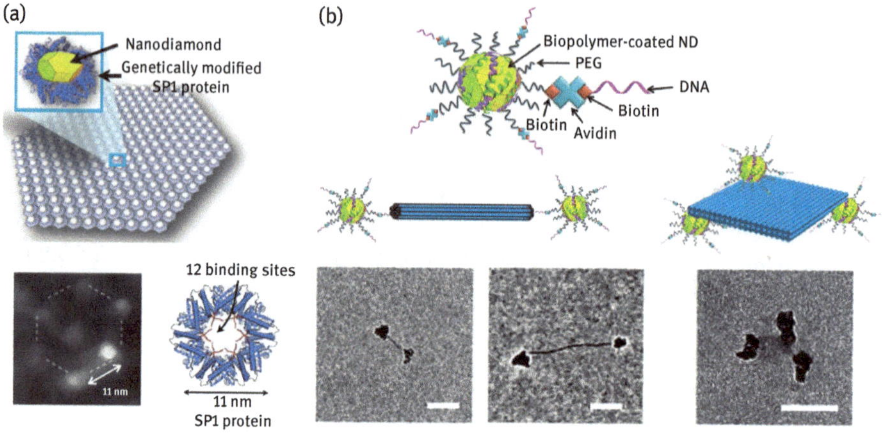

Figure 8.5: (a) ND arrays formed on genetic engineering SP1 proteins. The scheme above shows an ordered hexagonal array of SP1–ND hybrids consisting of an ND attached to the SP1 inner cavity. Here, the SP1-monolayer serves as a structural scaffold. The DF-STEM (dark-field scanning transmission electron microscopy) image shows a hexagonal structure formed of seven NDs (ND diameter ~5 nm). The SP1–protein ring consists of binding sites that are genetically modified to enable graphite-specific binding. (b) NDs assembled on DNA origami with 2D and 3D structures. The NDs are coated with biopolymer and the PEG side chains are functionalized with biotin to form biotin–avidin–biotin linkers for DNA conjugation. The DNA-modified NDs are stable without aggregation and therefore could be assembled on DNA origami nanostructures with defined geometries. A 2D assembly was demonstrated with two NDs fixed on two ends of a DNA origami tube with defined length. A 3D assembly was demonstrated with three NDs attached on a DNA origami square. The TEM images show the self-assembled structures. Scale bar = 100 nm. The figure is partially reprinted from Refs [31, 32] with permission.

information processing, quantum simulation and quantum sensing applications [32]. To achieve this goal, a novel approach was developed recently based on the self-assembling capabilities of biological systems. Techniques from bionanotechnology exploit the self-assembling behavior of special designed protein and DNA to produce periodical structures with nanometer spatial resolution. Through the conjugation techniques available for connecting NDs with protein and DNA building blocks, one could incorporate NDs onto these bionanostructures to achieve a defined nanoarray. One example is the formation of small ND arrays on SP1 protein variant (Stable Protein 1, Figure 8.5(a)), which could form numerous dimers and trimers along with large ordered structures such as a seven ND hexagon [32]. The emerging technique of DNA origami based on programmed folding of single-stranded DNA molecules allows creating even more complex structures ranging from periodic arrays to three-dimensional (3D) architectures [33] (Figure 8.5(b)). This offers an ideal platform to assemble ND arrays with flexible distance and sophisticated nanostructures. To assemble NDs onto DNA origami particularly requires DNA functionalized NDs with high colloidal stability even under high ionic strength. The biopolymer coating strategy discussed above provides an ideal method to achieve DNA functionalized NDs sufficient for preparing NDs on DNA origami. It has been demonstrated that with this method, NDs could be self-assembled on DNA origami in predefined one-dimensional, two-dimensional (2D) and 3D geometries and defined distance [31] (Figure 8.5(b)). These assemblies provide unique tools for studying self-assembled NV center spin lattices or plasmon-enhanced spin sensors as well as improving fluorescence labeling for bioimaging.

8.5 ND for Drug delivery

One major biological application of NDs is serving as drug delivery nanocarrier. ND-based drug delivery systems receive an increasing recognition due to their unique combination of promising biocompatibility and attractive optical properties. The highly stable emission of FNDs without bleaching and blinking enables long-time monitoring of the drug delivery process and drug release from the carrier. Therefore, using FNDs as nanomedicine platform for drug delivery into cancer cells and tissue would allow combining fluorescence-based tumor diagnostic and cancer therapy within one particle, which is essential for the evolving new concepts of "theranostic" (a portmanteau of therapeutics and diagnostics). The different strategies to achieve drug delivery with NDs will be discussed in this section.

8.5.1 Drug Delivery with Detonation ND Clusters

The first generation of ND–drug delivery systems has been prepared by simply absorbing mainly lipophilic drug molecules onto the ND surface [34]. These ND–drug complexes comprised of about 4–6 nm primary detonation ND particles clustered in solution into 100–200 nm aggregates, which can substantially absorb drug molecules and significantly enhance their blood circulation half-life and tumor retention [34, 35]. Dean Ho's group has done intensive studies with this strategy. They investigated the absorption of hydrophobic anticancer drugs such as DOX [34, 35], purvalanol A, 4-hydroxytamoxifen [36] and mitoxantrone (MTX) [37], anti-inflammatory drug dexamethasone [37] and the diabetes drug insulin [26] onto the ND complexes. Delivering anticancer drugs with ND complexes was found to be able to circumvent drug resistance. A major mechanism for drug resistance is that the cells pump drugs out through ATP-binding cassette (ABC) transporters. They have shown that ND–drug complex could overcome the drug efflux via an ABC transporter-independent pathway [35a]. Both ND–DOX complex [35a] and ND–MTX complex [37] have shown significant improved drug retention in respective drug-resistant tumor cell line. They have also demonstrated in vivo that ND–DOX complex could significantly increase apoptosis and inhibit tumor growth in both murine LT2-Myc liver tumor and 4T1 mammary tumor without affecting adjacent normal tissues [35a] (Figure 8.6). The reason for the advantages of ND–DOX complex over free DOX is proposed to be the gradual release of the drug from NDs, which allows for slow drug release over a long time to maintain the free DOX concentration below a toxic level to normal tissues.

The simple absorption strategy is high throughput and universal to different types of therapeutics independent of their structure. However, the complexes are not really stable inside blood plasma and the nonspecific drug release is always difficult to avoid. Therefore, covalent linking of drug molecules to the detonation ND

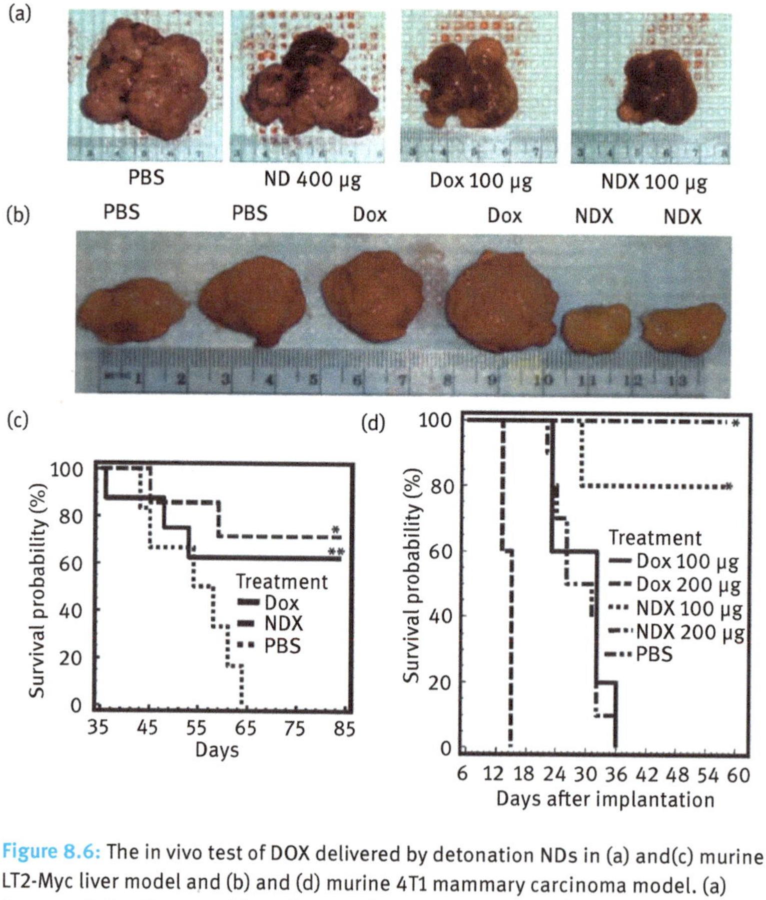

Figure 8.6: The in vivo test of DOX delivered by detonation NDs in (a) and(c) murine LT2-Myc liver model and (b) and (d) murine 4T1 mammary carcinoma model. (a) Representative images of livers/tumors from LT2-Myc mice. (b) Representative images of excised tumors from 4T1 mice. (c) Kaplan–Meier survival plot for LT2-Myc mice treated with PBS ($n = 5$), Dox (100 mg) ($n = 8$), or NDX (100 mg of Dox equivalent) ($n = 7$) by tail vein injection every 7 days. *$P < 0.03$; **$P < 0.06$. (d) Kaplan–Meier survival plot for 4T1 mice treated with PBS ($n = 7$), Dox (100 mg) ($n = 10$), NDX (100 mg of Dox equivalent) ($n = 10$), Dox (200 mg) ($n = 5$), or NDX (200 mg of Dox equivalent) ($n = 5$) by tail vein injection every 6 days. *$P < 0.003$ [35a]. Reprinted from Ref. [35a] with permission.

clusters was also investigated. By 1-Ethyl-3-(3-dimethylaminopropyl)-carbodiimide/*N*-Hydroxysuccinimide (EDC/NHS) coupling, DOX and cell penetration peptide TAT could be directly conjugated to the carboxylic acid groups on ND surface to enhance penetration of cell membrane [38]. PTX [39] has also been conjugated to the hydroxyl groups on NDs to avoid dissociation process. The activity of covalently linked PTX was maintained as shown by inducing mitotic arrest, apoptosis and anti-tumorigenesis of human lung cancer cells in xenograft severe combined immunodeficiency (SCID) mice. However, the mechanism of PTX release is unclear in this case. Dean Ho et al. [30] have reported a more sophisticated strategy to attach fluorescently labeled paclitaxel (PTX)–DNA conjugates and anti-EGFR antibody onto the ND surface. The PTX–DNA conjugates were covalently attached to ND surface via a disulfide linker, which allowed controlled drug release only under reducing condition. They have shown that these complexes could be efficiently taken up into epidermal growth factor receptor (EGFR) overexpressing cell lines and exhibited efficient therapeutic efficacy which was twofold increased comparing to nontargeted ND delivery system.

8.5.2 Drug Delivery with Polymer-Modified NDs

Although covalent linking of drug molecules could reduce drug leakage, the complex formation by random aggregation of the extremely small denotation ND particles is still an uncontrolled process, thus the clusters may not be stable during circulation. In addition, the NV centers in these small NDs are not stable; therefore, their optical property is not sufficient for high-quality bioimaging. Milling HPHT diamond could obtain more stable ND particles with sizes ranging from tens to hundreds of nanometers. Using single NDs as drug delivery carrier instead of their uncontrolled clusters is a strategy to achieve a well-defined drug delivery system. In addition, they could have multiple very stable NV centers, which are particularly attractive to develop as theranostic agents that combine drug therapy and diagnostic imaging in one particle. However, the bare NDs tend to form aggregations in the biological environment; therefore, single ND-based drug delivery normally requires a polymer coating on the ND surface as discussed in Section 8.3. For instance, polyglycerol-coated NDs with average size of 50 nm could be conjugated with Pt-based anticancer drug via pH-responsive linker [15c]. As discussed earlier, ND coated with polyglycerol exhibits good solubility in a physiological environment and also provides a large number of hydroxyl groups that could be conjugated with drugs as well as Arg-Gly-Asp (RGD) peptide for tumor cell targeting. In vitro experiments showed that the Pt-RGD-conjugated NDs could be taken up by U87MG cells and induced high toxicity, but no uptake and toxicity effect was observed with HeLa cells, revealing the high targeting efficacy. The stable NV centers in these NDs allow fluorescence imaging of the drug delivery process in cells by confocal microscopy. Notably, the Pt complex is a water-soluble anticancer drug that will not reduce the solubility of drug-loaded NDs. To achieve high colloidal stability of NDs even after loading hydrophobic drugs still remains a key challenge since lipophilic groups at the ND surface usually promote the formation of larger aggregates [36]. Uncontrolled aggregation of nanoparticles has been connected with unexpected cell toxicities and the prevention of clogging capillary blood vessels is a critical concern for clinical development of nanomaterials [40].

Recently, the Cigler [21a] and the Weil group [21b] prepared NDs with polymer [21a] and biopolymer [21b] coatings biopolymer, which displayed excellent colloidal stability as discussed in Section 8.3 (Figure 8.3). A defined number of DOX molecules could be conjugated onto the functional groups on the biopolymer before coating with NDs, thus the DOX-loading process could be precisely controlled and well characterized. Particularly, NDs coated with the DOX-loaded biopolymer are still highly stable without any aggregation under all physiological conditions tested (Figure 8.3(c)). The biopolymer-coated NDs showed efficient cell uptake and excellent biocompatibility. Since high numbers of DOX molecules were linked to NDs by an acid-cleavable linker, they could be efficiently transported and released inside cancer cells [21b] (Figure 8.7(e)). They have also demonstrated the theranostic potential of this system by using FNDs with NV centers. The intracellular distribution of NDs and DOX was resolved in live cells by recording the fluorescence spectra of NDs and DOX in different cellular compartments [21b] (Figure 8.7(a–c)). Significant amounts of DOX were released from ND carriers and distributed inside the cytoplasm and the nucleus, whereas the ND carriers remained entrapped inside endosomal [21b] (Figure 8.7(a–c)). The antitumor efficacy was shown with in vitro cellular experiment as well as with chicken embryo models. The DOX-loaded NDs were found to have similar IC_{50} in comparison to free DOX in cellular experiments, but shown more efficient tumor inhibition in chicken embryo models due to tumor accumulation after longer blood circulation time [21b] (Figure 8.7(d)). This strategy could be particularly attractive for developing multimodal NDs combining therapy and imaging potential in one particle for the emerging concept of “theranostics.”

8.6 NDs for Imaging and Biosensing

The biocompatibility evaluation of NDs and their drug delivery studies supports the promising potential of using NDs in living biological systems. In addition, the unique fluorescence properties and electromagnetic properties of NV centers make NDs particularly attractive for biosensing. The fluorescence from NV centers is in the near-infrared region with extremely high photostability and sufficiently long lifetime, which makes them one of the best fluorescence labels for high-resolution fluorescence imaging. In addition, the ground and the excited states of the NV center display zero-field magnetic resonances at 2.88 and 1.42 GHz, respectively. Such magnetic resonance occurs between the $m_s = 0$ and the $m_s = \pm 1$ magnetic states of an electronic spin triplet (Figure 8.8). This electron spin exhibits remarkably long coherence

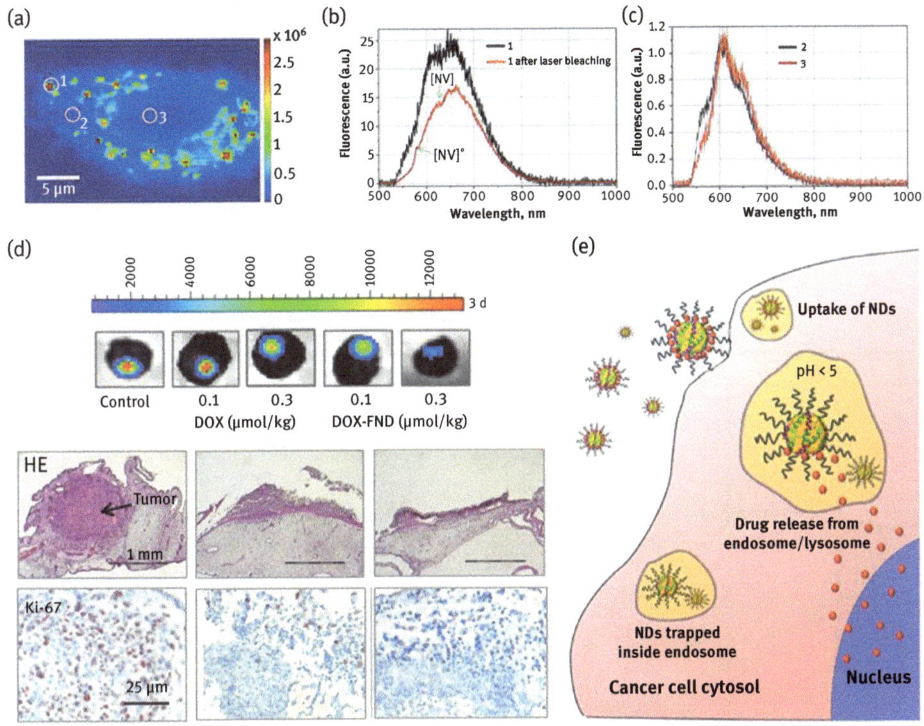

Figure 8.7: DOX delivery by biopolymer-coated NDs. (a) Fluorescence spectra inside living cells. A549 cells were treated with 55 µg/mL of fluorescent DOX-NDs for 24 h. Fluorescence spectra were recorded at positions highlighted by white circles. (b) Representative emission spectra inside cellular vesicles (line 1), e.g., circle 1, corresponds to the overlapping spectra of DOX and N–V centers. The emission spectrum after laser bleaching of DOX emission (line 1 after laser bleaching) yielding a typical emission spectrum of N–V centers with assignable peaks from [N–V]° and [N–V]⁻. (c) The representative emission spectra within the cytosol (line 2), e.g., circle 2, and inside the nucleus (line 3), e.g., circle 3, both revealing typical DOX spectra. (d) Antitumor effect of DOX–NDs on breast cancer xenografts in chorioallantoic membrane (CAM) model. Photograph on top showing the representative tumor luminescence in CAM model. Photograph below showing the immunohistochemical analysis of breast cancer xenografts. HE, hematoxylin and eosin staining of whole xenografts grown on CAM; Ki-67 antigen staining of tumor xenografts, brown-red nuclei are indicative of proliferating cells. (e) Illustration of cell uptake and drug-release process of DOX–ND. Reprinted from Ref. [21b] with permission.

(T_2) and relaxation times (T_1), which can reach milliseconds in ultrapure diamond [6]. With a concept of single-molecule optically detected magnetic resonance [41], the electronic spin state of even a single NV center can be detected by means of electron spin state-dependent light scattering that discriminates between the $m_s = 0$ and the $m_s = \pm1$ states [42]. These unique properties of NDs offer great potential to develop the smallest magnetosensing probes for detecting the tiniest of magnetic fields in vitro

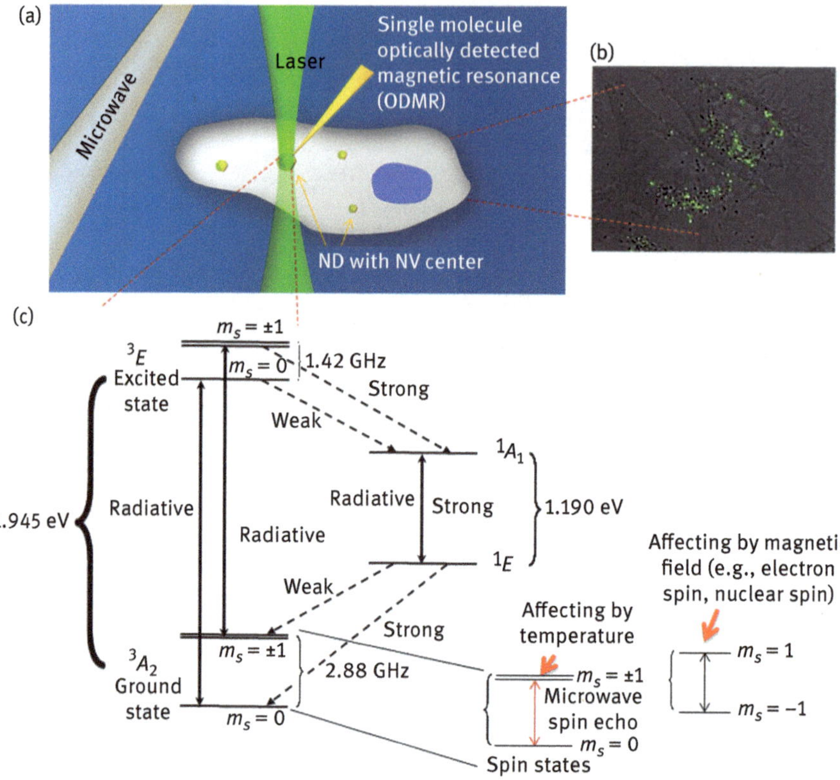

Figure 8.8: Illustration of ND quantum sensing in living cells. (a) The experimental setup for ND quantum sensing inside living cells. (b) Confocal fluorescence microscopy image of NDs with NV centers inside living cells (the bright dots). (c) The energy diagram of NV center. The dangling bonds form an electronic spin-1 system with an optical transition at 1.945 eV. Optical and infrared transition (solid arrows) and weak nonradiative transitions (dashed lines) allow for the observation of spin-dependent fluorescence and optical electron spin polarization. The spin states of ground state for temperature and magnetic sensing are amplified.

and in vivo. Moreover, the electron spin triplet to the $m_s = 0$ state could be hyperpolarized by optical pumping even at room temperature, and this hyperpolarization could be transferred to nuclear spins leading to potential applications in magnetic resonance imaging. Combining these advantages together with the great biocompatibility, NDs have great potential to serve as noninvasive nanosensors for studying the structure of biomolecules and monitoring biological reactions even inside living organisms. Although this field is just emerging, several successful examples have already been demonstrated and indicated the prospective of ND quantum sensors, which will be discussed in this section.

8.6.1 Fluorescence Imaging with NDs

NDs are emerging as efficient and safe candidates for fluorescent cellular imaging due to their bright fluorescence in the near-infrared region, high photostability with sufficiently long lifetime and excellent biocompatibility. Several studies have investigated FNDs as biomarker for in vitro and in vivo imaging. For instance, ND could serve as fluorescent marker to label biomolecules such as protein. ND-labeled transferrin was imaged by confocal fluorescence microscopy to confirm the receptor-mediated uptake into HeLa cells [43]. Actin and mitochondrial antibody-conjugated NDs were able to label actin and mitochondria in HeLa cells, suggesting the potential of NDs for imaging specific cellular structures [44]. The protein-conjugated NDs were found to exhibit negligible changes with respect to the fluorescence spectra and lifetimes [43]. Moreover, Mohan et al. [9b] have even imaged the bare FNDs and albumin-conjugated FNDs in *C. elegans* (Figure 8.9) by differential interference and wide-field epifluorescence microscopic images. The bare NDs were found in the lumen of the worm without intestinal absorption even after 12 h and were fast excreted when the worms were fed *Escherichia coli*. In contrast, albumin-conjugated NDs were taken up by the intestinal cells of the worm and remained there for 24 h, despite *E. coli* feeding. This result indicated that surface functionalization could significantly alter the in vivo fate of NDs. Albumin coating is known to improve the colloidal stability of NDs as discussed in Section 8.3. The enhanced dispersibility of the NDs was believed to be the reason for enhanced cellular uptake. In this study, no photobleaching or photoblinking of NDs was observed even after 48 h of continuous laser excitation, indicating the great potential of FNDs for long-term bioimaging in vivo.

Furthermore, the high-resolution fluorescence microscopy developed during the past two decades nowadays provides ultimate sensitivity of detection beyond the limits imposed by the diffraction of light. The fluorophores for superresolution imaging techniques are required to be completely photostable without bleaching after long time and strong illumination. Therefore, NDs are particularly attractive for single molecular imaging with these newly developed superresolution fluorescent microscopy. For instance, STED microscopy could image individual NV center in diamond crystals and map the 3D distribution of NV centers with resolution down to 6 nm [45]. Using STED microscopy, single NDs could be tracked in live cells with resolution down to 39 nm [19]. Notably, for such high-resolution fluorescence imaging, coated NDs with good dispersity without aggregation are particularly critical.

8.6.2 NDs as Nanoscale Magnetometer

The emission signal from an NV center of NDs is not only attractive for fluorescence imaging, but can also be developed as a magnetometer. As shown in Figure 8.8, the ground-state spin triplet of NV centers can be coherently manipulated using

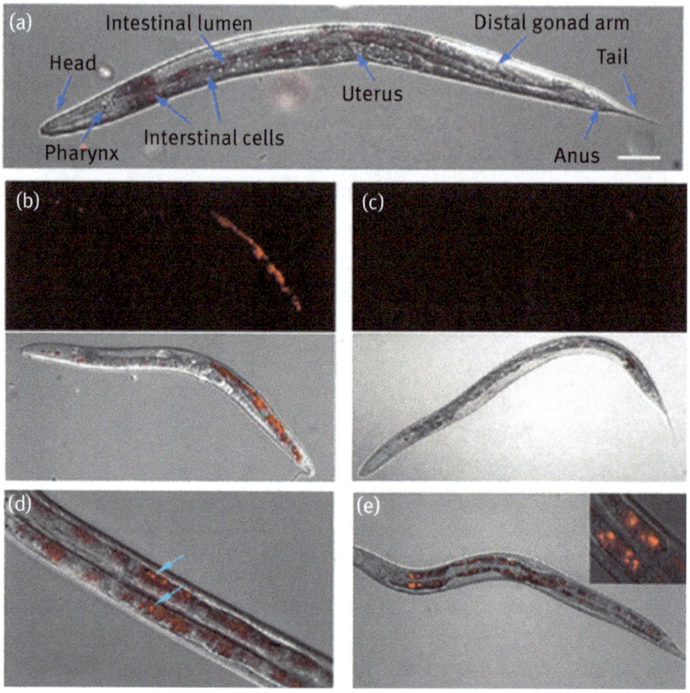

Figure 8.9: Epifluorescence and epifluorescence/DIC-merged images of wild-type *C. elegans*. (A) An untreated young adult. Scale bar: 50 µm. (B) Worms fed with bare FNDs for 12 h. The FNDs stayed inside the gut and were not excreted out when the worms were deprived of food. (C) Worms fed with bare FNDs for 2 h and recovered on to *E. coli* bacterial lawns for 40 min. Almost, if not all, FNDs were excreted out after feeding *E. coli* within 1 h. (D) Worms fed with BSA-coated FNDs for 3 h. FNDs can be seen to be localized within the intestinal cells (solid arrows). (E) Worms fed with BSA-coated FNDs for 3 h and recovered on to *E. coli* bacterial lawns for 1 h. The FNDs staying in the lumen are excreted out, whereas the ones localized in the cells retain. Insets: 100× magnified images of the FNDs within the intestinal cells. Anterior is left and dorsal is up in all the figures. Reprinted from Ref. [9b] with permission.

microwave pulses (spin echo) and efficiently initialized and detected by means of laser illumination. Thus, the energy of spin levels of individual NV centers could serve as fingerprints, allowing identification and tracking of ND particles with identical fluorescence. By this way, individual FNDs could be detected inside living cells with nanoscale precision with regard to their location, orientation, spin levels and spin coherence times [46].

In addition, the triplet ground state of the NV center is sensitive to the changes in the local environment, which could be detected by monitoring decoherence rates. This might provide unique insights into intracellular processes. For instance, ND could be developed as a sensitive nanoscale thermometry [47]. The transition frequency from the singlet ground state to triplet ground state (Figure 8.8) has a temperature dependence, which could be measured by the change of spin coherent time [47]. With this technique it is possible to noninvasively measure the local temperature in living cells at length scales as short as 200 nm [47].

Moreover, the triplet ground state of the NV center will also be modified by the presence of external spins, which could be recorded as shift of the fluorescence spectral lines or perturbance of the spin echo or changing of the relaxation time of NV centers (Figure 8.8). For instance, the fast fluctuating external spins generate noise to the relaxation of NV centers, which could be measured as a signal using decoherence microscopy. The sensitivity of this technique is able to reach the ultimate limit down to the detection of a single paramagnetic ion [48]. Therefore, the NDs could serve as a nanoscale magnetometer that could be delivered into cells and living organisms for electron spin sensing. If a biomolecule is labeled with paramagnetic ions, it could be detected even with single molecular sensitivity [49]. More straightforwardly, this technique is particularly interesting for studying magnetic proteins, which contain paramagnetic ions natively with special functions, such as ferritin. As the major iron reservoir in living organisms, ferritin is a protein cage that could store around 4,500 Fe^{3+} ions in one particle. Plasma ferritin concentration and the amount of iron inside ferritin is an important marker for many diseases such as anemia, hypothyroidism and celiac disease. The detection of ferritin with NDs is based on magnetic noise induced by the paramagnetic iron at the interior. When ferritins are absorbed on the surface of NDs, a significant reduction of both coherence (T_2) and relaxation time (T_1) could be obtained and the detection sensitivity is found to be able to reach the single molecule detection threshold [28, 50] (Figure 8.10). Since NDs are also ideal carriers to deliver biomolecules of interest into cells and living organisms as discussed earlier, one could envision a nanoscopic electron spin sensor for detection of electron spin containing biomolecules in vitro and in vivo. Furthermore, the remarkable sensitivity paired with the noninvasive character of NDs may also provide access to quantum

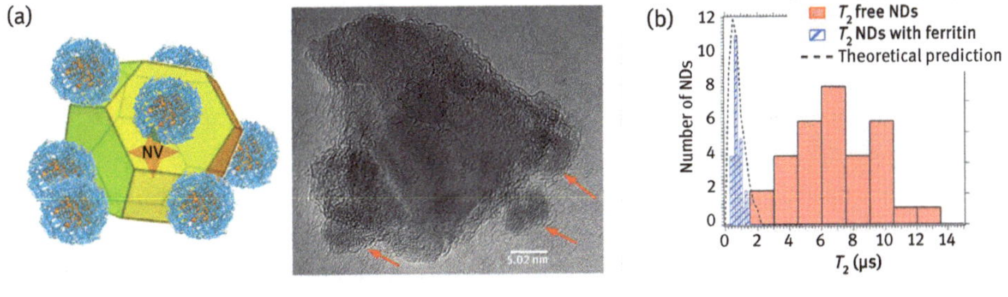

Figure 8.10: Detection of iron ions inside ferritin with fluorescent NDs. (a) Illustration of the ferritin adsorbed on fluorescent NDs and the TEM visualization. (b) The changing of T_2 time in response to iron ions in ferritin. Reprinted from Ref. [28] with permission.

mechanical studies in living organisms, which may have potential to resolve important electron transfer process and radical pair dynamics in cellular environments [51].

Similar like sensing of electron spin, the magnetic field from a nuclear spin could also induce changes on the triplet ground state of NV centers thus being detected optically (Figure 8.8). However, a single nuclear magnetic moment is approximately 1,000 times weaker than an electron spin. To detect such weak signal, the nuclear spin needs to be brought into close proximity to the NV center (a few nanometers) to produce sufficient field in the order of microtesla. In addition, using high order of spin echoes, for example, generated by microwave, the coherence time of NV defects in the presence of nuclear spins can be improved by orders of magnitude [52]. By this way, the NV centers could have great potential for developing nanoscale nuclear magnetic resonance (NMR) spectroscopy ultimately with even single molecular sensitivity [53]. Recently, the detection of proton NMR in only 5 nm^3 volume of liquid sample was achieved by directly placing the sample on single NV centers implanted 2.5–10 nm below the diamond surface [54]. By constructing a microfluidic device on diamond, picoliter-scale sample could be easily applied to allow detection of the nuclear spin in solution with below 500 nm spatial resolution [55] Single nuclear spin inside solid sample with long coherence time, such as Si29 in quartz and nuclear spins associated with hydrogen bound to diamond interface, could also be detected within a few seconds measurement time if the solid sample is directly placed on diamond surface [56]. Although the successful experiments are still based on NV centers implanted inside bulk diamonds, the same strategy should also be able to realize nuclear spin detection with NDs. However, it is much more challenging to manipulate a single NV center in ND and attach the molecule of interest within a few nanometer distance from the NV center. In addition, currently only location and intensity of the nuclear spin could be determined by the NV-based NMR technique. For more useful applications, such as the elucidation of biomolecule structures, the spectral resolution should be further improved in order to detect dipolar couplings and chemical shifts. Several strategies might be helpful in this respect, such as the development of new measurement protocols similar to 2D NMR [57], or new protocols to remove magnetic noise that are developed in the context of quantum computations (e.g., quantum error correction) [58].

8.6.3 Magnetic Resonance Imaging with Hyperpolarized NDs

Magnetic resonance imaging (MRI) is one of the most important diagnostic techniques allowing noninvasive imaging of tissue and organs inside the body. However, the sensitivity of MRI is only sufficient to detect large defects in tissue, which limits its applications for early diagnostic of important disease. This low sensitivity is partially related to low polarization of nuclear spins at low temperature. The electron spin native to the NV center can be initialized to a highly polarized quantum state by laser irradiation on microsecond timescale, which is known as hyperpolarization technique. This polarization can then be transferred to surrounding nuclear spins with the support of microwave radiation, thus to achieve hyperpolarized nuclear spins at room temperature [59]. This technique has emerged as a new approach for accessing highly sensitive MRI that might be sufficient for even single molecule detection [60]. The relaxation time of hyperpolarized NV centers in bulk diamond could reach hours. However, the polarization of diamond nanoparticles represents a more challenging task due to the more disordered orientation of NV centers. Advanced dynamic nuclear spin polarization protocols developed recently are able to address this problem [61]. With this method, it has been shown that the relaxation time of hyperpolarized NDs could also reach more than a minute, which is superior to any known MRI contrast agent. Furthermore, the polarization of NV centers can also be transferred to external nuclear spins if they are located close to the diamond surface [62]. Thus, by constructing flow channels with hyperpolarized diamond, the molecules in a tiny volume could be highly polarized, which might provide another mechanism for developing ultrasensitive NMR [63].

8.7 Conclusions

In summary, ND as one of the most biocompatible nanoparticles provides great potential for a broad range of biological applications, including bioimaging, drug delivery and quantum sensing. Compared to other carbon nanomaterials, such as carbon nanotubes, NDs were found to be biocompatible in vitro and in vivo. Although the strong aggregation inside biological systems represents one of the main limitations of NDs, many coating strategies including polymer and silica coatings have been already developed to achieve highly stable NDs with even better biocompatibility. Functional molecules, such as protein, DNA and drug molecules, could be conjugated to NDs via both physical adsorption and covalent conjugation chemistries. Based on these preparation methods, NDs are ready to be flexibly tailored as multifunctional drug delivery carrier. A large variety of drug molecules have been delivered with NDs and shown increased therapeutic efficiency and reduced systematic toxicity in vitro and in vivo. Particularly, with the excellent fluorescence and electromagnetic properties, NDs offer unique potential for bioimaging and sensing with different optical and quantum approaches. The highly stable near-infrared fluorescence of NV centers inside NDs is essential for advanced high-resolution fluorescence imaging in cells and tissue. The special electron spin states of NV centers in NDs enable unlimited potential for developing nanoscale quantum sensors for detection of the tiniest changes of magnetic fields, external electron and magnetic spins as well as temperature under biological environment. These techniques might eventually allow detection and analysis of structure, dynamics and function of single biomolecule under native conditions with high sensitivity and nanometer spatial resolution. Hyperpolarization of NV centers also offered great opportunity to develop high-resolution MRI that might be able to reach molecular detection sensitivity. The combination of these bioimaging techniques with drug delivery capability of NDs would be promising for next-generation theranostics to allow sensitive multimode diagnostic together with targeted drug therapy within one nanoparticle.

It should be stressed that several challenges still need to be addressed before these highly promising techniques could be really brought into clinical benefit. Particularly for the auspicious electromagnetic quantum sensing, a variety of challenges still need to be overcome. For instance, ND sensors with single molecular detection sensitivity will be easily perturbed by even weak noise such as nuclear and electron spins, charges and radicals in random motion, which are intrinsically present in biological systems. Furthermore, the ND sensor is not stationary. The fast movement and rotation leads to randomization of the signals. Several theoretical solutions for these challenges have been already developed, but need to be further verified experimentally in biological environment, even though current studies have already established the fundamental basis of these quantum-sensing techniques, which are highly promising to be brought into the application level in the near future.

Acknowledgments

Tanja Weil und Ihre Gruppe danken Frau Kalkhof-Rose für Ihr großes Interesse und Ihre freundliche Förderung unserer Forschung an Nanodiamanten für die biomedizinische Forschung. In addition, the authors thank Sean Harvey for proof-reading.

References

[1] Shenderova OA, Gruen DM. Ultrananocrystalline diamond: synthesis, properties and applications. William Andrew, New York, 2012.

[2] (a) Boudou JP, Curmi PA, Jelezko F, Wrachtrup J, Aubert P, Sennour M, et al. High yield fabrication of fluorescent nanodiamonds. Nanotechnology 2009;20(23): 235602; (b) Boudou J-P, Tisler J, Reuter R, Thorel A, Curmi PA, Jelezko F, et al. Fluorescent nanodiamonds derived from HPHT with a size of less than 10 nm. Diamond Relat Mater 2013;37:80–6.

[3] Zaitsev AM. Optical properties of diamond: a data handbook. Springer, New York, 2001.

[4] Doherty MW, Manson NB, Delaney P, Jelezko F, Wrachtrup J, Hollenberg LC. The nitrogen-vacancy colour centre in diamond. Phys Rep 2013;528(1):1–45.

[5] Fu C-C, Lee H-Y, Chen K, Lim T-S, Wu H-Y, Lin P-K, et al. Characterization and application of single fluorescent nanodiamonds as cellular biomarkers. Eur Spine J 2010;19(5):754–9.

[6] Balasubramanian G, Neumann P, Twitchen D, Markham M, Kolesov R, Mizuochi N, et al. Ultralong spin coherence time in isotopically engineered diamond. Nat Mater 2009;8(5):383–7.

[7] Schrand AM, Huang H, Carlson C, Schlager JJ, Ōsawa E, Hussain SM, et al. Are diamond nanoparticles cytotoxic? J Phys Chem B 2007;111(1):2–7.

[8] (a) Yuzhou W, Fedor J, Martin BP, Tanja W. Diamond quantum sensing in biology. Angew Chem Int Ed 2016; 55(23):6586–98; (b) Vaijayanthimala V, Chang HC. Functionalized fluorescent

nanodiamonds for biomedical applications. Nanomedicine 2008;4(1):47–55; (c) Mochalin VN, Shenderova O, Ho D, Gogotsi Y. The properties and applications of nanodiamonds. Nat Nano 2012;7(1):11–23.

[9] (a) Vaijayanthimala V, Cheng P-Y, Yeh S-H, Liu K-K, Hsiao C-H, Chao J-I, et al. The long-term stability and biocompatibility of fluorescent nanodiamond as an in vivo contrast agent. Biomaterials 2012;33(31):7794–802; (b) Mohan N, Chen C-S Hsieh H-H, Wu YC, Chang H-C. In vivo imaging and toxicity assessments of fluorescent nanodiamonds in Caenorhabditis elegans. Nano Lett 2010;10(9):3692–9; (c) Liu K-K, Cheng C-L, Chang C-C, Chao J. Biocompatible and detectable carboxylated nanodiamond on human cell. Nanotechnology 2007;18(32):325102; (d) Schrand AM, Dai L, Schlager JJ, Hussain SM, Osawa E. Differential biocompatibility of carbon nanotubes and nanodiamonds. Diamond Relat Mater 2007;16(12):2118–23.

[10] (a) Xing Y, Xiong W, Zhu L, Ōsawa E, Hussin S, Dai L. DNA damage in embryonic stem cells caused by nanodiamonds. ACS Nano 2011;5(3):2376–84; (b) Yuan Y, Chen Y, Liu J-H, Wang H, Liu Y. Biodistribution and fate of nanodiamonds in vivo. Diamond Relat Mater 2009;18(1):95–100.

[11] (a) Moore L, Grobarova V, Shen H, Man HB, Micova J, Ledvina M, et al. Comprehensive interrogation of the cellular response to fluorescent, detonation and functionalized nanodiamonds. Nanoscale 2014;6(20):11712–21; (b) Batsanov SS, Gavrilkin SM, Batsanov AS, Poyarkov KB, Kulakova II, Johnson DW, et al. Giant dielectric permittivity of detonation-produced nanodiamond is caused by water. J Mater Chem 2012;22(22):11166–72.

[12] Liu K-K, Wang C-C, Cheng C-L, Chao J-I. Endocytic carboxylated nanodiamond for the labeling and tracking of cell division and differentiation in cancer and stem cells. Biomaterials 2009;30(26):4249–59.

[13] Krueger A, Lang D. Functionality is key: recent progress in the surface modification of nanodiamond. Adv Funct Mater 2012;22(5):890–906.

[14] Price ME, Cornelius RM, Brash JL. Protein adsorption to polyethylene glycol modified liposomes from fibrinogen solution and from plasma. Biochim Biophys Acta Biomembr 2001;1512(2): 191–205.

[15] (a) Zhang X, Fu C, Feng L, Ji Y, Tao L, Huang Q, Li S, Wei Y. PEGylation and polyPEGylation of nanodiamond. Polymer 2012;53(15):3178–84; (b) Wang D, Tong Y, Li Y, Tian Z, Cao R, Yang B. PEGylated nanodiamond for chemotherapeutic drug delivery. Diamond Relat Mater 2013;36:26–34; (c) Zhao L, Xu Y-H, Qin H, Abe S, Akasaka T, Chano T, et al. Platinum on nanodiamond: a promising prodrug conjugated with stealth polyglycerol, targeting peptide and acid-responsive antitumor drug. Adv Funct Mater 2014;24(34):5348–57; (d) Rehor I, Mackova H, Filippov SK, Kucka J, Proks V, Slegerova J, Turner S, Van Tendeloo G, Ledvina M, Hruby M, Cigler P. Fluorescent Nanodiamonds with Bioorthogonally Reactive Protein-Resistant Polymeric Coatings. ChemPlusChem 2014;79(1):21–4.

[16] Zhao L, Takimoto T, Ito M, Kitagawa N, Kimura T, Komatsu N. Chromatographic separation of highly soluble diamond nanoparticles prepared by polyglycerol grafting. Angew Chem Int Ed 2011;50(6):1388–92.

[17] Barras A, Lyskawa J, Szunerits S, Woisel P, Boukherroub R. Direct functionalization of nanodiamond particles using dopamine derivatives. Langmuir 2011;27(20):12451–7.

[18] Kong XL, Huang LCL, Hsu CM, Chen WH, Han CC, Chang HC. High-affinity capture of proteins by diamond nanoparticles for mass spectrometric analysis. Anal Chem 2005;77(1) 259–65.

[19] Tzeng Y-K, Faklaris O, Chang B-M, Kuo Y, Hsu J-H, Chang H-C Superresolution imaging of albumin-conjugated fluorescent nanodiamonds in cells by stimulated emission depletion. Angew Chem Int Ed 2011;50(10):2262–5.

[20] Zhang X-Q, Chen M, Lam R, Xu X, Osawa E, Ho D. Polymer-functionalized nanodiamond platforms as vehicles for gene delivery. ACS Nano 2009;3(9):2609–16.

[21] (a) Slegerova J, Hajek M, Rehor I, Sedlak F, Stursa J, Hrubyd M, Cigler P. Designing the nanobiointerface of fluorescent nanodiamonds: highly selective targeting of glioma cancer cells. Nanoscale 2015;7(2):415–20; (b) Wu Y, Ermakova A, Liu W, Pramanik G, Vu TM, Kurz A, et al. Programmable biopolymers for advancing biomedical applications of fluorescent nanodiamonds. Adv Funct Mater 2015;25(42):6576–85.
[22] (a) Wu Y, Chakrabortty S, Gropeanu RA, Wilhelmi J, Xu Y, Er KS, et al. pH-responsive quantum dots via an albumin polymer surface coating. J Am Chem Soc 2010;132(14):5012–14; (b) Wu Y, Pramanik G, Eisele K, Weil T. Convenient approach to polypeptide copolymers derived from native proteins. Biomacromolecules 2012;13(6):1890–8.
[23] Bumb A, Sarkar SK, Billington N, Brechbiel MW, Neuman KC. Silica encapsulation of fluorescent nanodiamonds for colloidal stability and facile surface functionalization. J Am Chem Soc 2013;135(21):7815–18.
[24] Rehor I, Slegerova J, Kucka J, Proks V, Petrakova V, Adam M-P, et al. Fluorescent nanodiamonds embedded in biocompatible translucent shells. Small 2014;10(6):1106–15.
[25] Rehor I, Mackova H, Filippov SK, Kucka J, Proks V, Slegerova J, et al. Fluorescent nanodiamonds with bioorthogonally reactive protein-resistant polymeric coatings. ChemPlusChem 2014;79(1):21–4.
[26] Shimkunas RA, Robinson E, Lam R, Lu S, Xu X, Zhang X-Q, et al. Nanodiamond–insulin complexes as pH-dependent protein delivery vehicles. Biomaterials 2009;30(29):5720–8.
[27] Chang B-M, Lin H-H, Su L-J, Lin W-D, Lin R-J, Tzeng Y-K, et al. Highly fluorescent nanodiamonds protein-functionalized for cell labeling and targeting. Adv Funct Mater 2013;23(46): 5737–45.
[28] Ermakova A, Pramanik G, Cai JM, Algara-Siller G, Kaiser U, Weil T, et al. Detection of a few metallo-protein molecules using color centers in nanodiamonds. Nano Lett 2013;13(7): 3305–9.
[29] Krueger A. The structure and reactivity of nanoscale diamond. J Mater Chem 2008;18(13): 1485–92.
[30] Zhang X-Q, Lam R, Xu X, Chow EK, Kim H-J, Ho D. Multimodal nanodiamond drug delivery carriers for selective targeting imaging and enhanced chemotherapeutic efficacy. Adv Mater 2011;23(41):4770–5.
[31] Zhang T, Neumann A, Lindlau J, Wu Y, Pramanik G, Naydenov B, et al. DNA-based self-assembly of fluorescent nanodiamonds. J Am Chem Soc 2015;137(31):9776–9.
[32] Albrecht A, Koplovitz G, Retzker A, Jelezko F, Yochelis S, Porath D, et al. Self-assembling hybrid diamond–biological quantum devices. New J Phys 2014;16(9):093002.
[33] Rothemund PW. Folding DNA to create nanoscale shapes and patterns. Nature 2006;440(7082):297–302.
[34] Huang H, Pierstorff E, Osawa E, Ho D. Active nanodiamond hydrogels for chemotherapeutic delivery. Nano Lett 2007;7(11):3305–14.
[35] (a) Chow EK, Zhang X-Q, Chen M, Lam R, Robinson E, Huang H, et al. Nanodiamond therapeutic delivery agents mediate enhanced chemoresistant tumor treatment. Sci Transl Med 2011;3(73):73ra21; (b) Xiao J, Duan X, Yin Q, Zhang Z, Yu H, Li Y. Nanodiamonds-mediated doxorubicin nuclear delivery to inhibit lung metastasis of breast cancer. Biomaterials 2013;34(37):9648–56.
[36] Chen M, Pierstorff ED, Lam R, Li SY, Huang H, Osawa E, et al.Nanodiamond-mediated delivery of water-insoluble therapeutics. ACS Nano 2009;3(7):2016–22.
[37] Toh TB, Lee DK, Hou W, Abdullah LN, Nguyen J, Ho D, Chow EK. Nanodiamond-mitoxantrone complexes enhance drug retention in chemoresistant breast cancer cells. Mol Pharm 2014;11(8):2683–91.
[38] Li X, Shao J, Qin Y, Shao C, Zheng T, Ye L. TAT-conjugated nanodiamond for the enhanced delivery of doxorubicin. J Mater Chem 2011;21(22):7966–73.

[39] Liu KK, Zheng WW, Wang CC, Chiu YC, Cheng CL, Lo YS, et al. Covalent linkage of nanodiamond-paclitaxel for drug delivery and cancer therapy. Nanotechnology 2010;21(31):315106.
[40] Azad N, Rojanasakul Y. Nanobiotechnology in drug delivery. Am J Drug Delivery 2012;4(2):79–88.
[41] Wrachtrup J, von Borczyskowski C, Bernard J, Orritt M, Brown R. Optical detection of magnetic resonance in a single molecule. Nature 1993;363(6426):244–5.
[42] Gruber A, Dräbenstedt A, Tietz C, Fleury L, Wrachtrup J, Borczyskowski Cv. Scanning confocal optical microscopy and magnetic resonance on single defect centers. Science 1997;276(5321):2012–14.
[43] Weng M-F, Chiang S-Y, Wang N-S, Niu H. Fluorescent nanodiamonds for specifically targeted bioimaging: application to the interaction of transferrin with transferrin receptor. Diamond Relat Mater 2009;18(2–3):587–91.
[44] Mkandawire M, Pohl A, Gubarevich T, Lapina V, Appelhans D, Rödel G, et al. Selective targeting of green fluorescent nanodiamond conjugates to mitochondria in HeLa cells. J Biophoton 2009;2(10):596–606.
[45] (a) Rittweger E, Han KY, Irvine SE, Eggeling C, Hell SW. STED microscopy reveals crystal colour centres with nanometric resolution. Nat Photon 2009;3(3):144–7; (b) Han KY, Willig KI, Rittweger E, Jelezko F, Eggeling C, Hell SW. Three-Dimensional Stimulated Emission Depletion Microscopy of Nitrogen-Vacancy Centers in Diamond Using Continuous-Wave Light. Nano Letters 2009 9 (9) 3323–9; (c) Arroyo-Camejo S, Adam M. P, Besbes M, Hugonin J. P, Jacques V, Greffet J. J, Roch J. F Hell S. W Treussart F. Stimulated Emission Depletion Microscopy resolves individual nitrogen vacancy centers in diamond nanocrystals. ACS Nano 2013;7(12):10912–19.
[46] Maze JR, Stanwix PL, Hodges JS, Hong S, Taylor JM, Cappellaro P, et al. Nanoscale magnetic sensing with an individual electronic spin in diamond. Nature 2008;455(7213):644–U41.
[47] Kucsko G, Maurer PC, Yao NY, Kubo M, Noh HJ, Lo PK, et al. Nanometre-scale thermometry in a living cell. Nature 2013;500(7460):54–8.
[48] Sushkov AO, Chisholm N, Lovchinsky I, Kubo M, Lo PK, Bennett SD, et al. All-optical sensing of a single-molecule electron spin. Nano Lett 2014;14(11):6443–8.
[49] Shi FZ, Zhang Q, Wang PF, Sun HB, Wang JR, Rong X, et al. Single-protein spin resonance spectroscopy under ambient conditions. Science 2015;347(6226):1135–8.
[50] Schafer-Nolte E, Schlipf L, Ternes M, Reinhard F, Kern K, Wrachtrup J. Tracking temperature-dependent relaxation times of ferritin nanomagnets with a wideband quantum spectrometer. Phys Rev Lett 2014;113(21).
[51] Huelga SF, Plenio MB. Vibrations, quanta and biology. Contemp Phys 2013,54:181–207.
[52] Romach Y, Muller C, Unden T, Rogers LJ, Isoda T, Itoh KM, et al. Spectroscopy of surface-induced noise using shallow spins in diamond. Phys Rev Lett 2015;114(1):017601.
[53] Lovchinsky I, Sushkov AO, Urbach E, de Leon NP, Choi S, De Greve K, et al. Nuclear magnetic resonance detection and spectroscopy of single proteins using quantum logic. Science 2016;351(6275):836–41.
[54] (a) Staudacher T, Shi F, Pezzagna S, Meijer J, Du J, Meriles CA, et al. Nuclear magnetic resonance spectroscopy on a (5-nanometer)3 sample volume. Science 2013;339(6119):561–3; (b) Mamin HJ, Kim M, Sherwood MH, Rettner CT, Ohno K, Awschalom DD, Rugar D. Nanoscale nuclear magnetic resonance with a nitrogen-vacancy spin sensor. Science 2013;339(6119):557–60.
[55] Steinert S, Ziem F, Hall LT, Zappe A, Schweikert M, Götz N, et al. Magnetic spin imaging under ambient conditions with sub-cellular resolution. Nat Commun 2013;4:1607.
[56] Muller C, Kong X, Cai JM, Melentijevic K, Stacey A, Markham M, et al. Nuclear magnetic resonance spectroscopy with single spin sensitivity. Nat Commun 2014;5:4703.
[57] Ajoy A, Bissbort U, Lukin MD, Walsworth RL, Cappellaro P Atomic-scale nuclear spin imaging using quantum-assisted sensors in diamond Phys Rev X 2015;5:011001; (b) Kost M, Cai J, Plenio MB. Resolving single molecule structures with Nitrogen-vacancy centers in diamond. Sci Rep 2015;5:11007.

[58] (a) Kessler EM, Lovchinsky I, Sushkov AO, Lukin MD. Quantum error correction for metrology. Phys Rev Lett 2014;112(15):150802; (b) Arrad G, Vinkler Y, Aharonov D, Retzker A. Increasing sensing resolution with error correction. Physical Rev Lett 2014;112(15):150801.
[59] (a) Scheuer J, Schwartz I, Chen Q, Schulzesünninghausen D, Carl P, Höfer P, et al. Optically induced dynamic nuclear spin polarisation in diamond. New J Phys 2016;18(1):013040; (b) London P, Scheuer J, Cai JM, Schwarz I, Retzker A, Plenio MB, et al. Detecting and polarizing nuclear spins with double resonance on a single electron spin. Phys Rev Lett 2013;111(6):067601; (c) King JP, Jeong K, Vassiliou CC, Shin CS, Page RH, Avalos CE, et al. Room-temperature in situ nuclear spin hyperpolarization from optically pumped nitrogen vacancy centres in diamond. Nat Commun 2015;6:8965; (d) Fischer R, Bretschneider CO, London P, Budker D, Gershoni D, Frydman L. Bulk nuclear polarization enhanced at room temperature by optical pumping. Phys Rev Lett 2013;111(5).
[60] (a) Dutta P, Martinez GV, Gillies RJ. Nanodiamond as a new hyperpolarizing agent and its C-13 MRS. J Phys Chem Lett 2014;5(3):597–600; (b) Chen Q, Schwarz I, Jelezko F, Retzker A, Plenio MB. Optical hyperpolarization of 13C nuclear spins in nanodiamond ensembles. Phys Rev B 2015;92(18):184420; (c) Jochen S, Ilai S, Qiong C, David S-S, Patrick C, Peter H, et al. Optically induced dynamic nuclear spin polarisation in diamond. New J Phys 2016;18(1):013040.
[61] Rej E, Gaebel T, Boele T, Waddington DE, Reilly DJ. Hyperpolarized nanodiamond with long spin-relaxation times. Nat Commun 2015;6:8459.
[62] Abrams D, Trusheim ME, Englund DR, Shattuck MD, Meriles CA. Dynamic nuclear spin polarization of liquids and gases in contact with nanostructured diamond. Nano Lett 2014;14(5):2471–8.
[63] Chen Q, Schwarz I, Jelezko F, Retzker A, Plenio MB. Resonance-inclined optical nuclear spin polarization of liquids in diamond structures. Phys Rev B 2016;93(6):060408.

Soumyajit Das and Jishan Wu

9 Polycyclic Hydrocarbons with an Open-Shell Ground State

Abstract: Organic π-conjugated polycyclic hydrocarbons (PHs) with an open-shell ground state have been the subject of intensive study for a long time due to their unusual electronic structure, which makes them unique in terms of electronic, magnetic and photophysical properties. Incorporation of quinodimethane subunit is the most fundamental way to generate PHs with "diradical character" in the ground state. Thanks to the development of chemistry that evolved in the past two decades, such PHs are now synthetically accessible and will be discussed in this chapter.

9.1 Introduction

π-Conjugated polycyclic hydrocarbons (PHs) can either accommodate π-electrons in the bonding orbitals to form a closed-shell ground state or show open-shell ground state due to the existence of one or more unpaired electrons [1]. Monoradical PH is comprised of an unpaired electron in neutral molecule in the ground state [2]. Existence of two unpaired electrons (radicals) in a molecule can be classified by either diradical or biradical [3]. In the spin-restricted (R) molecular orbital (MO) representation [4], diradical character tends to increase as the highest occupied molecular orbital (HOMO)– lowest unoccupied molecular orbital (LUMO) gap gets reduced, causing a variation in bond nature from the stable bond regime to the bond dissociation regime through increasing the weight of doubly excited configuration from HOMO to LUMO. In a stable bond limit, i.e., closed-shell, no diradical character exists while in the bond dissociation limit, the perfect biradical or a pure open-shell configuration emerges. Any intermediate state is referred to as "diradicaloid" (diradical-like). Diradicals are even-electron molecules that have one bond less than the number permitted by the standard rule of valence and the two electrons occupy two near-degenerate MOs. The spin multiplicity (2S + 1) of monoradicals is doublet, whereas for diradicals, two spins can orient in parallel or antiparallel fashion, producing triplet biradical or singlet diradical species. Singlet diradical, as a molecular species, has all the electrons paired but a pair of these electrons, being weakly coupled through anti-ferromagnetic interaction, occupies different parts of the space with a small shared region. PHs with moderate to strong singlet diradical character generally show a broad electron spin resonance (ESR) signal due to the reduced singlet–triplet energy gap, which is believed to increase the population of the magnetically active triplet species. However, the ground state of such species is still singlet as at least one Kekulé structural formula can be drawn (e.g., *p*-quinodimethane (*p*-QDM)- and *o*-quinodimethane (*o*-QDM)-based PHs, Figure 9.1(a)). On the other hand, a pure

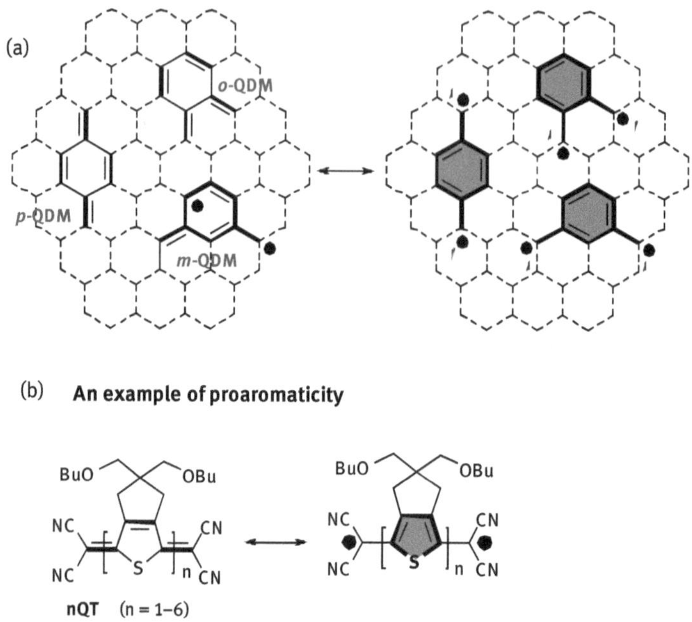

Figure 9.1: (a) Structures of *p*-QDM, *o*-QDM and their singlet diradical resonance forms (Kekulé form); *m*-QDM in its triplet biradical form. (b) The concept of proaromaticity is shown as an example of quinoidal oligothiophene.

biradical PH (e.g., *m*-quinodimethane (*m*-QDM)-based PHs, Figure 9.1(a)) is a molecular species with two electrons occupying two degenerate or nearly degenerate MOs, and there exists no Kekulé structural formula for this type of system [3a].

Aromaticity is a key concept for chemistry in the electronic ground state, and reactions in which aromaticity is gained are normally highly favorable. Antiaromatic molecules like parent cyclobutadiene can be categorized as delocalized diradicals according to their MO diagram as two electrons occupy two degenerate non-bonding MOs which are "spin-up" unpaired, making them strongly reactive species and thus their synthesis is highly elusive. In fact, later it was found that cyclobutadiene derivatives prefer a rectangular shape in its singlet ground state instead of delocalized square geometry caused by the Jahn–Teller effect [5]. In the proaromatic concept (Figure 9.1(b)) for the formation of diradicals [5b], the driving force for such structure is the recovery of aromaticity from consecutive non-aromatic rings (quinoidal structure). A net energy minimization that can eventually surpass the energy required to break a C–C double bond is the key in generation of an open-shell diradical ground state. The smaller proaromatic tetracyano-quinoidal-thiophene oligomers are closed shell, whereas the ESR spectra of the 5QT and 6QT constitute the first reports on the unquestionable diradical character due to the population of thermally (room temperature (rt)) excited magnetically active species as a result of aromatization of the thiophene rings [6].

Acenes can be regarded as one-dimensional fragments of graphene and belong to a class of PHs consisting of linearly fused benzene rings. Bendikov et al. showed that the RB3LYP wave function becomes unstable for oligoacenes as small as hexacene and all higher oligoacenes, implying that the calculated energies for singlet states are unbelievably high. Re-optimization by using the broken-symmetry (BS) UB3LYP/6-31G* method, it was shown that acenes, larger than hexacene, possess a nonzero bandgap with a large amount of diradical character in a singlet open-shell ground state and the singly occupied molecular orbital (SOMO) is largely populated on the zigzag edges [7]. Later, Hachmann et al. also showed that the ground state of linear polyacenes is singlet for all chain lengths from naphthalene to dodecacene, and acenes larger than dodecacene were found to exhibit singlet polyradical character in their ground state [8]. Interestingly, experimental observations already established the closed-shell ground states (as evident from clear and sharp ^{1}H nuclear magnetic resonance (NMR) signal) of higher-order acenes except for the first isolated crystalline nonacene derivative [9] that apparently showed low HOMO–LUMO gap of 1.2 eV which is a criterion for the origin of diradical character in PH. Absence of NMR signal and the presence of ESR signal with $g_e = 2.0060$ for nonacene derivative were reported, although no clear conclusion regarding the ground state electronic structure was given in the report as the ESR signal may also come from some monoradical impurities.

Theoretically, diradical character is estimated by a complete active space self-consistent field (SCF) calculation using the restricted Hartree–Fock and two-configuration SCF calculations. A BS approach [10], later modified with spin-projection technique [11] to discard spin contamination in BS approach, using the unrestricted Hartree–Fock wave function [12] can define the occupation number of the lowest unoccupied natural orbital (LUNO) as the extent of diradical character. As simplified by Nakano [4], for a two-electron two-orbital model the ground state closed-shell molecule has the natural orbital (NO) occupation numbers 2 (occupied) or 0 (unoccupied). So the LUNO is 0 for closed shell, whereas in the open-shell singlet molecules, the NO occupation number can be an intermediate value between 0 and 2, depending on the diradical character; therefore, the occupation number of the LUNO should increase with the increase in the diradical character and eventually approaches 1 for a pure diradical. This implies that the diradical character, which is denoted by y ($0 \le y \le 1$ where $y = 0$ is closed shell and $y = 1$ is pure open-shell biradical), can be defined by the occupation number of the LUNO. Experimentally, diradical character can also be determined using the following equation [13]:

$$y = 1 - \sqrt{1 - \left(\frac{E_{S1u,S1g} - E_{T1u,S1g}}{E_{S2g,S1g}} \right)^2},$$

where $E_{S1u'S1g}$ and $E_{S2g,S1g}$ correspond to the energy of the lowest energy peaks in the one- and two-photon absorption (TPA) spectra, and $E_{T1u,S1g}$ corresponds to the energy gap between the triplet and the singlet ground state.

9.2 Quinodimethane-Based Open-Shell Polycyclic Hydrocarbons

The most basic, yet powerful, approach to generate a diradicaloid PH is to embed an *o*-QDM or *p*-QDM subunit into a π-conjugated framework. This approach has been a topic of intense interest due to their inherent diradical character arising from the recovery of the aromaticity of the central benzenoid ring of the respective quinoidal subunits, in the ground state. Comparatively, *o*-QDM derivatives such as pleiadene [14] are extremely reactive compared to *p*-QDM derivatives. Unlike *o*- and *p*-QDM, the *m*-QDM (i.e., *m*-xylylene biradical) cannot have a closed-shell quinoidal structure and thus is classified into non-Kekulé PH. Embedding of an *m*-xylylene biradical subunit into a planarized polycyclic benzenoid hydrocarbon generates open-shell π-conjugated triplet PH systems such as triangulenes (*vide infra*), with large spin densities at the edge sites.

9.2.1 *o*-QDM-Embedded Diradicaloids

Tobe et al. reported [15] the synthesis and crystallographic structure of an indeno-[2,1-*a*]fluorene (**1**) derivative **5** which could show diradical resonance contribution due to recovery of one additional Clar's sextet (highlighted in gray color) in the ground state, with radical sites mainly located on the five-membered rings (Figure 9.2). Mesityl groups were introduced to kinetically protect the reactive sites, by adding mesitylmagnesium bromide to the diketone **3** to afford the diol **4** which on treatment with tin(II)chloride gave **5** as a stable, purple-colored solid. The unaffected ^{1}H NMR spectra of **5** at higher temperature and a significant bond-length alternation in *o*-QDM core, as per crystal structure analysis, indicate a large singlet–triplet energy gap with small diradical contribution, which was also supported by theory. The singlet diradical character of **1** and **5** estimated by the Yamaguchi scheme was 0.33 and 0.21, respectively. Nucleus-independent chemical shift calculation further supports that **5** is weakly antiaromatic as a result of the *as*-indacene moiety. The vertical π-extension of *o*-QDM can lead to Kekulé structure **2** which can also have two diradical resonance forms and the spin density may distribute to the inner naphthalene ring. A similar synthetic approach was conducted to prepare the vertically extended *o*-QDM derivative **8** which was confirmed by X-ray crystallographic analysis [16]. This compound is reported to be quite reactive in air affording an endoperoxide **9** (Figure 9.2). Compound **8** showed sharp signals at rt and broadening of the signals at higher temperatures was observed due to thermally accessible triplet state because of a reduced

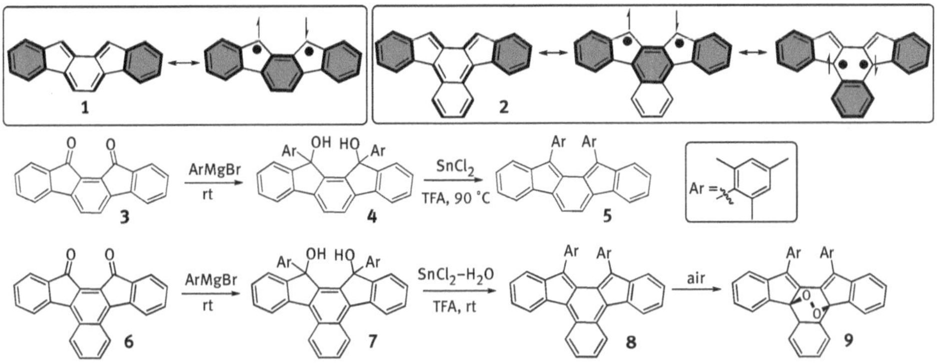

Figure 9.2: Structures and synthesis of Tobe's *o*-QDM derivatives.

singlet–triplet energy gap of –7.46 kcal mol^{-1}, indicating an enhanced diradical character in **8** compared to **5**. The index *y* was estimated to be 0.63 which was considerably larger when compared to that of **1** (0.33).

9.2.2 *m*-Xylylene-Based Systems

An unsuccessful attempt by Clar to synthesize an *m*-xylylene derivative, triangluene, was explained by the triplet ground state in neutral state [17]. Nakasuji et al. treated a dihydrotriangulene **10** precursor with *p*-chloranil to obtain a highly reactive triangulene derivative **11**, although being protected by bulky *tert*-butyl groups on the three vertexes (Figure 9.3) [18]. Variable temperature (VT) ESR measurement was conducted to monitor the chemical oxidation step and rapid freezing of the sample at intermediate stage gave a superposition of a doublet monoradical species and a fine-structure ESR spectrum of a typical triplet state which was attributed to compound **11**, a true hydrocarbon with three-fold rotation axis. **11** is highly reactive and tends to polymerize to give oligomer/polymer **12**.

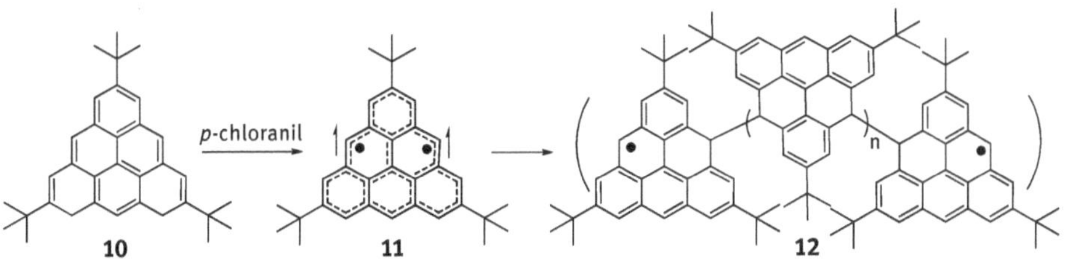

Figure 9.3: Synthesis of Nakasuji's tri-*tert*-butyl triangulene derivative.

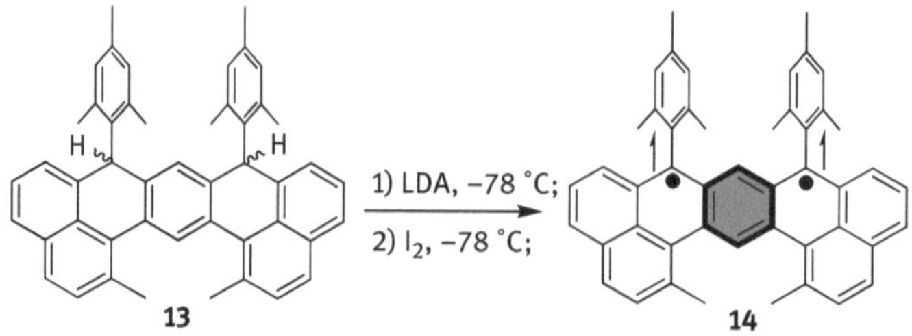

Figure 9.4: Synthesis of Wu's *m*-xylylene-embedded heptazethrene isomer.

Two naphthalene units bridged by an *m*-xylylene moiety cannot have a Kekulé structure and can only be drawn in its open-shell biradical form with two spins aligned parallel to each other leading to a reactive triplet biradical (Figure 9.4). A kinetically protected high-spin derivative **14** was synthesized by deprotonation of **13** using lithium diisopropylamide and subsequent oxidation of the dianion with iodine in dry 2-methyltetrahydrofuran at –78°C. This *in situ* generated triplet biradical was persistent at –78°C under nitrogen protection for at least one day [19].

An *m*-xylylene unit, incorporated into indeno[2,1-*b*]fluorene framework, can be drawn into both closed- and open-shell resonance forms (Figure 9.5). More resonance stabilization energy in the diradical resonance form of indeno[2,1-*b*]fluorene (1 sextet in closed shell vs. 3 aromatic sextets in open shell) compared to indeno-[2,1-*a*]fluorene (2 sextets in closed shell vs. 3 sextets in open shell) is supposed to increase its diradical character. Indeed, the mesityl-substituted isomer **17**, prepared by addition of mesitylmagnesium bromide to the diketone **15** and subsequent reduction of the diol **16**, turned out to be a singlet diradical in the ground state [20]. Absence of rt ^{1}H NMR signal and only broad NMR signals at low temperature down to –93°C indicate the existence of the thermally active triplet species. The ESR signal intensity decreased with decreasing temperature, indicating a singlet open-shell ground state. However, the singlet–triplet energy gap estimated experimentally, −4.2 kcal mol^{-1}, was relatively higher compared to the theoretical value of –2.46 kcal mol^{-1}. X-ray crystallographic analysis indeed showed the bond lengths of bond *a* (1.437 Å) and *b* (1.431 Å) are similar and lie between the lengths of the C(sp^2)–C(sp^2) bond in benzene (1.39 Å)

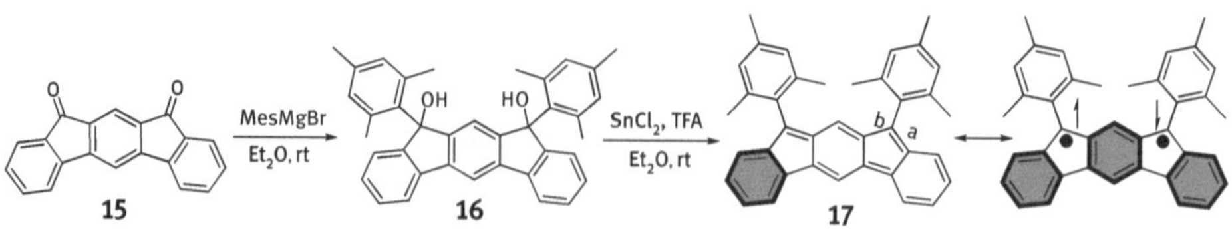

Figure 9.5: Synthesis of Tobe's open-shell indeno[2,1-*b*]fluorene.

and the C(sp^2)–C(sp^3) bond in fluorene (1.468 Å), supporting significant diradical contribution (y = 0.68) in ground state.

9.2.3 *p*-QDM-Embedded Systems and Its π-Extended Derivatives

Substitution of the terminal methylene sites in *p*-QDM by four phenyl groups resulted in the reasonably stable Thiele's hydrocarbon **18** due to a quinoidal closed-shell ground state with a strong bond-length alternation in the *p*-QDM core (1.346 and 1.449 Å) (Figure 9.6) [21]. Electron-withdrawing cyano (CN) groups, at the terminal methylene sites (radical centers), can help stabilize the *p*-QDM derivatives and, indeed, the tetracyanoquinodimethane (TCNQ, **21**) is stable and commercially available. Incorporation of a *p*-QDM moiety into a fused π-conjugated PH framework has been an efficient approach to generate singlet diradicaloids, and typical examples include the bis(phenalenyl)s like indacenodiphenalene (IDPL) and heptazethrenes, to be discussed later. *p*-QDM-embedded indenofluorene congeners have been reported as typical closed-shell antiaromatic systems with negligible diradical character [22]. Indeno-[1,2-*b*]fluorene **27** derivatives have shown their promise as electron-transporting materials, although no open-shell characteristics are found in the ground state [23]. Similarly, indeno[1,2-*c*]fluorene **28** derivative was also found to be closed shell in the ground state [24]. A fluoreno[4,3-*c*]fluorene (**29**) derivative, comprising of a laterally π-extended *p*-QDM architecture (2,6-naphthoquinodimethane), also exhibited no diradical characteristics in the ESR spectrum on heating or any line broadening in NMR spectrum as the triplet state was theoretically found to be 16.1 kcal mol^{-1} above the singlet ground state, thereby confirming a stable closed-shell ground state [25]. In contrast, linearly π-extended *p*-QDM derivative, the tetracyano-substituted diphenoquinodimethane **22**, had been reported to be very reactive and underwent instantaneous polymerization [26]. The bond distance in the central part of the π-extended *p*-QDM derivative **19** with four phenyl protected at the terminal methylene (Tschitschibabin's hydrocarbon) is intermediate between double- and single-bond values (1.420 and 1.372 Å), indicating significant diradical contribution in the ground state [21]. A more enhanced diradical character was observed in π-extended Müller's hydrocarbon **20**

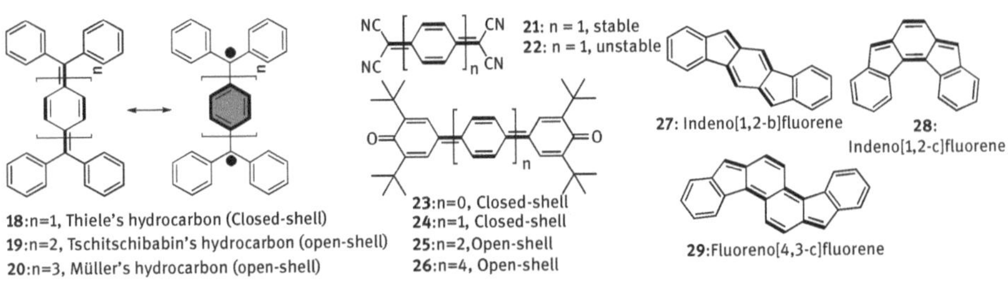

Figure 9.6: Examples of π-extended *p*-QDMs and *p*-QDM-embedded PHs.

(Figure 9.6) that showed an ESR signal typical for triplet species at 153 K [27]. The biphenoquinone **23** is a quinoidal closed-shell system, and terphenoquinone **24** also showed a strong bond length alternation in the ground state [28], whereas the more *para*-extended quaterphenoquinone **25** [29] and quinquephenoquinone **26** [30] are ESR active both in the solution and solid states, indicating significant diradical contribution in the ground state. These findings, indeed, lead to a new observation that the diradical character in π-conjugated planar PHs should emerge with the increasing number of proaromatic units, simply because more aromatic sextet rings can be recovered from a closed-shell quinoid to an open-shell diradical form.

Nakamura et al. used palladium catalyzed Takahashi coupling on the iodo-derivatives **30** and **32** to generate air and thermally stable tetracyano-substituted oligo-(*para*-phenylene vinylenes) **31** and **33**, respectively (Figure 9.7) [31]. Compound **33** displayed temperature-dependent broadening of its ^{1}H NMR spectrum and a quite small singlet–triplet gap of 2.12 kcal/mol was estimated by superconducting quantum interference device (SQUID) measurements, suggesting the population of thermally excited triplet species causing an NMR line broadening and an ESR signal at $g_e = 2.003$ at 340 K. In contrast, the closed-shell compound **31** exhibited sharp NMR signals and a relatively large HOMO–LUMO energy gap (1.58 eV) compared to **33** (1.02 eV).

A closed-shell diindeno[1,2-b:20,10-*n*]perylene **35**, consisting a Tschitschibabin's hydrocarbon framework, comprised of two indene groups being simultaneously fused at the *peri*- and *β*- position of a perylene core was synthesized from the corresponding diol precursor **34** by treating tin(II)chloride [32]. No extra aromatic sextet rings can be drawn in its diradical resonance form (Figure 9.8) and thus it behaves as a closed-shell quinoidal hydrocarbon and was used as semiconductor in ambipolar field effect transistors.

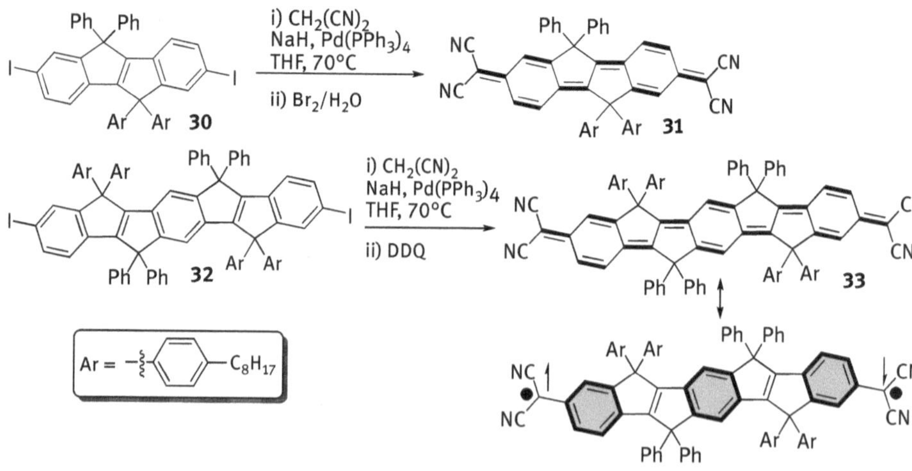

Figure 9.7: Synthesis of Nakamura's quinoidal oligo-(*para*-phenylene vinylene) diradicaloids.

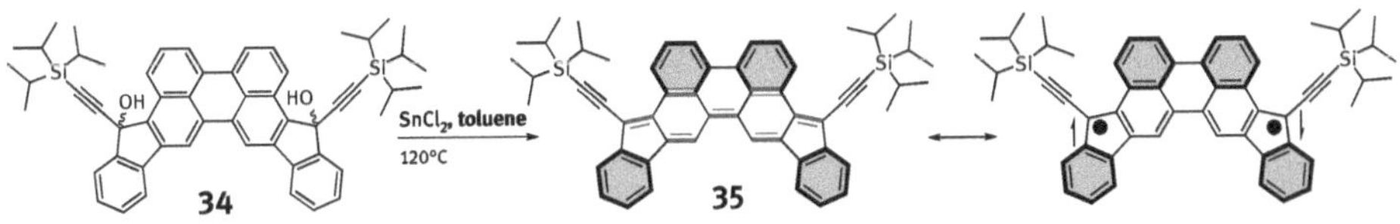

Figure 9.8: Synthesis of diindeno[1,2-b:20,10-*n*]perylene **35** and its open-shell form.

Wu et al. had developed a series of tetracyano-terminated, quinoidal *N*-annulated perylene (NP)-based diradicaloids **37** comprising of extended *p*-QDM framework (Figure 9.9) [33]. These oligomers were synthesized by Takahashi coupling from the dibromo-oligomers **36** followed by oxidation with *p*-chloranil. It was found that upon extension of the chain length, the monomer possesses a closed–shell quinoidal structure while the dimer, trimer and tetramer are open-shell singlet diradicals, and pentamer and hexamer become weakly coupled biradicals in the ground state. The enhanced diradical character in the higher-order oligomers is due to the large steric repulsion between the neighboring NPs causing a strain release from the rigid quinoidal structure to the more flexible diradical form. In addition, the recovery of aromaticity of the quinoidal perylene units in the diradical form helps the longer oligomers to gain more stability, and hence the diradical character approaches nearly 1 for the pentamer and hexamer.

To minimize the influence of strain release, planarized tetracyano-quaterylenequinodimehtane (**40**, $m = 0$) and hexarylenequinodimethane (**40**, $m = 1$) were prepared from respective dibromo-rylenes **38** (Figure 9.10) [34]. In contrast to the unfused perylene dimer **37** ($n = 2$), the fused dimer **40** ($m = 0$) is closed shell due to the strong intramolecular antiferromagnetic coupling of the two spins via a double-spin polarization mechanism. The fused trimer **40** ($m = 1$) has an open-shell singlet diradical ground state due to the stabilization through the recovery of six aromatic naphthalene units in the open-shell form, although the diradical character, theoretically, was found to be quite smaller ($y = 0.064$) with a large singlet–triplet energy gap of −4.21 kcal/mol. This comparison leads to a valuable conclusion that the dihedral angle between the rylene cores plays a critical role in fine-tuning the electronic configuration of quinoidal molecules.

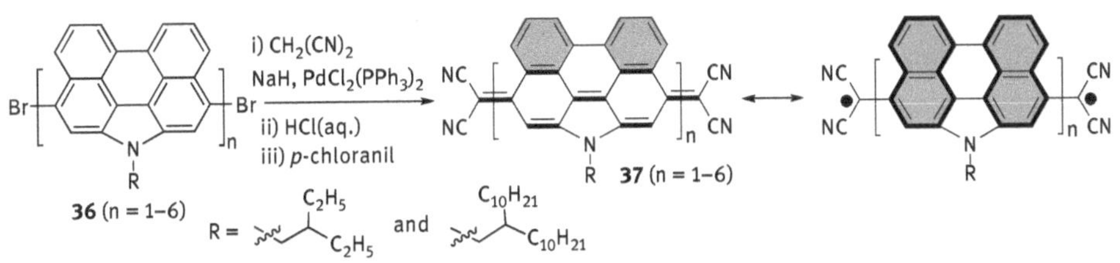

Figure 9.9: Synthesis and structures of tetracyano-quinoidal NP oligomers.

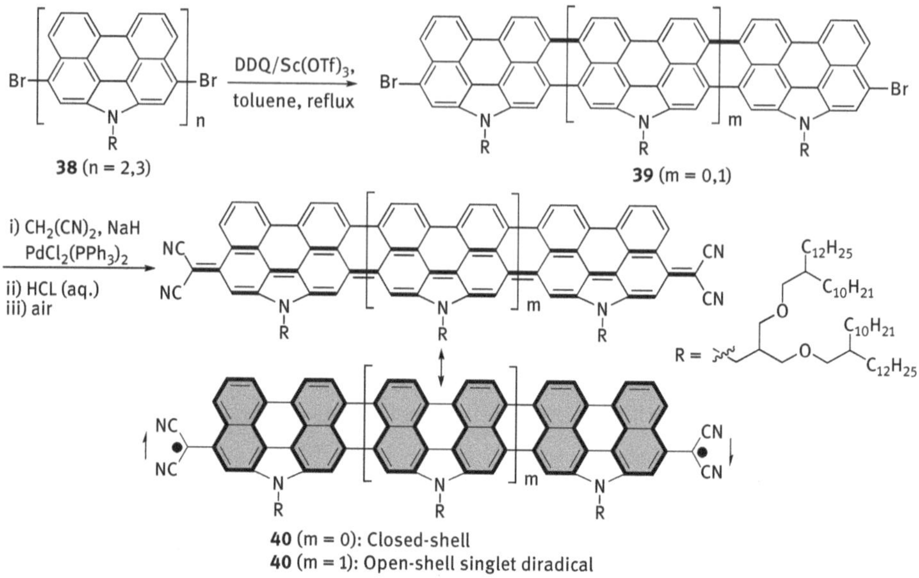

Figure 9.10: Synthesis and structures of fused tetracyano-quinoidal rylenes.

Incorporation of one thiophene unit between the rylene and the dicyanomethylene site in closed-shell **37** ($n = 1$) and **40** ($m = 0$) can readily turn on their diradical states, making **42** and **44** singlet diradical in the ground state with diradical character estimated as 0.81 and 0.93, respectively [35]. The larger diradical contribution for **43** is in agreement with the stronger ESR signal and small singlet–triplet energy gaps of −0.16 kcal/mol as compared to −4.71 kcal/mol for **42**, determined by SQUID measurements. Such a strong diradical contribution can be attributed to the recovery of two additional aromatic thiophene rings in the diradical resonance forms together with the conformational flexibility around the thiophene-rylene connections (Figure 9.11).

Nakano et al., theoretically, predicted that the asymmetric open-shell singlet systems with intermediate diradical characters could exhibit further enhancement of static first and second hyperpolarizabilities as compared to conventional asymmetric closed-shell systems or symmetric open-shell singlet systems with similar conjugation size [4]. In order to achieve such systems, Wu et al. reported the first push–pull type quinoidal NP oligomers with two terminal sites substituted by benzo-1,3-dithiol-2-ylidene (donor) and dicyanomethylene (acceptor), respectively (Figure 9.12) [36]. **46a** and **46b** represent the smallest derivatives with different substituents at the amine site, while **47** and **48** are the higher-order analogues. The bulky 2,6-dioctoxytolyl group in **46b** suppressed the strong aggregation observed in **46a**. Their synthesis was achieved by Pd-catalyzed Takahashi coupling from the corresponding bromo- and benzo-1, 3-dithiol-2-yl-substituted oligo- or (*N*-annulated perylenes) **45**, followed by *p*-chloranil treatment to ensure complete dehydrogenation. The ^{1}H and ^{13}C NMR spectrum of **46b** in $CDCl_3$ exhibited sharp peaks at rt and even at elevated

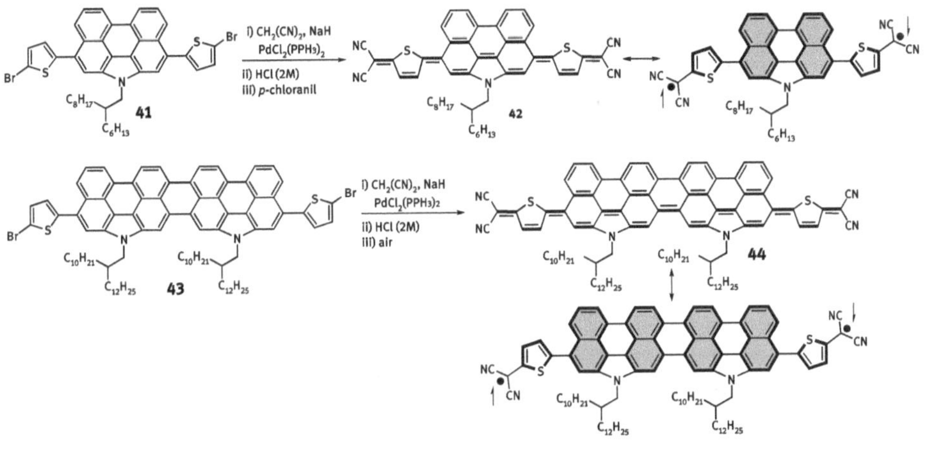

Figure 9.11: Synthesis and structures of tetracyano-thiophene capped rylene QDMs.

temperature, confirming its closed-shell ground state. In contrast, the higher series **47** and **48** showed NMR silence even at the low temperature (e.g., –100°C), implying their open-shell diradical ground state. Accordingly, ESR measurements on both displayed a broad spectrum both in solid and solution state with g_e = 2.0017. The VT ESR measurements on the powder disclosed that the ESR intensity decreased as the temperature decreased, indicating that both **47** and **48** have an open-shell singlet ground state. $\Delta E_{S\text{-}T}$ values were estimated by SQUID to be −0.348 kcal mol^{-1} and −0.114 kcal mol^{-1} for **47** and **48**, respectively, indicating an increased diradical character with increasing chain length. The contribution of the diradical form to the ground state in the push–pull type systems is slightly smaller than the pull–pull type framework **37** (n = 2) (−0.341 kcal mol^{-1}) and **37** (n = 3) (−0.107 kcal mol^{-1}) with the same conjugation length. The monomer **46b** gave a maximum TPA cross-section value of about 1,000 GM at 1,500 nm compared to the values of about 660 GM and 670 GM at 1,900 nm for **47** and **48**, respectively. This observation was reasoned for the enhancement of diradical character from closed-shell monomer (y = 0.123) to nearly

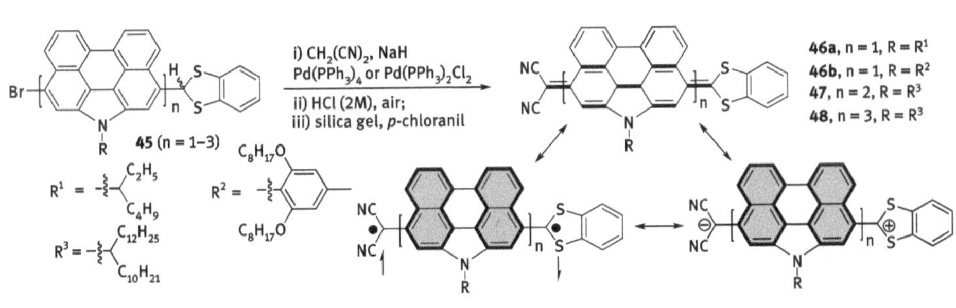

Figure 9.12: Synthesis and structures of push–pull type quinoidal NP oligomers.

pure open-shell **47** ($y = 0.899$) and **48** ($y = 0.995$) that does not benefit the non-linear optical (NLO) response. This abrupt increase of y from monomer to dimer is because of the reasonably large steric repulsion between the NP units in the rigid quinoidal structure, which tends to relax back to more flexible diradical/ionic form. To evaluate the intramolecular charge transfer character in the derivatives, the Hirshfeld charges of the central NP units and the terminal regions were calculated and it was found that with an extension of the chain length, the contribution of the zwitterion character decreases while the diradical character increases, and thus their physical properties become less solvent dependant. As a result, solvent polarity has a significant effect on the ground state and physical properties of the monomer only, but insignificant on the dimer and trimer.

An unorthodox approach to stabilize the Tschitschibabin's hydrocarbon is benzannulation of the central biphenyl unit, reported by Wu et al., to generate the tetrabenzannulated hydrocarbons **50** and **52** (Figure 9.13) which should favor a closed-shell ground state (**50-CS** and **52-CS**) as two aromatic Clar's sextets are lost in their respective diradical resonance forms [37]. However, the closed-shell forms of **50** and **52** adopt a contorted geometry with a larger steric hindrance among the anthryl *peri*-hydrogens, which could make the corresponding orthogonal diradical structures (**50-OS** and **52-OS**) more favorable. A thorough experimental investigation assisted with theoretical calculation supported a closed-shell quinoidal ground state for **50** while **52** can be regarded as a weakly coupled triplet biradical in the ground state. The better thermodynamic stability of the diradical ground state in **52** can be credited to the efficient spin delocalization on the fluorenyl units. 4-*tert*-Butylphenyl-substituted compound **50** favors a closed–shell ground state (i.e., **50-CS**) and a diradical excited state which was chemically obtained by reduction of the diol precursor **49**, and the diradical excited state gradually decayed to the closed-shell ground state **50-CS** with a half-life time of 495 min overcoming the large energy barrier (22.7 kcal mol^{-1}) for the transition from the orthogonal diradical form to a butterfly-like quinoidal form. On the other hand, the triplet biradical ground state of **52** (i.e., **52-OS**) was confirmed by the absence of NMR signal and presence of strong ESR signal with a singlet–triplet energy gap of 0.334 kcal mol^{-1} from SQUID measurements. An attempt to isolate the

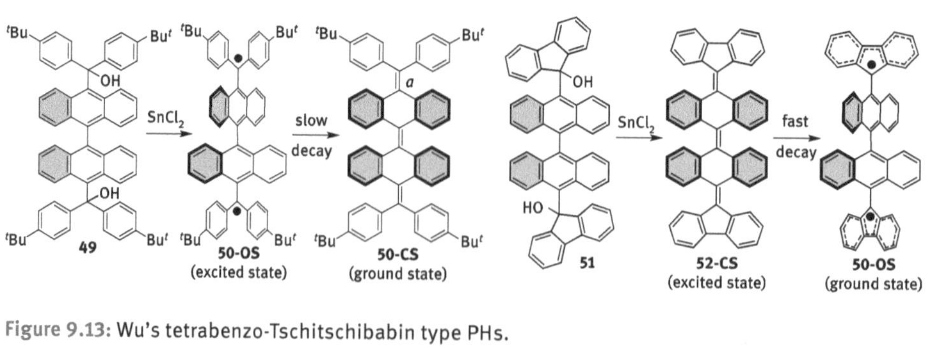

Figure 9.13: Wu's tetrabenzo-Tschitschibabin type PHs.

closed–shell form of **52** (i.e., **52-CS**) was failed as the intermediate quickly relaxed back to the more stable open-shell form.

9.3 Open-Shell Anthenes and *Peri*-Fused Acenes

Anthenes and *peri*-fused acenes (periacenes) are rectangular polycyclic aromatic hydrocarbons (PAHs) comprising both zigzag and armchair edges with extended zigzag edges (Figure 9.14). As the conjugation is extended to a certain point ($n > 0$ for anthenes and $m > 2$ for periacenes), a remarkable open-shell diradical character may emerge which originates from a narrow bandgap and stabilization through recovery of Clar's sextet in the diradical resonance form. Anthracene units fused together by three single bonds between neighboring anthryls generates extended graphene-like molecules called bisanthene ($n = 0$, $m = 1$), teranthene ($n = 1$, $m = 1$), quarteranthene ($n = 2$, $m = 1$) and so on. Closed-shell bisanthene is an unstable material and stabilization of closed-shell bisanthene was achieved by introduction of electron-withdrawing imide groups onto the zigzag edges or by substitution at the *meso*-positions with aryl or alkyne groups [38].

9.3.1 Anthenes

Teranthene [39] and quarteranthene [40] derivatives have been prepared and isolated in the crystalline form by Kubo's group, thus allowing a detailed investigation on their ground-state molecular structure, chemical behavior and physical properties (Figure 9.15). Treatment of **53** with 2-mesitylmagnesium bromide in the presence of $CeCl_3$ followed by NaI/NaH_2PO_2 generated a partially cyclized hydrocarbon **55** which

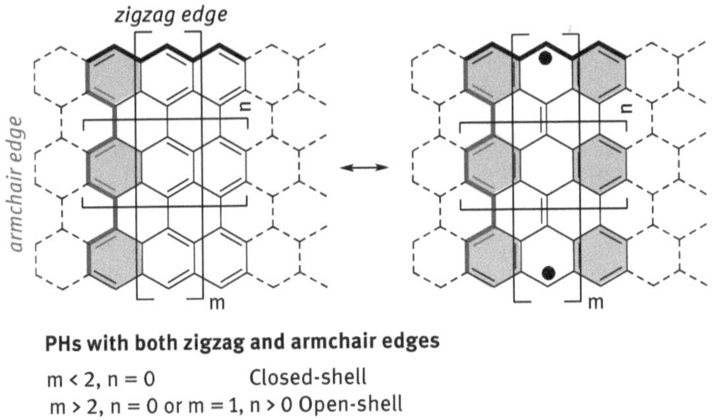

Figure 9.14: PAHs with both zigzag and armchair edges and their diradical form.

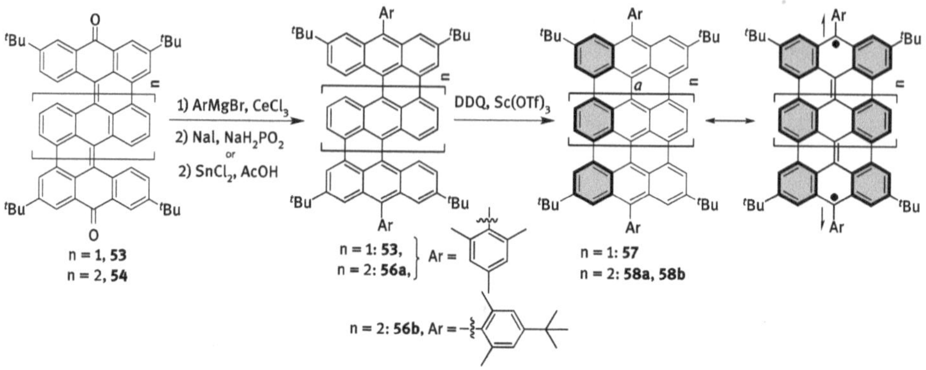

Figure 9.15: Synthesis of Kubo's teranthene (n = 1) and quarteranthene (n = 2) derivatives.

upon ring-closure reaction by 2,3-Dichloro-5,6-dicyano-1,4-benzoquinone (DDQ)/$Sc(OTf)_3$ afforded teranthene **57** [39]. The quarteranthene was synthesized using an identical synthetic approach [40]. Treatment of **54** with respective arylmagnesium bromides in the presence of $CeCl_3$ and subsequent reductive aromatization with $SnCl_2$/acetic acid afforded the partially ring-closed hydrocarbons **56a** and **56b**. Upon DDQ/$Sc(OTf)_3$ treatment, **56a/56b** gave the desired quarteranthenes **58a/58b**. The CD_2Cl_2 solution of **57** was ^{1}H NMR silent at rt while progressive line sharpening was observed as the temperature was lowered which is attributed to the lower population of triplet species at low temperature. A small singlet–triplet energy gap of –3.81 kcal mol^{-1} was estimated from SQUID. In contrast, the NMR baseline of **58a** remained flat even when the temperature was lowered to 183 K. The absence of NMR signals for **58a** accounts for the large population of thermally accessible triplet species. The singlet–triplet energy gap for **58a** was found to be 0.689 kcal mol^{-1} from SQUID which indicates that **58a** was easily activated to a triplet state with nearly 50 % populated triplet species at rt. The crystallographic data gave further information about the bond length alternation as *a* bond in **58a** is 1.412 Å, much shorter than teranthene **57** (1.424 Å), which is considerably shorter than length of a C(sp^2)–C(sp^2) single bond (1.467 Å) and the corresponding one in bisanthene (1.447 Å), resulting from the significant diradical contribution. In addition, the highest harmonic oscillator model of aromaticity (HOMA) values of outer six-membered rings in quarteranthene indicates more benzenoid character at the peripheral rings and, hence, a larger diradical character compared to bisanthene. According to the desnsity functional theory (DFT) calculation, the singlet diradical character values are estimated to be 0.07 for bisanthene, 0.54 for teranthene and 0.91 for quarteranthene [40]. The diradical character of these molecules in terms of the energy balance between the formal loss of a double bond and the aromatic stabilization was well explained by Kubo. The aromatic stabilization energy of benzene, based on the homodesmic stabilization energy, is ~21.5 kcal mol^{-1} which is around three times less than the C–C π-bond energy of ~64.53 kcal mol^{-1}. Upon transformation to the diradical form, bisanthene contains only two additional sextets; therefore, the destabilization energy due to the π-bond cleavage cannot be fully compensated, so electron pairing is favored. In contrast, teranthene and quarteranthene both include three and four additional Clar sextets, respectively, in its diradical resonance form; hence the diradical forms contribute a lot to their ground state.

9.3.2 Periacenes

Perylene (**59**) and bisanthene (**60**) can be subcategorized as another class of PAHs, *peri*-fused acenes or *periacenes* (Figure 9.16), because they look like two naphthalenes or anthracenes have been *peri*-fused, respectively. The synthesis of the higher-order periacenes, namely peritetracene (**61**) and peripentacene (**62**), is quite challenging due to the extremely high reactivity, presumably originating from the more diradical/polyradical contribution (Figure 9.16), and the lack of proper synthetic methodology. Theoretical calculations conducted by Jiang et al. pointed out a crossover from non-magnetic phase to antiferromagnetic phase for both periacenes and anthenes starting from bisanthene onward, and an open-shell ground state is therefore expected for these types of graphene-like molecules [41]. Synthesis of peritetracene derivative is yet to be reported, and only one potential precursor, i.e., monobromo-tetracene dicarboximide, was reported by Wu's group [42]. The only experimental evidence of peripentacene was the mass spectroscopic peak of gas-phase disproportionation products [43].

A partially fused tetracene dimer was very recently reported by Müllen et al. The intermediate compound **63** was first prepared in 8 steps and was subjected to Grignard treatment followed by a ring fusion and oxidative dehydrogenation to afford the tetrabenzo[*a*,*f*,*j*,*o*]perylene (a bistetracene analogue) derivative **65** as green powder (Figure 9.17) [44]. The diradical contribution to the ground state, arising from the 5 Clar sextets in open-shell form compared to 2 sextets in closed shell, was derived by the VT NMR measurement where progressive line broadening from 213 K to 298 K was

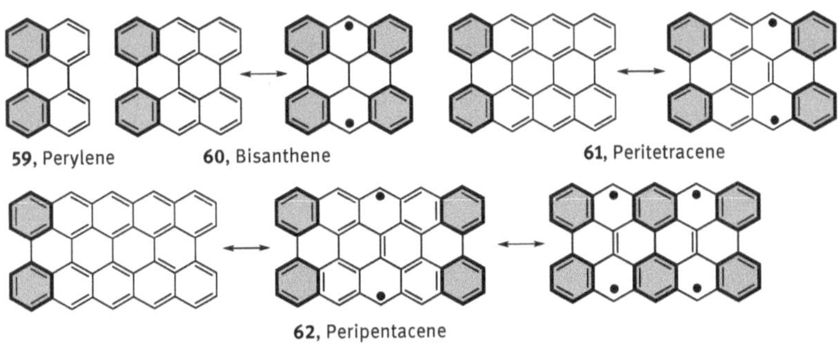

Figure 9.16: Periacenes and their open-shell resonance forms.

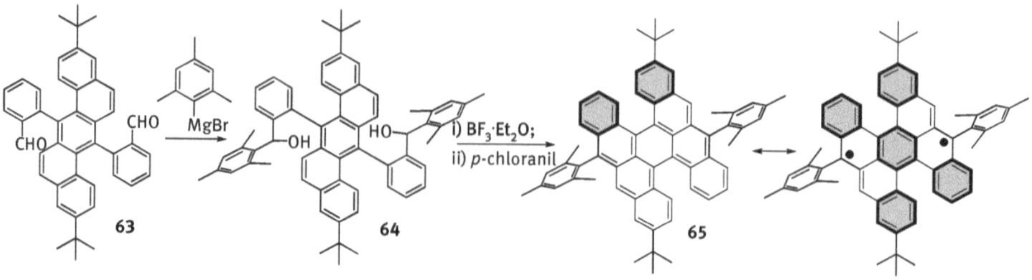

Figure 9.17: Synthesis of Müllen's diradicaloid bistetracene.

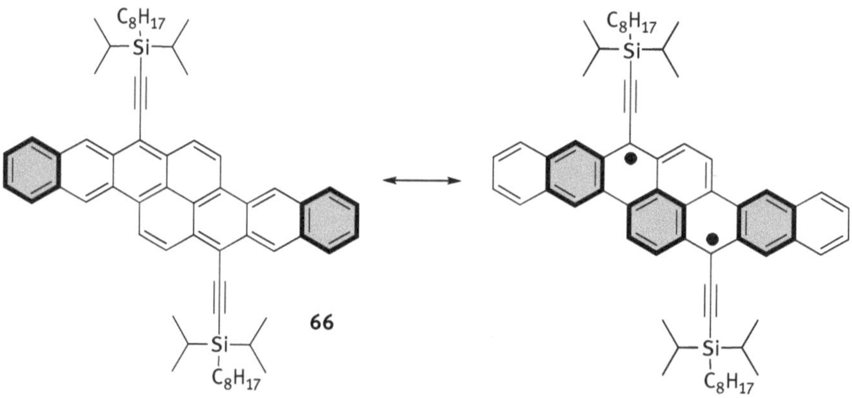

Figure 9.18: Briseno's bistetracene derivative.

attributed to the population of the thermally excited triplet species which was –3.4 kcal mol^{-1} (derived from VT ESR) higher than the singlet diradical ground state. Theoretically (at B3LYP/6-31G(d,p) level), the compound has been shown to have $y = 0.615$, even though a large singlet–triplet energy gap of –6.7 kcal mol^{-1} was deduced.

In contrast, a bistetracene derivative **66** (Figure 9.18) with high-charge carrier mobility of 6.1 $cm^2\ V^{-1}\ s^{-1}$ was reported [45] to exhibit closed-shell ground state as in the open-shell form only one extra Clar's sextet was recovered which may not be high enough to compensate the breaking of one C–C double bond energy.

There are a couple of synthetic approaches toward peripentacene reported in literatures; however, none of them succeeded (Figure 9.19(a)). One of them, contributed by Wu's group, involves the oxidative photocyclization of **67** to obtain partially fused bis-pentacenequinone **68** [46]. However, the subsequent nucleophilic reaction did not generate the desired 1, 2-addition adduct; instead an unexpected 1, 4-Michael addition product **69** was obtained when compound **68** was treated with excess Grignard reagent of 1-bromo-3,5-di-*tert*-butylbenzene in anhydrous tetrahydrofuran (THF) followed by acidification in air. Further treatment of **69** with excess Grignard reagent followed by acidification in air gave the tetraaryl-substituted fused bispentacenequinone **70**. Crystallographic analysis revealed an *α*, *β*-unsaturated ketone structure

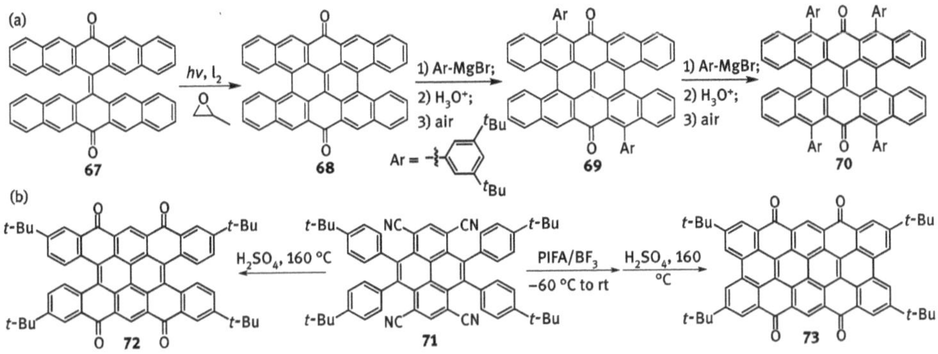

Figure 9.19: Attempted synthesis of peripentacene, reported by (a) Wu and (b) Müllen.

in the fused bispentacenequinones **68** and **69**, which may account for the unusual Michael addition reactions. Another synthetic approach, documented by Müllen et al. (Figure 9.19(b)), involves the cyclization of tetracyano pyrene derivative **71** to afford the tetraketone **72** which unfortunately didn't undergo further photocyclization [47]. Adopting an alternative path by oxidative cyclodehydrogenation of **71** in the presence of PIFA/BF_3 (PIFA: phenyliodine bis(trifluoroacetate)) and subsequent acid-promoted acylation afforded the peripentacenetetraketone **73** in trace amount with very poor solubility and no further reactions were conducted. It concludes that periacenes beyond bisanthene are still elusive.

A tetrabenzo-peripentacene derivative **76** [48] was recently reported by Yamada et al., utilizing another pyrene derivative, in 17 % yield by oxidation of **74** with $FeCl_3$ following a quick separation and further oxidation with DDQ/Sc(OTf)$_3$ in a microwave at 130°C [49]. This compound is better be described as a tetrabenzo-peripentacene rather than a "peripentacene" derivative because HOMA analysis revealed localized aromaticity as defined by the eight aromatic sextet rings in the core, with a closed-shell ground state (Figure 9.20). The structure was established by mass spectrum and single-crystal X-ray crystallographic analysis. The rigid framework of the compound is well reflected by the fine and sharp UV-vis absorption structure and a small stokes shift of 10 cm^{-1}.

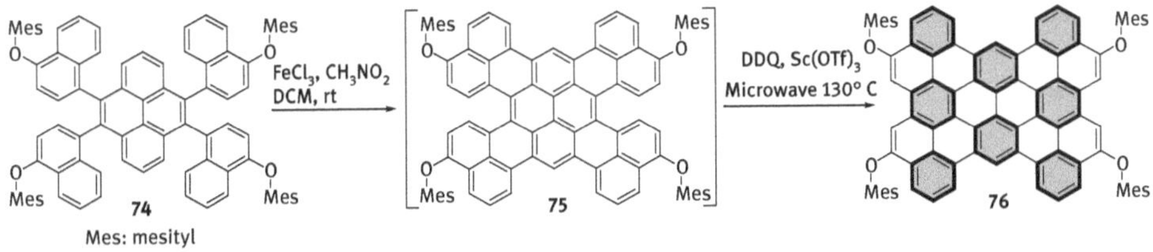

Figure 9.20: Synthesis of Yamada's tetrabenzo-peripentacene derivative.

9.4 Phenalenyl-Based Open-Shell PHs

Phenalenyl (**77**) is regarded as rigid π-conjugated neutral radical composed of a triangular fusion of three benzene rings and is also the smallest "open-shell graphene fragment" (Figure 9.21). Research on phenalenyls could lead to a better understanding of the intriguing electronic and magnetic properties of nanographene [50]. The planar structure of phenalenyl with the radical spin delocalized over the entire molecule, and high redox amphotericity grabbed attention of chemists, and earlier, Haddon claimed its potential use as single-component molecular electroconductors [51]. The kinetic instability of phenalenyl stimulated a great deal of synthetic efforts for the stabilization of the phenalenyl by chemical modifications to suppress its immediate dimerization by intermolecular σ-bond formation as well as reactivity toward oxygen. In the following sections, the significant progress in neutral monoradicals and diradicaloids based on the phenalenyl system is discussed.

9.4.1 Phenalenyl-Based Monoradicals

Introduction of three *tert*-butyl groups at the *β*-positions of phenaleny helped Nakasuji with the isolation of **82** in the solid state in air in 1999 (Figure 9.22) [52]. Formylation on 4-bromo-2,7-di-*tert*-butylnaphthalene (**78**) by lithiation and then Reformatsky reaction followed by a reductive elimination of hydroxyl group, subsequent hydrolysis and Friedel–Crafts acylation, gave the phenalanone derivative **80.** The radical precursor **81** was obtained as pale yellow crystals by reduction of **80** with $LiAlH_4$ and subsequent dehydration in *p*-TsOH. Oxidation of the phenalene **81** by *p*-chloranil in degassed toluene afforded phenalenyl **82** as deep blue crystals. The bulky substituents not only successfully suppress the σ-bond dimerization, but are also possessed with a minimal perturbation to the electronic structure of the parent phenalenyl in that the connection positions were β positions with negligible spin densities. Crystal structure showed the formation of π-dimer in a herringbone packing motif with interplanar distance ranging from 3.201 to 3.323 Å, much shorter than the sum of the Van der Waals radius of the carbon atoms. A strong antiferromagnetic interaction within the π-dimer as evidenced by a large antiferromagnetic intermolecular exchange interaction ($2J/k_B$ = −2,000 K) from SQUID measurements corroborates well with such

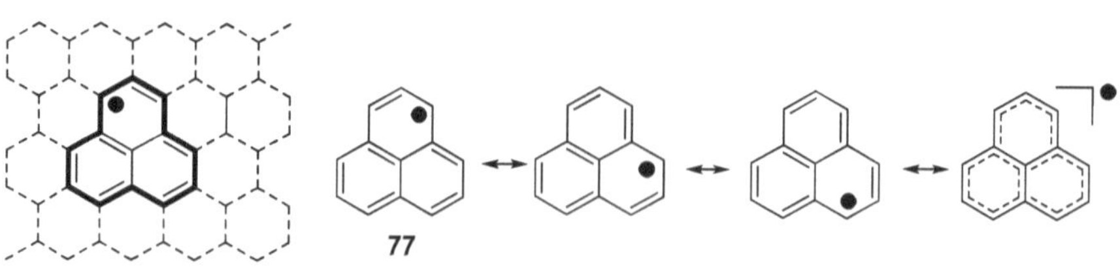

77

Figure 9.21: Phenalenyl as an open-shell graphene fragment and its resonance forms.

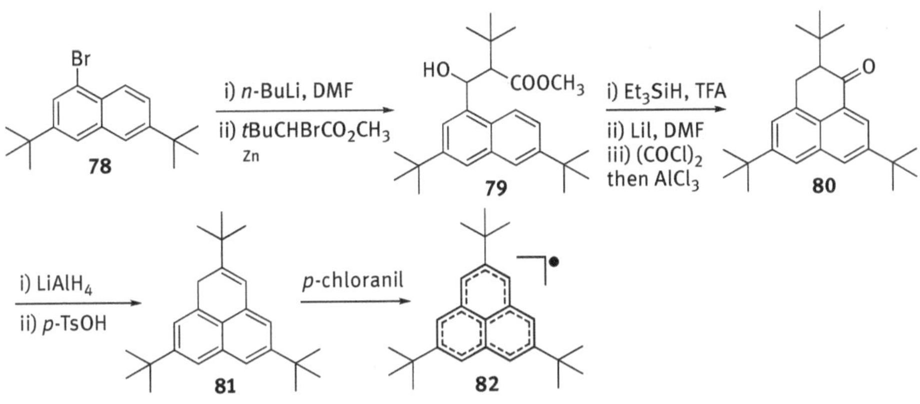

Figure 9.22: Synthesis of Nakasuji's *tert*-butyl-protected phenalenyl monoradical.

short distance that eventually opened the area of multicenter bonding [53]. Interestingly, spectroscopic study showed the same dimeric behavior (pancake bonding) in solution (similar to crystal) as the red purple solution of **82** gradually turned to blue upon cooling which is in accordance with the increase of absorbance in 530–670 nm region [54]. More definite evidence was provided by low-temperature ^{1}H and ^{13}C NMR measurements. The molecular weight of the π-dimer was also detected by cold-spray ionization mass spectrometry (MS) which allows substance ionized at lower temperatures, and this result represented the first detection of a radical dimer with NMR and MS techniques.

Perchlorophenalenyl radical **88** with all α- and β-positions substituted by chlorine atom was first prepared by Haddon et al. [55], and the X-ray crystallographic structure (Figure 9.23) was obtained in 2001 [56]. The key starting material

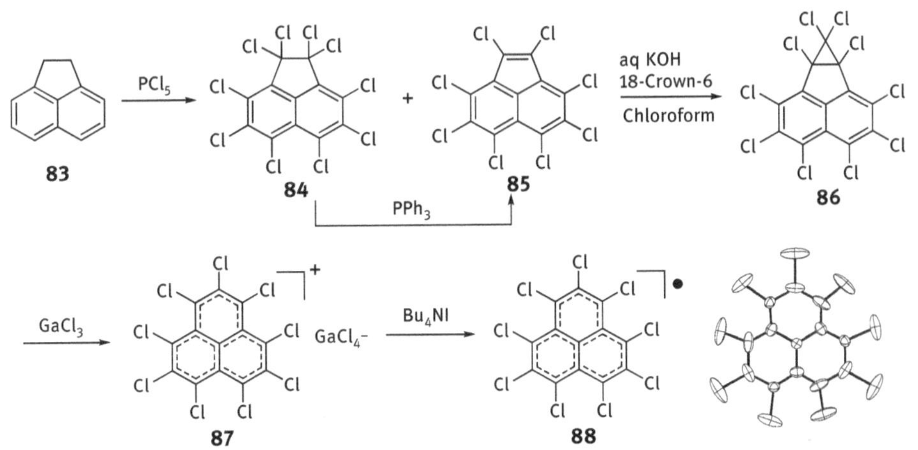

Figure 9.23: Synthesis of Haddon's perchlorophenalenyl radical. (Adapted with permission from Ref. [56]. Copyright 2001, American Chemical Society.)

perchloroacenaphthylene **85** was obtained as mixture with **84** from chlorination of acenaphthene **83**. Conversion of **85** from **84** also was realized by thermolysis or treatment with PPh_3. Addition of dichlorocarbene to **85** afforded **86** which underwent subsequent allyl rearrangement in the presence of Lewis acid to yield phenalenyl cation **87**. Reduction of **87** afforded the air stable **88** as a shiny black hexagonal crystal. X-ray analysis revealed a propeller-shaped **88** due to the bulky chlorine atom-induced deformation from the planar structure, and nonuniform stacking was also observed. This molecule stacked in columns separated by 3.78 Å, in sharp contrast to the *tert*-butyl phenalenyl **82** which was dimeric in the solid state with a much closer intermolecular distance. The reason for the long intermolecular distance of **88** was due to the non-planarity which inhibited the formation of 60°-rotated stacking motif. The magnetic susceptibility measurements disclosed that the solid **88** exhibited Curie paramagnetism at 100–380 K and antiferromagnetic ordering below 100 K. A rt conductivity of 10^{-10} S cm^{-1} was reasoned for the isolation of individual molecules, making **88** a Mott–Hubbard insulator.

In 2014, Kubo et al. tried to create an ideal 1D chain by utilizing the strong electrostatic interaction of pentafluorophenyl groups that were introduced at the β-positions of phenalenyl. A hydro precursor **89** was prepared in ten steps from a commercially available 2,8-dibromonaphthalene and dehydrogenation with DDQ in final step resulted in the formation of a σ-dimer (**90**) of pentafluorophenyl-substituted phenalenyl radical, and not the desired 2,5,8-tris(pentafluorophenyl)phenalenyl radical **91** [57]. The single-crystal X-ray analysis is the first structural characterization of a σ-dimer of an electronically unperturbed phenalenyl compound. The σ-dimer featured a long s-bond 1.636(7) Å between two phenalenyl rings, thereby implying the weakness of the σ bond. Indeed, a solution of the σ-dimer in toluene at rt showed a well-resolved multiline ESR spectrum that corresponded to compound **91**; therefore, compound **91** and its s-dimer existed as an equilibrium mixture in solution. The ESR signals decreased in intensity with decreasing temperature and almost disappeared at 230 K. The σ-dimer in solid form was further heated at 300°C in a degassed tube. The heating resulted in melting to a liquid accompanied by a color change from yellow to purple and, surprisingly, cooling the liquid afforded relatively large X-ray quality needles that at 10 K showed the distinctive feature that **91** forms a 1D chain with equidistant stacking of the molecule. It exhibited bimodal (σ and π) association modes. In the π mode, a consecutive multicenter bonding interaction led to electron delocalization in a uniform 1D stack of the radicals and a covalent bonding interaction of about –600 K ($2J/k_B$). High thermodynamic stability of the phenalenyl moiety decreased the stability of the s-dimer, whilst its planar structure adequately stabilized the p-association, owing to a multicenter bonding interaction, which resulted in a small energy difference between the s- and p-modes (Figure 9.24).

Bis(phenalenyl)s linked through a five-membered ring to produce a monoradical system **93a**, obtained as black powder by the reduction of precursor **92** with decamethylferrocene, was reported by Kubo et al. [58] to be highly stable in the solid

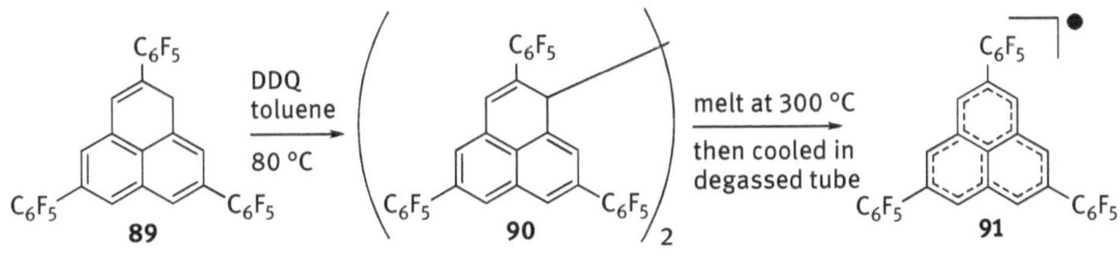

Figure 9.24: Kubo's pentafluorphenyl-substituted phenalenyl radical.

state sans steric protection due to extensive delocalization (thermodynamically stabilized) (Figure 9.25). The extensive delocalization of unpaired electron also suppressed the formation of σ-dimer as evident from the unchanged multiline ESR signal intensity from rt to –90°C. The X-ray crystallographic analysis of the *n*-butyl-substituted compound **93b** indicated the formation of π-dimers in the solid state with multicenter bonding, and the strength of antiferromagnetic coupling interaction was measured as –1,600 K ($2J/k_B$) within the π-dimer, which would split the SOMO of **93b** into bonding and antibonding MOs within the π-dimer. The half-life in air determined at rt was almost 60 h, which is in contrast to the rapid reaction of the phenalenyl radical with oxygen, suggesting that spin delocalization is very important for the stabilization of organic radicals. Fusion of three phenalenyl units onto one benzene ring was supposed to lead to a highly delocalized monoradical system (Figure 9.25), but the very low solubility of the synthetic intermediates and, furthermore, the oxygen-sensitive trihydro precursor of **94a** hampered its characterization [59]. *Tert*-butyl groups were then introduced to improve the solubility and stability, and subsequent reactions eventually afforded a hexa-*tert*-butyl-tribenzodecacyclenyl derivative **94b** which showed six reversible one-electron redox waves, providing evidence for formation of stable mono-, di- and tri-valent species, making **94b** one of the rare examples of compounds with six-stage amphoteric redox behavior.

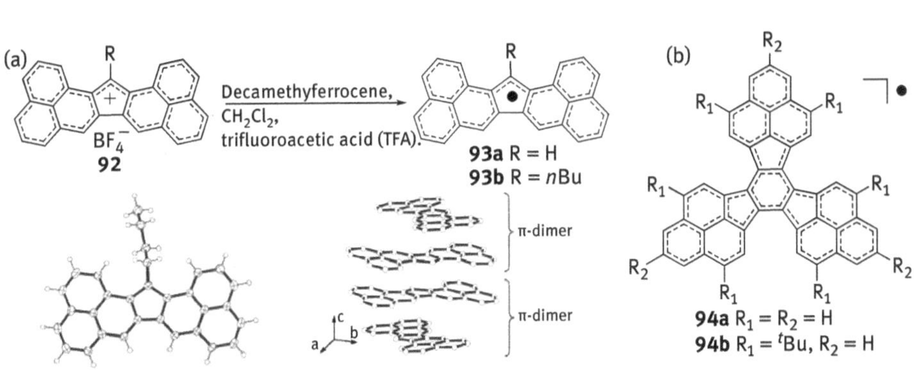

Figure 9.25: Kubo's (a) bis(phenalenyl) and (b) tribenzodecacyclenyl-derived stable monoradicals. (Adapted with permission from Ref. [58]. Copyright 2011, American Chemical Society.)

With the aim to obtain stable neutral radicals with less steric hindrance and extended delocalization, Haddon et al. designed a series of phenalenyl radicals with a disulfide bridge across two neighboring active positions (Figure 9.26). The dithiophenalenyl **98** was firstly prepared in 1978 by reduction of the corresponding cationic species **97** and it survived in solid state in air for up to 24 h but decomposed quickly in solution [60]. In the crystalline phase, **98** stacked in a sandwich herringbone motif of face-to-face π-dimers with 180° rotation between two radicals (Figure 9.26). The intermolecular distance in this π-dimer (3.13–3.22 Å) is even shorter than that of **82**, indicating a stronger intermolecular interaction. Notably, in spite of the absence of bulky substituents, no σ-dimerization was formed in the solid state as the spin is largely delocalized even on the two sulfur atoms stabilizing the neutral radical thermodynamically. An almost complete superposition of the α carbon atoms can be observed from a perpendicular view, maximizing the overlap of SOMOs. Other contacts in the lattice, such as S–S interactions between different π-dimers, were also observed, but the dimers remain isolated with electrons trapped within dimer pairs, which was further demonstrated by the conductivity measurements revealing it as insulator [61]. The attempt to prepare a more spin-delocalized radical analogue of tetrathiophenalenyl was performed; however, the reduction of the cation **100** led to formation of a closed-shell dimer **101** with a S–S σ-bond (Figure 9.26), and only a weak ESR signal can be observed in solution [62].

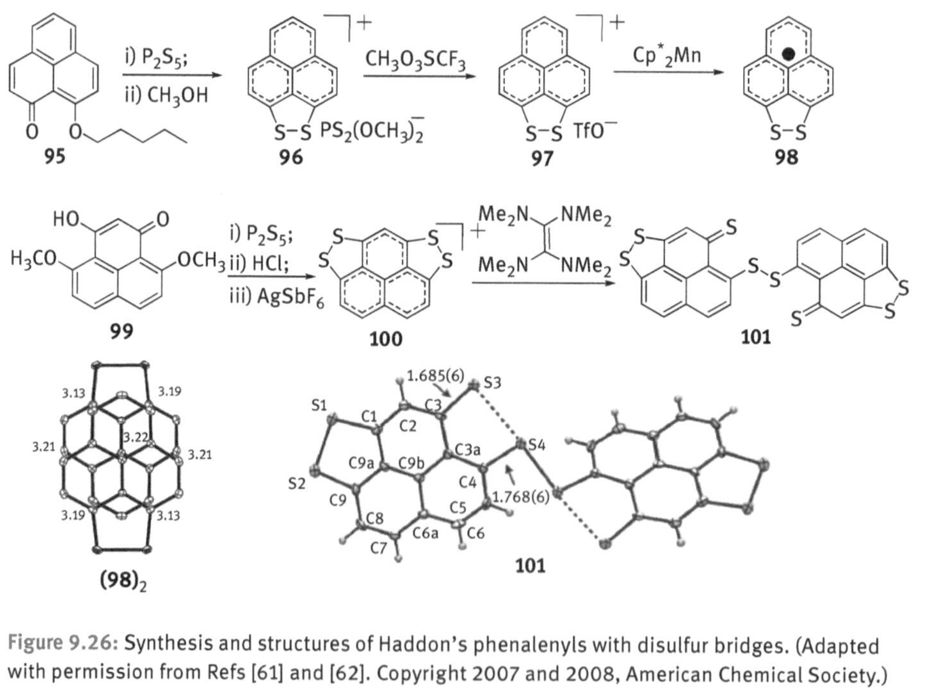

Figure 9.26: Synthesis and structures of Haddon's phenalenyls with disulfur bridges. (Adapted with permission from Refs [61] and [62]. Copyright 2007 and 2008, American Chemical Society.)

Modifications within the phenalenyl moiety, instead of around the phenalenyl unit, were also investigated to stabilize the open-shell system; for example, incorporation of heteroatoms such as nitrogen atom is an effective way to stabilize phenalenyl radical [63], and a chiral diazaphenalenyl having [4]helicene structure was also reported as stable radical [64].

9.4.2 Phenalenyl-Based Diradicaloids

9.4.2.1 Bis(phenalenyl)s

Pentalenodiphenalene **102** [65], a pentalene subunit bridging two phenalenyl moieties, was first reported by Nakasuji et al. (Figure 9.27) and showed interesting amphoteric multistage redox properties. However, the neutral state was substantially destabilized by the electronic contribution from 8π-electron antiaromatic pentalene subunit. Afterward, an IDPL **103a** with one benzenoid ring fused in the center was developed, and various substituents were introduced to improve the solubility (Figure 9.27) [66]. The line broadening in the ^1H NMR spectra at elevated temperature as well as emergence of sharp peaks at lower temperature together with solid-state ESR peak clearly indicated a thermally accessible triplet species at rt, and the energy separation between ground singlet state to excited triplet state was further determined as ~4.87 kcal mol^{-1}. The crystallographic structure of **103d** demonstrated one-dimensional chains in staggered stacking mode with an average π–π distance of 3.137 Å, which is significantly shorter than the van der Waals contact of carbon atoms (3.4 Å). This packing mode will maximize the SOMO–SOMO overlapping between the

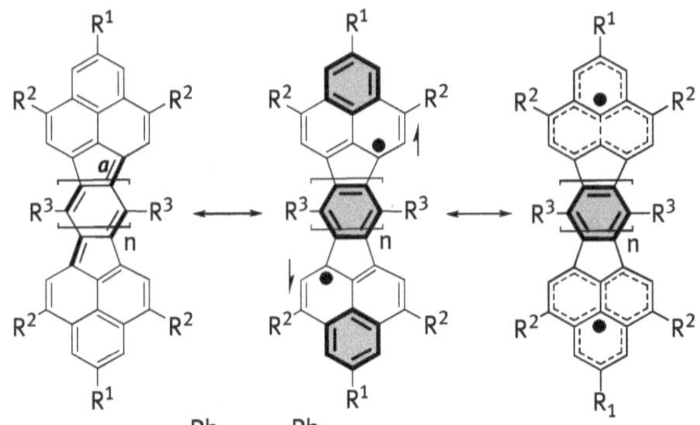

102: n = 0, $R^1 = R^2 = R^3 = H$
103a: n = 1, $R^1 = R^2 = R^3 = H$
103b: n = 1, $R^1 = R^2 = H$, $R^3 = {}^iPr$
103c: n = 1, $R^1 = R^3 = H$, $R^2 = {}^tBu$
103d: n = 1, $R^1 = R^2 = H$, $R^3 = Ph$
103e: n = 1, $R^1 = Me$, $R^2 = H$, $R^3 = Ph$
104a: n = 2, $R^1 = {}^tBu$, $R^2 = H$, $R^3 = Ph$
104b: n = 2, $R^1 = R^2 = H$, $R^3 = Ph$

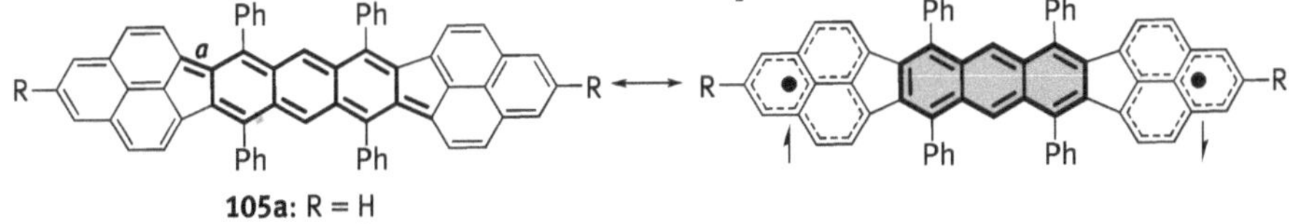

105a: R = H
105b: R = ^tBu

Figure 9.27: Bis(phenalenyl)s bridged by pentalene and different aromatic linkers.

radicals, leading to stabilized intermolecular orbitals corresponding to intermolecular covalency [67]. A naphthalene-linked bis(phenalenyl) **104a** with enhanced diradical character than **103** was also reported [68] to feature a smaller HOMO–LUMO and singlet–triplet gap of 1.04 eV and –3.77 kcal mol^{-1} (by SQUID), respectively, than **103d** (1.15 eV and $\Delta E_{S\text{-}T}$ = –4.37 kcal mol^{-1}). The crystal packing of **104b** adopted similar stepped mode to **103d** in the one-dimensional chain, and the intermolecular bonding was stronger than the intramolecular one due to the spin-localized nature on the phenalenyl moieties, which can be more adequately described as multicenter bonding. Bis(phenalenyl)s linked by anthracene unit **105** (Figure 9.27) were also reported by Kubo et al. [69] and have even larger diradical contribution (y = 0.68) compared to their naphthalene (y = 0.50) and benzene (y = 0.30) counterparts. The enhanced diradical contribution of **105** was further supported by crystallographic analysis where the bond a in **105b** is more elongated (1.467(3) Å) compared to that of **104a** (1.465(7) Å) and **103e** (1.457(2) Å). A similar intermolecular covalent bonding interaction in **105a** with a distance of 3.122 Å, which is considerably stronger between molecules compared to within a molecule in a one-dimensional stack, was observed. The significant diradical contribution of **105** was attributed to the high aromatic stabilization energy of the anthracene bridge. These compounds are featured by a singlet diradical ground state as two factors play major role in stabilization of the systems: (1) the intrinsic delocalization of phenalenyl moiety and (2) the aromatic stabilization through the recovery of additional aromatic ring from quinoidal resonance form to the diradical resonance form.

A thiophene-fused bis(phenalenyl) **106** (Figure 9.28) was reported to show y = 0.35 and DFT calculation at UB3LYP/6-31G(d,p) level predicted a singlet diradical ground state which lies 1.7 kcal mol^{-1} below the singlet closed-shell state. The single crystal analysis revealed two kinds of dimeric pairs with substantially short nonbonding contacts of about 3.1 Å between each thiophene ring [70]. Accommodation of a doubly

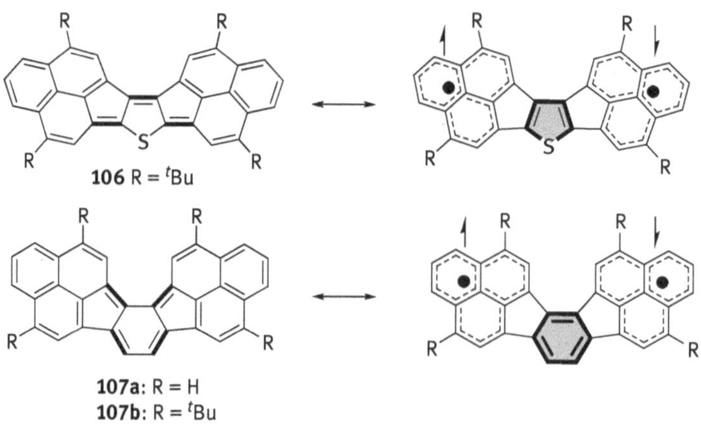

106 R = ^{t}Bu

107a: R = H
107b: R = ^{t}Bu

Figure 9.28: Bis(phenalenyl)s linked with thiophene and benzene spacer.

excited configuration into the ground state configuration stabilized this system by suppressing four-electron repulsion arising from interaction between fully occupied orbitals. Fusion of phenalenyl to *ortho*- positions of phenyl ring led to compounds **107a** and **107b** (Figure 9.28), and **107b** with *tert*-butyl substituents was reported to be extremely air-sensitive and decomposed in the air compared to several week long stable **103c** [71]. The absence of ^{1}H NMR signals gave no useful information about **107b**, however, dissolution of **107b** in D_2SO_4 gave clear spectra assignable to the dicationic species **107b**$^{2+}$, thereby supporting the formation of **107b**. The singlet open-shell form was stabilized by 8.45 kcal mol^{-1} than the singlet closed shell, supporting a diradical ground state [72].

9.4.2.2 Zethrenes

A head-to-head fusion of two phenalenyl moieties will generate a Z-shape graphene fragment, called zethrene **108**, which looks like two naphthalene units are fixed by a *trans*-1,3-butadiene unit (Figure 9.29) [73]. The diradical form of zethrene cannot be stabilized by recovery of additional aromatic sextet ring, therefore, they could show closed-shell ground state, although, remarkable open-shell diradical character y = 0.407 was predicted by Nakano et al., based on the occupancy numbers of spin-unrestricted Hartree–Fock natural orbitals (UNOs) [74]. However, synthesis and experimental investigations of 7, 14-bis(phenylethynyl)zethrene **109** by Tobe et al. [75], the phenyl-substituted zethrene derivative **110** by Y. -T. Wu et al. [76], a zethrene bis-carboximide derivative **111** by Wu et al. [77], and a parent zethrene derivative **108** by Miao et al. [78] were in favor of a closed-shell ground state. A diphenyl-substituted 1,2:8,9-dibenzozethrene **114** [79], prepared by adopting a similar approach used by

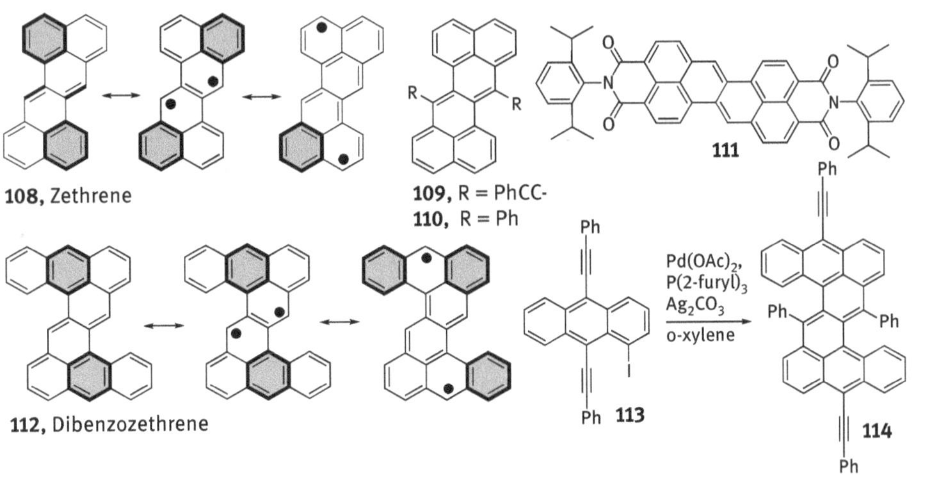

Figure 9.29: Zethrene, dibenzozethrene, and their derivatives.

Y. -T. Wu [76], exhibits larger diradical character index of 0.210 compared to that of 0.128 for unsubstituted 1,2:8,9-dibenzozethrene, most probably due to the distorted geometry and extended delocalization. However, the large singlet–triplet gap (−7.32 kcal mol^{-1}) didn't allow any thermally promoted magnetic activity, making it behave like a typical closed-shell PH.

Extension of the butadiene unit gives higher-order zethrenes such as heptazethrene **115**, octazethrene **116**, and nonazethrene **117** (not reported yet) in which two naphthalene units are bridged by a *p*-QDM, 2,6-naphthoquinodimethane, and 2,6-anthraquinodimethane unit, respectively (Figure 9.30). Unlike zethrene, diradical resonance form of heptazethrene (y = 0.537) and higher homologues (octazethrene, y = 0.628) are stabilized by the recovery of aromaticity of the aromatic linker ring in the center together with the thermodynamic stabilization through radical delocalization on entire hydrocarbon backbone. Vertical fusion of two benzenoid rings onto the heptazethrene core, in different modes, can lead to two isomeric structures, 1,2:9,10-dibenzoheptazethrene **118** and 5,6:13,14-dibenzoheptazethrene **119** (Figure 9.30).

Theoretical prediction of a larger diradical character in heptazethrene stimulated our group to synthesize the first relatively stable heptazethrene derivative, heptazethrenebis(dicarboximide) **121** (Figure 9.31(a)). This compound was synthesized by using a transannular cyclization approach from **120** involving simultaneous cyclization of an octadehydronaphthoannulene intermediate [80]. ^1H NMR spectrum showed a line broadening at rt due to the existence of thermally accessible triplet species and progressive line sharpening upon cooling due to the shift of the equilibrium to the singlet state. On the basis of DFT calculations, the energy of singlet diradical state of **121** was located 5.8 and 7.9 kcal/mol lower than the closed-shell quinoidal state and open-shell triplet biradical state, respectively, supporting an open-shell singlet

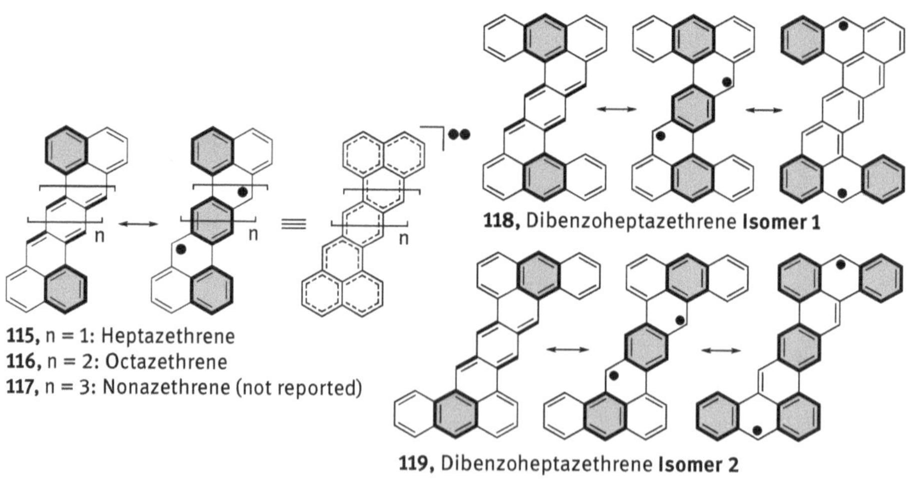

Figure 9.30: Higher-order zethrenes and two dibenzoheptazethrene isomers.

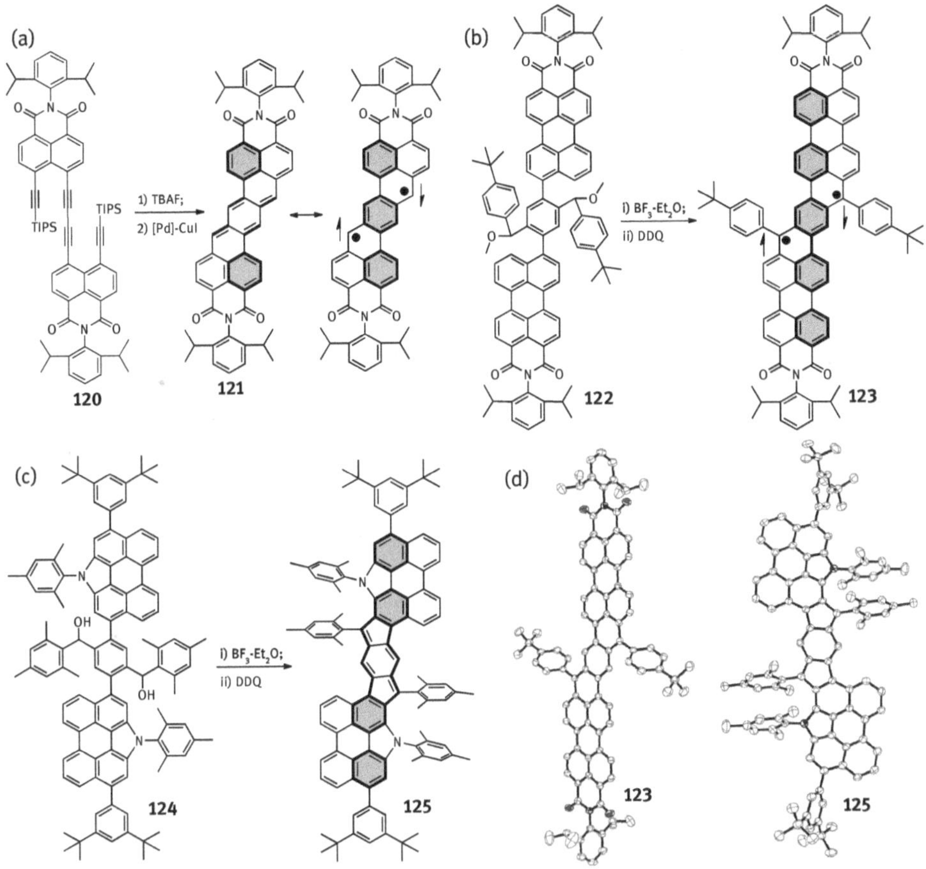

Figure 9.31: Synthesis and structures of (a) heptazethrenebis(dicarboximide), (b) *p*-QDM-bridged perylene monoimide dimer or extended heptazethrenebis(dicarboximide), (c) *s*-indacene-bridged NP dimer, (d) X-ray crystallographic structures of **123** and **125**.

diradical ground state with a homogeneously distributed spin density on the entire hydrocarbon backbone. A low electrochemical bandgap of 0.99 eV associated with lower-energy UV-vis-NIR absorption band that originated from admixing of the doubly excited electronic configuration (H,H-L,L) into the ground state further supports the open-shell character of **121**. Compound **121** showed reasonable photostability in solution; however, the material slowly decomposed during storage either as solution or in solid state. In order to improve the stability without compromising the solubility, kinetic blocking at the most reactive radical sites was necessary. With the aim to improve the stability and tune the bandgap and diradical character, an extended-heptazethrenebis(dicarboximide) (a *p*-QDM-bridged perylene monoimide dimer) **123** was synthesized which was kinetically blocked by *t*-butyl phenyl groups at the most reactive radical sites (Figure 9.31(b)) [81]. BF_3·OEt_2-mediated intramolecular ring-cyclization of the dimethoxy derivative **122** followed by oxidative dehydrogenation by DDQ gave the target compound **123**, which is soluble in common organic solvents and stable under ambient conditions. An absorption spectrum characteristic to admixture of a doubly excited electronic configuration with low HOMO–LUMO gap of 0.99 eV, absence of NMR signal at rt and presence of broad ESR signal due to thermally populated triplet species ($\Delta E_{S\text{-}T}$ = –2.97 kcal mol^{-1}, from SQUID) were in favor of its singlet diradical character (y = 0.465) in ground state. It is worth to mention that a different regio-selectivity for the ring cyclization reaction was observed with NP derivative **124** leading to eventual isolation of a closed-shell *s*-indacene-bridged perylene dimer **125** with negligible diradical character (y = 0.009), instead of the extended heptazethrene derivative (Figure 9.31(c)) [81]. This observation was attributed to the activation of the β-position of electron-rich NP unit (in **124**) along with formation of two five-membered rings rather than six-membered rings which is thermodynamically more favorable. The structures of **123** and **125** are both confirmed by X-ray crystallographic analysis (Figure 9.31(d)).

To improve the stability of the higher-order zethrenes, bulky triisopropylsilylethynyl (TIPS)-blocked heptazethrene **126** and octazethrene **127** were prepared from the corresponding diketone precursors (Figure 9.32) [82]. Compound **126** showed a typical *p*-band similar to closed-shell PHs, while **127** displayed an absorption pattern similar to the **121** and **123** indicating a probable open-shell character for **127**. The open-shell nature of **127** was further supported by the broadened ^{1}H NMR spectrum and the appearance of ESR signal at rt due to presence of thermally excited triplet species as a consequence of small singlet–triplet energy gap (–3.87 kcal/mol based on SQUID). Both **126** and **127** were packed into a 1D infinite chain via intermolecular π–π interactions, with an average π-stacking distance of 3.38 and 3.35 Å, respectively, which is larger than intermolecular covalent π-bonding interaction seen in Kubo's bis(phenalenyl)-based diradicaloids.

Using a similar synthetic strategy, nucleophilic addition of TIPS-magnesium chloride to the corresponding diketone precursor **128** following reduction with $SnCl_2$ provided a dibenzoheptazethrene derivative **129** (a derivative of **119**) which appeared to have an open-shell singlet diradical ground state (Figure 9.33). Another dibenzoheptazethrene isomer **131** (a derivative of **118**) was also obtained using a

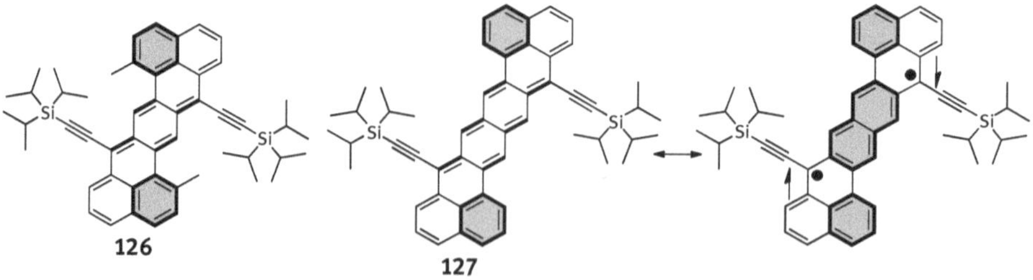

126 **127**

Figure 9.32: Wu's TIPS-blocked heptazethrene and octazethrene derivatives.

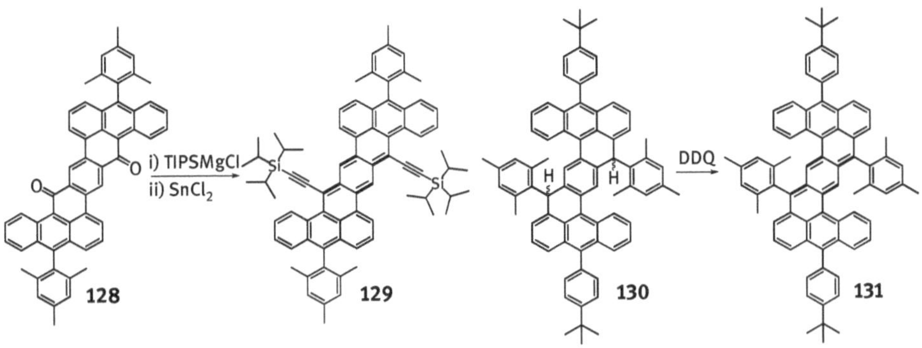

Figure 9.33: Wu's dibenzoheptazethrene isomers.

DDQ-mediated oxidative dehydrogenation of a dihydro precursor **130**. Experimental and theoretical investigation revealed **129** to possess a larger diradical contribution (y = 0.576) in the ground state compared to isomeric **131** (0.309) [83]. This difference in diradical character between two isomeric structures was explained by the different number of aromatic sextet rings in the respective diradical resonance forms (Figure 9.30). Hence, *benzenoid PHs with the same chemical composition, the molecule with more aromatic sextet rings in the diradical resonance forms, exhibits greater diradical character.* This can be regarded as an extension of Clar's aromatic sextet rule in the benzenoid PH-based singlet diradicaloids.

9.5 Miscellaneous Open-Shell PHs

A potential tetraradicaloid hydrocarbon **132** (Figure 9.34), tetracyclopenta-[*def,jkl,pqr,vwx*]tetraphenylene, was successfully prepared by Tobe et al., which on NMR timescale rapidly equilibrated between its two D_{2h} structures through valence tautomerization in solution [84]. It exhibited an intense absorption maximum at 475 nm together with a broad band centered at 909 nm with the lowest energy absorption band shifted to much longer wavelengths because of the extended conjugation. Using a long-range-corrected LC-UBLYP/6-311+G(d,p) method, the diradical and

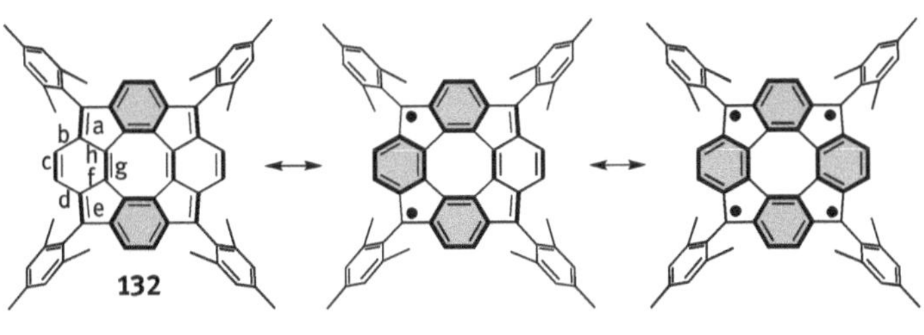

Figure 9.34: Tobe's tetracyclopenta-[*def,jkl,pqr,vwx*]tetraphenylene derivative.

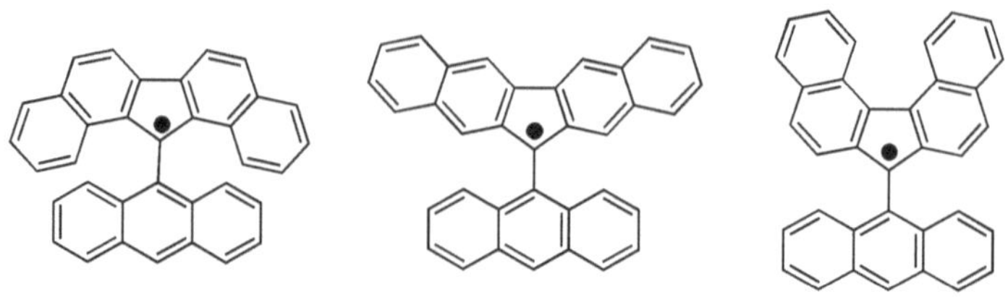

133 134 135

Figure 9.35: Kubo's extended fluorenyl monoradicals.

tetraradical characters were measured where (y_0, y_1) = (1, 0) indicated a pure diradical character while (y_0, y_1) = (1, 1) indicated pure tetraradical character. Theoretical calculation, on three different models with D_{2h}, D_{4h} symmetry and the crystal structure, predicted the (y_0, y_1) as (0.095, 0.032) for D_{2h} structure, (1.00, 0.166) for D_{4h} structure and (0.258, 0.085) for the crystallographic structure. However, a clear and sharp ^{1}H NMR can't experimentally validate the open-shell ground state which also corroborates the strong bond length alternation from crystallographic bond length analysis as the relatively short bonds *a* (1.371 Å), *c* (1.349 Å), *e* (1.373 Å) and *g* (1.359 Å), and longer bonds *b* (1.433 Å) and *d* (1.440 Å) bonds were in favor of the presence of a *p*-QDM subunit. Notably, ^{1}H NMR signal broadening at 50°C was observed which was indicative of thermally accessible triplet state, although vertical singlet–triplet energy gap was theoretically found to be quite large. So, this compound is better regarded as a singlet diradicaloid with a moderate diradical character and minor tetraradical character.

Stable fluorenyl-based monoradicals were recently reported by Kubo (Figure 9.35) [85]. The attachment of the 9-anthryl group at the most reactive site of the fluorenyl moiety efficiently stabilizes the radicals as anthracene can effectively protect the spin centers [37] with large spin density from σ-dimerization and oxygen attack, and benzannulation of the fluorenyl center increases the stability of the radicals through extended delocalization. All three radicals showed significant stabilities with the half lives of 7, 3.5 and 43 days for **133**, **134** and **135**, respectively. Compound **134** was isolated as an unassociated monoradical in crystalline form, whereas **133** gave crystalline σ-dimers and **135** existed in both unassociated monoradical and σ-dimer forms in its solid state. The dimerization in solution for **133** was evident from decrease in ESR signals at low temperature, whereas **134** and **135** did not show such ESR behavior. Solid UV measurements disclosed mechanochromic properties of crystals of $\mathbf{133}_2$ and $\mathbf{135}_2$ (σ-dimers) that indicated the dissociation of σ-dimer to monoradical could be a mechanochemical process.

9.6 Conclusion

As said by Lambert [86], "the future of these biradical PAHs clearly lies in materials science" and therefore efforts have to be made to realize their importance in materials sciences. Open-shell systems with moderate diradical contribution, theoretically, showed impressive third-order NLO properties [4, 87] which was justified experimentally (large TPA cross section was observed) by several groups including ours [3e]. Their role in future photovoltaic devices had also been investigated in terms of singlet-fission process [4, 88] and lately, promising results for organic light emitting device material have also been reported [89]. Furthermore, investigations of molecular-sized open-shell graphene fragments showed promise in molecule-based batteries [90], field-effect transistors [91], and as electroconductor [92]. Such advancement is believed to make path for the open-shell PHs as the next-generation spintronic materials [93] given that realistic synthetic approaches evolve from a clear understanding on the stabilization issues of such systems in the molecular level.

Acknowledgments

The project at Singapore was financially supported by AcRF MOE Tier 2 grants (MOE2014-T2-1-080), Tier 3 grant (MOE2014-T3-1-004) and A*STAR JCO grant (1431AFG100). We also would like to say special thanks to other major collaborators, Professor D. Kim in Yonsei University, Professor J. Casado and Professor J. T. Lopéz Navarrete in University of Malaga, Professor J. Ding in National University of Singapore, Professor K. -W. Huang in KAUST, Professor K. Furukawa in Niigata University, Professor Masayoshi Nakano in Osaka University, and Professor R. D. Webster in Nanyang Technological University for their valuable contributions.

References

[1] Sun Z, Wu J. Closed-shell and open-shell 2D nanographenes, in Polyarenes I: topics in current chemistry. Siegel JS, Wu YT, editors. Springer-Verlag Berlin Heidelberg, 2014:197–248.

[2] Morita Y, Nishida S. Phenalenyls, cyclopentadienyls, and other carbon-centered radicals, in Stable radicals: fundamentals and applied aspects of odd-electron compounds. Hicks RG, , editor. Germany, 2010:81–145.

[3] (a) Abe M. Diradicals. Chem Rev 2013;113:7011–88. (b) Sun Z, Wu J. Open-shell polycyclic aromatic hydrocarbons. J Mater Chem 2012;22:4151–60. (c) Sun Z, Ye Q, Chi C, Wu J. Low bandgap polycyclic hydrocarbons: from closed-shell near infrared dyes and semiconductors to open-shell radicals. Chem Soc Rev 2012;41:7857–89. (d) Sun Z, Zeng Z, Wu J. Benzenoid polycyclic hydrocarbons with an open-shell biradical ground state. Chem Asian J 2013;8:2894–904. (e) Sun Z, Zeng Z, Wu J. Zethrenes, extended *p*-QDMs, and periacenes with a singlet biradical ground state. Acc Chem Res 2014; 47:2582–91. (f) Kubo T. Phenalenyl-based open-shell polycyclic aromatic hydrocarbons. Chem Rec 2015;15:218–32. (g) Kubo T. Recent progress in quinoidal singlet biradical molecules. Chem Lett 2015;44:111–122. (h) Das S, Wu J. Open-shell benzenoid polycyclic hydrocarbons, in: Polycyclic arenes and heteroarenes: synthesis, properties, and applications. Qian M, editor. John Wiley & Sons Inc., 2015:1–36.

[4] Nakano M. Excitation energies and properties of open-shell singlet molecules: applications to a new class of molecules for nonlinear optics and singlet fission, Springer International Publishing, 2014. DOI: 10.1007/978–3–319–08120–5.

[5] (a) Senn P. A simple quantum mechanical model that illustrates the Jahn–Teller effect. J Chem Educ 1992;69:819. (b) Zeng Z, Shi X, Chi C, López Navarrete JT, Casado J, Wu J. Pro-aromatic and anti-aromatic π-conjugated molecules: an irresistible wish to be diradicals. Chem Soc Rev 2015;44:6578–96.
[6] Takahashi T, Matsuoka K, Takimiya K, Otsubo T, Aso Y. Extensive Quinoidal oligothiophenes with dicyanomethylene groups at terminal positions as highly amphoteric redox molecules. J Am Chem Soc 2005;127:8928.
[7] Bendikov M, Duong HM, Starkey K, Houk KN, Carter EA, Wudl F. Oligoacenes: theoretical prediction of open-shell singlet diradical ground states. J Am Chem Soc 2004;126:7416–17.
[8] Hachmann J, Dorando JJ, Avilés M, Chan GK. The radical character of the acenes: A density matrix renormalization group study. J Chem Phys 2007;127:134309.
[9] Purushothaman B, Bruzek M, Parkin SR, Miller AF, Anthony J E. Synthesis and structural characterization of crystalline nonacenes. Angew Chem Int Ed 2011;50:7013.
[10] Noodleman L. Valence bond description of antiferromagnetic coupling in transition metal dimers. J Chem Phys 1981;74:5737–5743.
[11] (a) Yamaguchi K. Self-consistent field: theory and applications. Carbo R, Klobukowski M, editors. Amsterdam: Elsevier, 1990:727–823. (b) Yamanaka S. Okumura M, Nakano M, Yamaguchi K. J Mol Struct: THEOCHEM 1994;310:205–218.
[12] Döhnert D, Koutecký J. Occupation numbers of natural orbitals as a criterion for biradical character. Different kinds of biradicals. J Am Chem Soc 1980;102:1789–96.
[13] Kamada K, Ohta K, Shimizu A, et al. Singlet diradical character from experiment. J Phys Chem Lett 2010;1:937–40.
[14] (a) Kolc J, Michl J. Photochemical synthesis of matrix-isolated pleiadene. J Am Chem Soc 1970;92:4147–8. (b) Kolc J, Michl J. π,π-Biradicaloid hydrocarbons. Pleiadene family. I. Photochemical preparation from cyclobutene precursors. J Am Chem Soc 1973;95:7391–401.
[15] Shimizu A, Tobe Y. Indeno[2,1-a]fluorene: an air-stable ortho-quinodimethane derivative. Angew Chem Int Ed 2011;50:6906–6910.
[16] Miyoshi H, Nobusue S, Shimizu A, Hisaki I, Miyatab M, Tobe Y. Benz[c]indeno[2,1-a]fluorene: a 2,3- naphthoquinodimethane incorporated into an indenofluorene frame. Chem Sci 2014;5:163.
[17] Clar E, Stewart DG. Aromatic hydrocarbons. LXV. Triangulene derivatives. J Am Chem Soc 1953;75:2667–72.
[18] Inoue J. Fukui K, Kubo T, et al. The first detection of a Clar's hydrocarbon, 2,6,10-tri-*tert*-butyltriangulene: a ground-state triplet of non-Kekulé polynuclear benzenoid hydrocarbon. J Am Chem Soc 2001;123:12702–03.
[19] Li Y, Huang KW, Sun Z, et al. A kinetically blocked 1,14:11,12-dibenzopentacene: a persistent triplet diradical of a non-Kekule polycyclic benzenoid hydrocarbon. Chem Sci 2014;5:1908–14.
[20] Shimizu A, Kishi R, Nakano M, et al. Indeno[2,1-b]fluorene: A 20-π-electron hydrocarbon with very low-energy light absorption. Angew Chem Int Ed 2013;52:6076.
[21] Montgomery LK, Huffman JC, Jurczak EA, Grendze MP. The molecular structures of Thiele's and Chichibabin's hydrocarbons. J Am Chem Soc 1986;108:6004.
[22] Shimizu A, Nobusue S, Miyoshi H, Tobe Y. Indenofluorene congeners: biradicaloids and beyond. Pure Appl Chem 2014;86:517–28.
[23] (a) Chase DT, Rose BD, McClintock SP, Zakharov LN, Haley MM. Indeno[1,2-b]fluorenes: fully conjugated antiaromatic analogues of acenes. Angew Chem Int Ed 2011;50:1127. (b) Chase DT, Fix AG, Rose BD et al. Electron-accepting 6,12-diethynylindeno[1,2-b]fluorenes: synthesis, crystal structures, and photophysical properties. Angew Chem Int Ed 2011;50:11103 (c) Chase DT, Fix AG, Kang SJ et al. 6,12-Diarylindeno[1,2-b]fluorenes: syntheses, photophysics, and ambipolar OFETs. J Am Chem Soc 2012;134:10349. (d) Nishida J, Tsukaguchi S, Yamashita Y. Synthesis, crystal structures, and properties of 6,12-diaryl-substituted indeno[1,2-b]fluorenes. Chem Eur J 2012;18:8964.

[24] Fix AG, Deal PE, Vonnegut CL, Rose BD, Zakharov LN, Haley MM. Indeno[2,1-c]fluorene: a new electron-accepting scaffold for organic electronics. Org Lett 2013;15:1362.

[25] Rose BD, Vonnegut CL, Zakharov LN, Haley MM. Fluoreno[4,3-c]fluorene: A closed-shell, fully conjugated hydrocarbon. Org Lett 2012;14:2426.

[26] (a) Hartzler HD. Synthesis and decomposition of 3-diazo-6-dicyanomethylene-1,4-cyclohexadiene. Arylmalononitriles. J Am Chem Soc 1964;86:2174–5. (b) Addison AW, Dalal NS, Hoyano Y, Huizinga S, Weiler L. The chemistry of anions derived from tetracyanodiphenoquinodimethane (TCNDQ). Can J Chem 1977;55:4191–9.

[27] Schmidt R, Brauer HD. Energetic positions of the lowest singlet and triplet state of the Schlenk and of the Mueller hydrocarbon. Angew Chem Int Ed 1971;10:506.

[28] West R, Jorgenson JA, Stearley KL, Calabrese JC. Synthesis, structure and semiconductivity of a p-terphenoquinone. J Chem Soc Chem Commun 1991:1234.

[29] Rebmann A, Zhou J, Schuler P, Stegmann HB, Rieker A. Synthesis, EPR spectroscopy and voltammetry of a *p*-quaterphenyl biradical/Quinone. J Chem Res Synop 1996;318.

[30] Rebmann A, Zhou J, Schuler P, Rieker A, Stegmann HB. Synthesis of two novel *para*-extended bisaroxyls and characterization of their triplet spin states. J Chem Soc Perkin Trans 1997; 2:1615.

[31] Zhu X, Tsuji H, Nakabayashi K, Ohkoshi S, Nakamura E. Air- and heat-stable planar tri-p-quinodimethane with distinct biradical characteristics. J Am Chem Soc 2011;133:16342–5.

[32] Sbargoud K, Mamada M, Marrot J, Tokito S, Yassar A, Frigoli M. Diindeno[1,2-b:20,10-n] perylene: a closed shell related Chichibabin's hydrocarbon, the synthesis, molecular packing, electronic and charge transport properties. Chem Sci 2015;6:3402.

[33] Zeng Z, Ishida M, Zafra, JL, et al. Pushing extended p-quinodimethanes to the limit: stable tetracyano-oligo(*N*-annulated perylene)quinodimethanes with tunable ground states. J Am Chem Soc 2013;135:6363–71.

[34] Zeng Z, Lee S, Zafra JL, et al. Tetracyanoquaterrylene and tetracyanohexarylenequinodimethanes with tunable ground states and strong near-infrared absorption. Angew Chem Int Ed 2013;52:8561–5.

[35] Zeng, Z, Lee S, Zafra JL, et al. Turning on the biradical state of tetracyano-perylene and quaterrylenequinodimethanes by incorporation of additional thiophene rings. Chem Sci 2014;5:3072–80.

[36] Zeng, Z, Lee S, Son M, et al. Push–pull type oligo(*N*-annulated perylene)quinodimethanes: chain length and solvent-dependent ground states and physical properties. J Am Chem Soc 2015;137:8572–83.

[37] Zeng Z, Sung YM, Bao N, et al. Stable tetrabenzo-chichibabin's hydrocarbons: tunable ground state and unusual transition between their closed-shell and open-shell resonance forms. J Am Chem Soc 2012;134:14513–25.

[38] (a) Yao JH, Chi C, Wu J, Loh KP. Bisanthracene bis(dicarboxylic imide)s as soluble and stable NIR dyes. Chem Eur J 2009;15:9299–302. (b) Li J, Zhang K, Zhang X, Chi C, Wu J. meso-Substituted bisanthenes as soluble and stable near-infrared dyes. J Org Chem 2010;75:856–63. (c) Zhang K, Huang K, Li J, Chi C, Wu J. A soluble and stable quinoidal bisanthene with NIR absorption and amphoteric redox behavior. Org Lett 2009;11:4854–7. (d) Hirao Y, Konishi A, Matsumoto K, Kurata H, Kubo T. Synthesis and electronic structure of bisanthene: a small molecular-sized graphene with zigzag edges. AIP Conf Proc 2012;1504:863–6.

[39] Konishi A, Hirao Y, Nakano M, et al. Synthesis and characterization of teranthene: A singlet biradical polycyclic aromatic hydrocarbon having kekule structures. J Am Chem Soc 2010;132:11021–3.

[40] Konishi A, Hirao Y, Matsumoto K, et al. Synthesis and characterization of quarteranthene: elucidating the characteristics of the edge state of graphene nanoribbons at the molecular level. J Am Chem Soc 2013;135:1430–7.

[41] Jiang D, Sumpter BG, Dai S. First principles study of magnetism in nanographenes. J Chem Phys 2007;127:124703.
[42] Yin J, Zhang K, Jiao C, Li J, Chi C, Wu J. Synthesis of functionalized tetracene dicarboxylic imides. Tetrahedron Lett 2010;51:6313–15.
[43] Roberson LB, Kowalik J, Tolbert L, et al. Pentacene disproportionation during sublimation for field-effect transistors. J Am Chem Soc 2005;127:3069–75.
[44] Liu J, Ravat P, Wagner M, Baumgarten M, Feng X, Müllen K. Tetrabenzo[a,f,j,o]perylene: a polycyclic aromatic hydrocarbon with an open-shell singlet biradical ground state. Angew Chem Int Ed 2015. DOI: 10.1002/anie.201502657.
[45] Zhang L, Fonari A, Liu Y, et al. Bistetracene: an air-stable, high-mobility organic semiconductor with extended conjugation. J Am Chem Soc 2014;136:9248–51.
[46] Zhang X, Li J, Qu H, Chi C, Wu J. Fused bispentacenequinone and its unexpected Michael addition. Org Lett 2010;12:3946–3949.
[47] Zöphel L, Berger R, Gao P, et al. Toward the *peri*-pentacene framework. Chem – Eur J 2013;19:17821–6.
[48] Matsumoto A, Suzuki M, Kuzuhara D, Hayashi H, Aratani N, Yamada H. Tetrabenzoper-ipentacene: stable five-electron donating ability and a discrete triple-layered β-graphite form in the solid state. Angew Chem Int Ed 2015;127:8293–6.
[49] Matsumoto A, Suzuki M, Kuzuhara D, et al. A kinetically protected pyrene: molecular design, bright blue emission in the crystalline state and aromaticity relocation in its dicationic species. Chem Commun 2014;50:10956–8.
[50] Enoki T, Kobayashi Y, Fukui K. Electronic structures of graphene edges and nanographene. Int Rev Phys Chem 2007;26:609–45.
[51] Haddon RC. Design of organic metals and superconductors. Nature 1975;256:394–6.
[52] Goto K, Kubo T, Yamamoto K, et al. A stable neutral hydrocarbon radical: synthesis, crystal structure, and physical properties of 2,5,8-tri-*tert*-butylphenalenyl. J Am Chem Soc 1999;121:1619–20.
[53] Mota F, Miller JS, Novoa JJ. Comparative analysis of the multicenter, long bond in [TCNE]- and phenalenyl radical dimers: a unified description of multicenter, long bonds. J Am Chem Soc 2009;131:7699–707.
[54] Suzuki S, Morita Y, Fukui K, et al. Aromaticity on the pancake-bonded dimer of neutral phenalenyl radical as studied by MS and NMR spectroscopies and NICS Analysis. J Am Chem Soc 2006;128:2530–1.
[55] Haddon RC, Chichester SV, Stein SM, Marshall JH, Mujsce AM. Perchloro-*7H*-cycloprop[*a*]acenaphthylene and the perchlorophenalenyl system. J Org Chem 1987;52:711–12.
[56] Koutentis PA, Chen Y, Cao Y, et al. Perchlorophenalenyl radical. J Am Chem Soc 2001;123:3864–3871.
[57] Uchida K, Hirao Y, Kurata H, Kubo T, Hatano S, Inoue K. Dual association modes of the 2,5,8-tris(pentafluorophenyl)phenalenyl radical. Chem Asian J 2014;9:1823–9.
[58] Kubo T, Katada Y, Shimizu A, et al. Synthesis, Crystal Structure, and Physical Properties of Sterically Unprotected Hydrocarbon Radicals. J Am Chem Soc 2011;133:14240–3.
[59] (a) Kubo T, Yamamoto K, Nakasuji K, Takui T, Murata I. Hexa-tert-butyltribenzodecacyclenyl: a six-stage amphoteric redox system. Angew Chem Int Ed 1996;35:439. (b) Kubo T, Yamamoto K, Nakasuji K, Takui T, Murata I. 4,7,11,14,18,21-hexa-t-butyltribenzodecacyclenyl radical: a six-stage amphoteric redox system. Bull Chem Soc Jpn 2001;74:1999.
[60] Haddon RC, Wudl F, Kaplan ML, Marshall JH, Cais RE, Bramwell FB. 1,9- Dithiophenalenyl system. J Am Chem Soc 1978;100:7629–33
[61] Beer L, Mandal SK, Reed RW, et al. The first electronically stabilized phenalenyl radical: effect of substituents on solution chemistry and solid-state structure. Cryst Growth Des 2007;7: 802–9.

[62] Beer L, Reed RW, Robertson CM, Oakley RT, Tham FS, Haddon RC. Tetrathiophenalenyl radical and its disulfide-bridged dimer. Org Lett 2008;10:3121–3123.

[63] Morita Y, Aoki T, Fukui K, et al. A new trend in phenalenyl chemistry: a persistent neutral radical, 2,5,8-tri-*tert*-butyl-1,3-diazaphenalenyl, and the excited triplet state of the gable syn-dimer in the crystal of column motif. Angew Chem Int Ed 2002;41:1973–6.

[64] Ueda A, Wasa H, Suzuki S, et al. Chiral stable phenalenyl radical: synthesis, electronic-spin structure, and optical properties of [4]helicene-structured diazaphenalenyl. Angew Chem Int Ed 2012;51:6691–5.

[65] Nakasuji K, Yoshida K, Murata I. Design and synthesis of a highly amphoteric condensed hydrocarbon with the highest reduction potential. Pentaleno[1,2,3-cd:4,5,6-c'd']diphenalene. J Am Chem Soc 1983;105:5136–7.

[66] Ohashi K, Kubo T, Masui T et al. 4, 8, 12,16-Tetra-tert-butyl-s-indaceno[1,2,3-cd:5,6,7-c'd'] diphenalene: a four-stage amphoteric redox system. J Am Chem Soc 1998;120:2018–27.

[67] Kubo T, Shimizu A, Sakamoto M, et al. Synthesis, intermolecular interaction, and semiconductive behavior of a delocalized singlet biradical hydrocarbon. Angew Chem Int Ed 2005;44:6564–8.

[68] Kubo T, Shimizu A, Uruichi M, et al. Singlet biradical character of phenalenyl-based kekule hydrocarbon with naphthoquinoid structure. Org Lett 2007;9:81–4.

[69] Shimizu A, Hirao Y, Matsumoto K et al. Aromaticity and π-bond covalency: prominent intermolecular covalent bonding interaction of a Kekule hydrocarbon with very significant singlet biradical character. Chem Commun 2012;48:5629.

[70] Kubo T, Sakamoto M, Akabane M, et al. Four-stage amphoteric redox properties and biradicaloid character of tetra-tert-butyldicyclopenta[b;d]thieno[1,2,3-cd;5,6,7-c'd']diphenalene. Angew Chem Int Ed 2004;43:6474–9.

[71] Kubo T, Yamamoto K, Nakasuji K, Takui T. Tetra-tert-butyl-as-indaceno[1,2,3-cd:6,7,8-c'd'] diphenalene: a four-stage amphoteric redox system. Tetrahedron Lett 2001;42:7997–8001.

[72] Kubo T, Sakamoto M, Nakasuji K. Biradicaloid character of phenalenyl-based aromatic compounds with a small HOMO-LUMO gap. Polyhedron 2005;24:2522–7.

[73] Clar E, Lang KF, Schulz-Kiesow H, Aromatic hydrocarbons. LXX. Zethrene (1,12;6,7-dibenzotetracene). Chem Ber 1955;88:1520–7.

[74] Nakano M, Kishi R, Takebe A, et al. Second hyperpolarizability of zethrenes. Comput Lett 2007;3:333–8.

[75] Umeda R, Hibi D, Miki K, Tobe Y. Tetradehydrodinaphtho[10]annulene: a hitherto unknown dehydroannulene and a viable precursor to stable zethrene derivatives. Org Lett 2009;11:4104–6.

[76] Wu TC, Chen CH, Hibi D, Shimizu A, Tobe Y, Wu YT. Synthesis, structure, and photophysical properties of dibenzo[de,mn]naphthacenes. Angew Chem Int Ed 2010;49:7059–62.

[77] Sun Z, Huang KW, Wu J. Soluble and stable zethrenebis(dicarboximide) and Its quinone. Org Lett 2010;12:4690–3.

[78] Shan L, Liang ZX, Xu XM, Tang Q, Miao Q. Revisiting zethrene: synthesis, reactivity and semiconductor properties. Chem Sci 2013;4:3294–7.

[79] Sun Z, Zheng B, Hu P, Huang KW, Wu J. Highly twisted 1,2:8,9-Dibenzozethrenes: synthesis, ground state, and physical properties. ChemPlusChem 2014;79:1549–53.

[80] Sun Z, Huang KW, Wu J. Soluble and stable heptazethrenebis(dicarboximide) with a singlet open-shell ground state. J Am Chem Soc 2011;133:11896–9.

[81] Das S, Lee S, Son M, et al. *para*-Quinodimethane-bridged perylene dimers and pericondensed quaterrylenes: the effect of the fusion mode on the ground states and physical properties. Chem -Eur J. 2014;20:11410–20.

[82] Li Y, Heng WK, Lee BS, et al. Kinetically blocked stable heptazethrene and octazethrene: closed-shell or open-shell in the ground state? J Am Chem Soc 2012;134:14913–22.

[83] Sun Z, Lee S, Park K, et al. Dibenzoheptazethrene isomers with different biradical characters: an exercise of Clar's aromatic sextet rule in singlet biradicaloids. J Am Chem Soc 2013;135:18229–36.
[84] Nobusue S, Miyoshi H, Shimizu A, et al. Tetracyclopenta[*def,jkl,pqr,vwx*]tetraphenylene: a potential tetraradicaloid hydrocarbon. Angew Chem Int Ed 2015;54:2090–4.
[85] Tian Y, Uchida K, Kurata H, Hirao Y, Nishiuchi T, Kubo T. Design and synthesis of new stable fluorenyl-based radicals. J Am Chem Soc 2014;136:12784–93.
[86] Lambert C. Towards polycyclic aromatic hydrocarbons with a singlet open-shell ground state. Angew Chem Int Ed 2011;50:1756.
[87] Nakano M, Champagne B. Theoretical design of open-shell singlet molecular systems for nonlinear optics. J Phy Chem Lett 2015;6:3236–56.
[88] (a) Smith MB, Michl J. Singlet Fission. Chem Rev 2010, 110, 6891–6936. (b) Minami T, Nakano M. Diradical character view of singlet fission. J Phys Chem Lett 2012;3:145–50.
[89] Peng Q, Obolda A, Zhang M, Li F. Organic light-emitting diodes using a neutral π radical as emitter: the emission from a doublet. Angew Chem Int Ed 2015;54:7091–5.
[90] Morita Y, Nishida S, Murata T, et al. Organic tailored batteries materials using stable open-shell molecules with degenerate frontier orbitals. Nat Mater 2011;10:947–51.
[91] (a) Chikamatsu M, Mikami T, Chisaka J, Yoshida Y, Azumi R, Yase K. Ambipolar organic field-effect transistors based on a low bandgap semiconductor with balanced hole and electron mobilities. Appl Phys Lett 2007;91:043506/1–/3. (b) Koike H, Chikamatsu M, Azumi R et al. Stable Delocalized Singlet Biradical Hydrocarbon for Organic Field-Effect Transistors. Adv Funct Mater 2016;26:277–83.
[92] Koike H, Kubo T, Uchida K et al. Direct observation of energy band development in a one-dimensional biradical molecular chain by ultraviolet photoemission spectroscopy. Appl Phys Lett 2013;102:134103/1–/4.
[93] Kim WY, Kim KS. Prediction of very large values of magnetoresistance in a graphene nanoribbon device. Nat Nanotechnol 2008;3:408–12.

Bjoern Schulte, Stephen Schrettl and Holger Frauenrath

10 Synthesis and Use of Reactive Molecular Precursors for the Preparation of Carbon Nanomaterials

Abstract: The use of reactive molecular carbon precursors is required if the preparation of carbon nanostructures and nanomaterials is to be achieved under conditions that are sufficiently benign to control their nanoscopic morphology and tailor their chemical functionalization. Recently, oligoyne precursors have been explored for this purpose, as they are sufficiently stable to be available in tangible quantities but readily rearrange in reactions that yield other forms of carbon. In this chapter, we briefly discuss available synthetic routes toward higher oligoynes that mostly rely on transition metal-mediated coupling reactions. Thereafter, a comprehensive overview of the use of oligoyne derivatives as precursors for carbon nanostructures and nanomaterials is given. While the non-templated conversion of simple oligoynes into carbonaceous matter exemplifies their potential as metastable carbon precursors, the more recent attempts to use functionalized oligoynes in host–guest complexes, self-assembled aggregates, thin films, colloids or other types of supramolecular structures have paved the way toward a new generation of carbon nanomaterials with predictable nanoscopic morphology and chemical functionalization.

10.1 Introduction

Fullerenes, carbon nanotubes, graphene sheets, porous carbons, carbon nanospheres and carbon nanofibers are just a few representative examples of the many different carbon nanostructures and nanomaterials that have become available [1–5]. The diverse range of accessible properties in combination with the high natural abundance of carbon, low specific weight as well as an exceptional chemical and thermal robustness distinguish carbon nanostructures as attractive components in organic electronic devices, polymer composites or energy storage systems [2–4, 6, 7]. The respective mechanical properties of the different nanostructures as well as their electrical or thermal conductivities are generally the result of the connectivity and hybridization of the constituting carbon atoms. The two main naturally occurring carbon allotropes, diamond and graphite, may serve as a basis to illustrate the broad range of accessible properties on the macroscopic scale. Thus, the carbon atoms in diamond are sp^3-hybridized, and their tetrahedral coordination geometry results in a three-dimensional network of covalent bonds. The resulting bulk material is transparent, extremely hard and electrically insulating, but it displays a high thermal conductivity [2, 8–10]. Graphite, on the other hand, is formed from stacks of carbon sheets that comprise a honeycomb lattice of sp^2-hybridized carbon atoms. Unlike diamond, the comparably soft graphite has a highly reflecting black appearance and is an electrical (semi)conductor; its properties drastically change with the direction of measurement, reflecting its anisotropic structure [2, 8, 9]. A corresponding variety of properties is typically observed on the nanoscale. However, it remains difficult to prepare pure, well-defined carbon nanomaterials in tangible quantities, since they feature dimensions that are only accessible through demanding, multistep organic syntheses ("bottom-up" approaches) or through delicate physical processing ("top-down" approaches) [11–14]. Typical physical processes used in top-down approaches are mechanical exfoliation, arc discharge or laser ablation of adequate bulk precursors. Bottom-up approaches are generally based on the wet-chemical synthesis of precursor molecules and their subsequent conversion into carbon nanomaterials at elevated temperatures. In this context, hard or soft templates can be used to control the nanoscopic morphology and microstructure of the resulting carbon materials [15–17]. It is important to point out that all chemical functionality is usually lost during the final carbonization of the molecular precursors at elevated temperatures. Thus, a post-synthetic functionalization of the carbon nanostructures is required when a defined chemical functionalization is desired. However, this typically results in an inhomogeneous distribution of the chemical functional groups and accompanying structural defects in the carbon nanostructures [18, 19].

The only way to circumvent this issue is the use of alternative molecular precursors that are sufficiently reactive to undergo carbonization under benign conditions, preferably close to room temperature. This would allow for the preservation of (most) chemical functional groups introduced with the precursors throughout the carbonization process. Along these lines, functional molecules containing segments rich in *sp*-hybridized carbons are an attractive choice. The inherent reactivity of such segments and the absence of other elements that would have to be eliminated in the course of carbonization facilitate a conversion of the *sp*-carbon-rich precursors into carbon nanostructures with predominantly sp^2-hybridized carbons under mild conditions. Moreover, the chemical functional groups that are introduced with the precursors and remain covalently linked throughout the conversion necessarily lead to sp^3-hybridized carbon atoms in the produced carbon nanostructures. The deliberate introduction of such sp^3 carbon "defects" allows for a control over the chemical functionality as well as its distribution in the obtained carbon materials and, consequently, the properties and processability.

In this context, the present chapter aims to give a concise introduction into the chemistry of functional molecules comprising chains of *sp*-hybridized carbon atoms and will then provide a comprehensive overview over the different strategies for their conversion into defined carbon nanostructures and nanomaterials.

10.2 Reactivity of Oligoynes

The extended chains of carbon–carbon triple bonds in conjugated oligoynes are thermodynamically unstable and exhibit an inherent reactivity toward a rearrangement to other, more stable carbon allotropes. Accordingly, all different series of oligoyne derivatives R–(C≡C)$_n$–R display a drastically decreasing stability with an increasing number n of triple bonds [20]. This is readily illustrated on the basis of the parent series H–(C≡C)$_n$–H of unsubstituted compounds. Thus, butadiyne may be distilled at low temperatures to furnish the pure compound but begins to polymerize above 0°C [21]. Hexatriyne slowly polymerizes even at temperatures below –50°C, and samples that were exposed to air reportedly exploded spontaneously leading to the formation of "carbonaceous matter" [22]. Similar accounts that describe the rapid decomposition of compounds containing conjugated acetylenes under formation of insoluble black material are abundant, especially in the literature concerning the preparation of higher oligoynes [23, 24]. Whereas the carbon materials produced by the (in most cases unwanted) decomposition of reactive carbon-rich molecules were rarely investigated in more detail, Kroto et al. [25] discussed whether oligoynes were the "small carbon clusters" involved in the formation of fullerenes by the arc discharge method. This hypothesis found strong support in reports by Grösser and Hirsch [26], who isolated a homologous series of dicyanide end-capped oligoynes NC–(C≡C)$_n$–CN as the main product when the vaporization of graphite was carried out in the presence of cyanogen CN_2. Moreover, Cataldo [27] reported the formation of the unsubstituted derivatives H–(C≡C)$_n$–H when an arc discharge experiment was performed with graphite electrodes submerged in organic solvents (Figure 10.1).

Despite this apparent role of carbon-rich molecules that feature extended chains of carbon–carbon triple bonds in the formation of common carbon

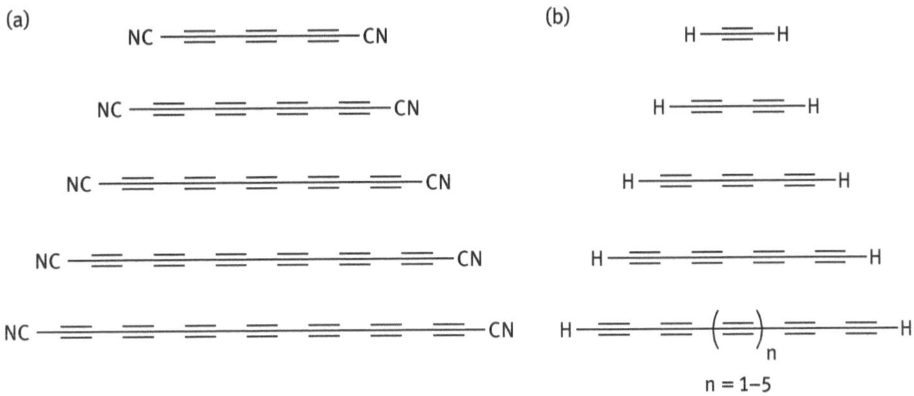

Figure 10.1: (a) Oligoynes produced by vaporization of graphite in the presence of cyanogen CN_2 and (b) with graphite electrodes submerged in organic solvents such as methanol [26, 27].

nanostructures, the deliberate utilization of oligoynes as molecular precursors for carbon-rich materials has rarely been addressed to date. Most investigations have rather aimed at the preparation of stable oligoyne moieties as model compounds toward the linear, fully *sp*-hybridized carbon allotrope carbyne $(C\equiv C)_\infty$ [28] or, alternatively, at the use of oligoynes as molecular wires for an efficient charge transport [29–31]. To these ends, thermodynamically unstable oligoyne derivatives have been kinetically stabilized by terminal substituents R with a high steric demand. The latter act in the sense of protecting groups that prevent an intermolecular cross-linking of the triple bonds, and a large number of oligoyne derivatives have been prepared employing this general strategy [20, 32, 33]. However, it needs to be pointed out that even such sterically shielded oligoynes beyond the decaynes become increasingly unstable as neat compounds and easily react in a manner that converts the molecules into carbonaceous matter. Besides the covalent introduction of protective groups, more recent supramolecular approaches relied on the use of rotaxanes in which an oligoyne segment as the guest was stabilized using a macrocyclic host [34–37].

10.3 Synthetic Approaches toward Oligoynes

Despite the difficulties associated with the stability of oligoynes, several approaches have been developed for their synthesis. Most of these strategies rely on copper-mediated coupling reactions between *sp*-hybridized carbons of terminal alkynes. Prominent examples are coupling reactions such as the Glaser [38], Hay [39, 40], Eglinton–Galbraith [41, 42], Sonogashira [43, 44] or Cadiot–Chodkiewicz [45] reactions (Figure 10.2) [20, 32, 46, 47]. Furthermore, procedures based on the Negishi coupling reaction have been developed that also allow to couple two different alkyne moieties to each other [48–51]. Besides these transition-metal-mediated approaches, the Fritsch–Buttenberg–Wiechell (FBW) rearrangement of 1,1-dibromovinylidene derivatives [52–54] has also been very successfully exploited for the preparation of higher oligoynes [55].

Among the chemical methods, the Glaser coupling represents the oldest coupling method for terminal acetylenes. Although the mechanism of the Glaser reaction is still under debate with regard to the oxidation state of the catalytically active copper species, it is clear that the presence of a copper salt, molecular oxygen and a base are crucial for the reaction to proceed [56]. Often encountered issues with the solubility of copper salts in common organic solvents were resolved by the use of the copper(I) chloride tetramethylethylenediamine (TMEDA) complex according to the Hay coupling conditions [57]. Closely related to Glaser and Hay reactions is a procedure first described by Eglinton and Galbraith [41, 42]. Instead of catalytic amounts of a copper(I) salt, stoichiometric amounts of a copper(II) salt are used for the coupling of the acetylene derivatives, obviating the need for oxygen as the oxidizing agent.

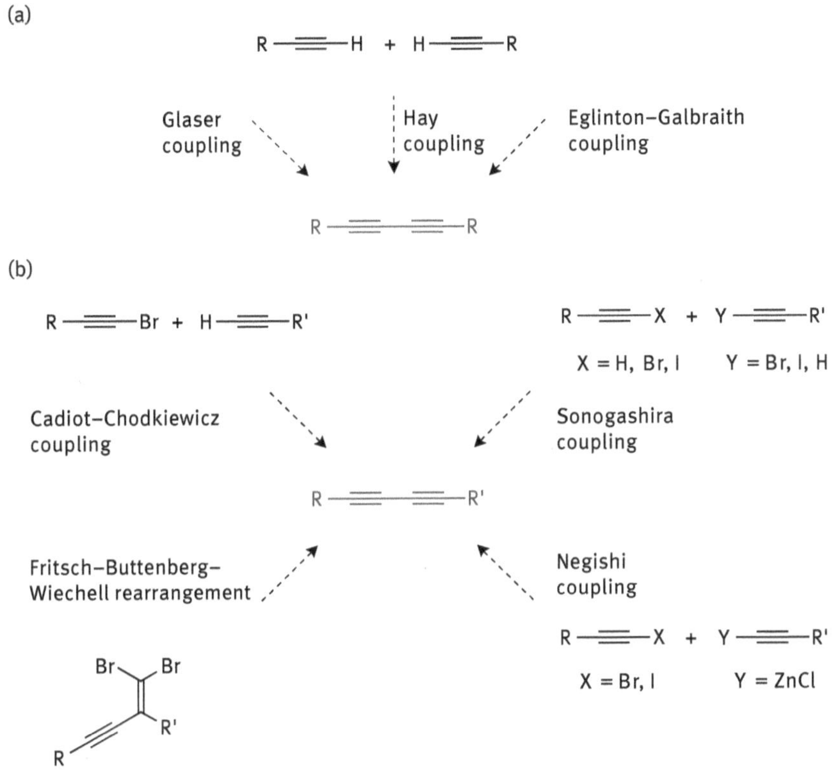

Figure 10.2: Overview of chemical methods that allow for the preparation of oligoynes with extended *sp*-hybridized segments. (a) Synthesis of symmetric oligoynes by Glaser coupling, Hay coupling or Eglinton–Galbraith coupling and (b) synthesis of asymmetric oligoynes by Cadiot–Chodkiewicz, Sonogashira, Negishi coupling or FBW rearrangement.

Whereas symmetric oligoynes can be obtained with all of the preceding procedures, the Cadiot–Chodkiewicz coupling allows for the synthesis of asymmetric oligoynes [58]. Here, a copper(I) halogenide, a base, as well as a halogenated and a nonhalogenated acetylene species are used as reactants. However, in spite of the vast number of investigations describing variations of the reaction conditions for the simple coupling of a terminal acetylene with a haloacetylene to obtain unsymmetrical diacetylenes [59, 60], the practical use of the Cadiot–Chodkiewicz coupling has remained limited, in particular, for the synthesis of higher oligoyne homologues. One possible reason is the seemingly inevitable formation of the symmetric side products resulting from homocoupling of the acetylenes and self-coupling of the haloacetylenes that often result in statistic mixtures and render isolation of the pure product difficult [61, 62]. This selectivity problem is mitigated using the established palladium-mediated cross-coupling reactions, such as Sonogashira or Negishi reactions, in which a haloacetylene is coupled to an in situ generated copper or zinc acetylide, respectively, using a palladium(0) complex as the catalytically active species [44, 63]. Despite a limited number of examples to date, it seems as if the extent of homocoupling side reaction products in Negishi couplings is smaller than under Sonogashira conditions [64].

In contrast to all of the previously described metal-mediated transformations that serve to directly build up the oligoyne segment, the FBW route first seeks to synthesize compounds containing one or more 1,1-dibromovinylidene groups that are then subjected to a base-induced rearrangement to yield the final oligoynes [65, 66].

Among the many notable examples of molecules prepared following the above-mentioned wet-chemical procedures are derivatives such as the methyl-substituted Me–$(C{\equiv}C)_6$–Me [67], pentafluorophenyl-substituted Pfp–$(C{\equiv}C)_8$–Pfp [68], *tert*-butyl-substituted *t*-Bu–$(C{\equiv}C)_{10}$–*t*-Bu [69, 70] or triisopropylsilyl-substituted derivatives TIPS–$(C{\equiv}C)_{10}$–TIPS [71]. Hirsch and coworkers [72] investigated the use of Frechet-type dendrons of the third generation to maximize the steric shielding and prepared G_3–$(C{\equiv}C)_{10}$–G_3. Zheng and Gladysz [73] explored the stabilizing effect of organometallic complexes at the termini with derivatives such as the platinum analogues L_nPt–$(C{\equiv}C)_{14}$–PtL_n (Figure 10.3).

Despite the increasing challenges encountered with the stability of higher oligoynes, Chalifoux and Tykwinski [76] were able to even prepare a derivative with 44 *sp*-hybridized carbon atoms (or 22 acetylene units). Their overall synthetic approach relied on the use of the very bulky tris(3,5-di-*tert*-butylphenyl)methyl ("supertrityl",

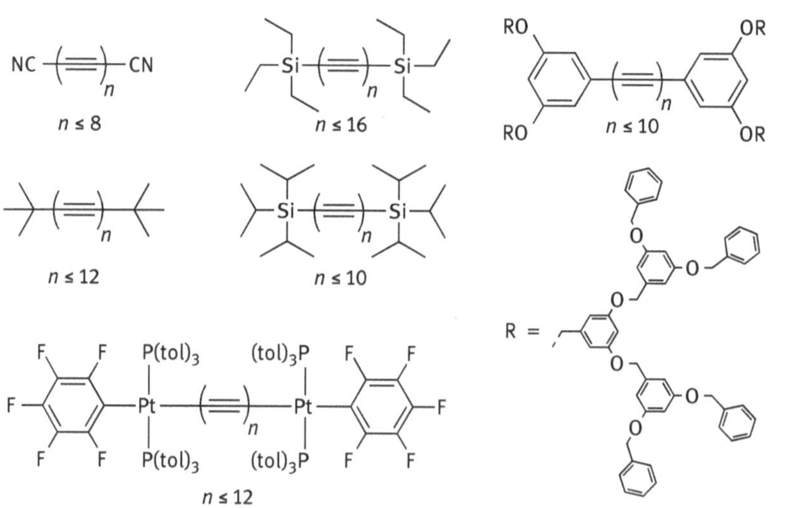

Figure 10.3: Different series of oligoynes have been synthesized so far as model compounds for the *sp*-hybridized carbon allotrope carbyne. Notable examples comprise derivatives such as the nitrile-capped series [74], *tert*-butyl-substituted derivatives [69, 70] or silylated compounds [71, 75].

Tr*) end-capping groups. Using a variant of the reliable Eglinton–Galbraith homocoupling protocol but adding a silyl-protected tetrayne in excess in the presence of copper acetate and 2,6-lutidine as a base, they obtained an asymmetric Tr*-protected heptayne (Figure 10.4). After desilylation, they elongated this heptayne applying the same reaction conditions. The obtained undecayne was eventually desilylated in situ in the presence of cesium fluoride, and the symmetric Tr*-protected compound with 22 acetylenes was obtained in a dimerization reaction in the presence of $Cu(OAc)_2 \cdot H_2O$ and 2,6-lutidine. The use of 2,6-lutidine as a base was essential, as attempts with other, more nucleophilic bases such as pyridine failed. The obtained products exclusively consisted of acetylenic carbons, apart from the terminal Tr* groups that were crucial to secure the stability of the final oligoyne derivatives.

These efforts toward the preparation of oligoynes of ever-increasing length demonstrate that the chemical synthesis of such reactive derivatives is feasible and delivers quantities that are sufficient to use oligoynes as precursors for carbon materials. At the same time, experiments with the isolated compounds served to further illustrate the reactive nature of such compounds as, for example, the differential scanning calorimetry traces of the triisopropyl-terminated derivatives TIPS–$(C{\equiv}C)_n$–TIPS revealed reversible melting for compounds with two to six acetylene units, and only the diyne did not exhibit an irreversible exothermic transition at higher temperatures [71].

Finally, physical methods have been used to prepare oligoynes, such as arc discharge at graphite electrodes in organic solvents, solution-spray flash vacuum pyrolysis of 3,4-dialkynyl-3-cyclobutene-1,2-diones, electrochemical reduction of poly(tetrafluoroethylene) or laser ablation of graphite [27, 77–80]. In an interesting

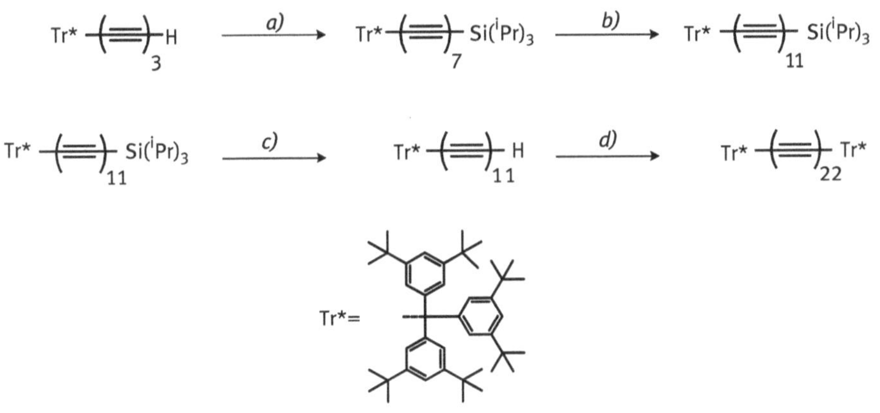

Figure 10.4: Reagents and conditions: (a) TMS–$(C{\equiv}C)_4$–TIPS, $Cu(OAc)_2 \bullet H_2O$, K_2CO_3, 2,6-lutidine/MeOH/CH_2Cl_2 (1:1:1); (b) (i) CsF in tetrahydrofuran/H2O (5:1), room temperature (ii) TMS–$(C{\equiv}C)_4$–TIPS, $Cu(OAc)_2 \bullet H_2O$, K_2CO_3, 2,6-lutidine/MeOH/CH_2Cl_2; (c) CsF in tetrahydrofuran/H_2O (5:1), room temperature; and (d) $Cu(OAc)_2 \bullet H_2O$ (excess), 2,6-lutidine, CH_2Cl_2. The preparation of an oligoyne derivative with 44 *sp*-hybridized carbon atoms is based on a modified Eglinton–Galbraith reaction, employing tris(3,5-di-*t*-butylphenyl) methyl moieties as end-capping groups [76].

approach, Shinohara and coworkers obtained a mixture of oligoynes H–$(C{\equiv}C)_n$–H of different lengths (n = 4–6) through laser ablation of graphite microparticles from which the pure decayne was isolated by high-performance liquid chromatography. Subsequently, they used double-walled carbon nanotubes (DWCNT) as hosts for the one-dimensional assembly of the decaynes [80]. The inclusion compounds were formed upon dipping of open-ended DWCNTs into a methanol solution containing decayne. Molecules that adhered to the outer surface of the DWCNT were removed by rinsing and drying of the inclusion complexes at 80°C. Thereafter, a thermal annealing step at temperatures of up to 1,000°C was carried out that initiated a "fusion reaction" between the decayne molecules leading to the formation of *sp*-hybridized carbon chains with more than 50 carbon atoms. While this constitutes an intriguing approach, the possibility to exploit the inherent reactivity of the decaynes to carbonize them under mild conditions was left unexplored. In general, the presented physical methods cannot be used to prepare oligoynes that can be isolated in sufficient quantities to subsequently use them as precursors for the tailored preparation of carbon nanostructures, in contrast to the wet-chemical approaches described earlier.

10.4 Carbonization of Non-Preorganized Oligoynes

In experiments that were originally aimed at the preparation of the elusive carbon allotrope carbyne $(C{\equiv}C)_\infty$, Lagow and coworkers [28] reported the unintended use of oligoynes as molecular precursors for carbonaceous materials. Thus, when investigating the stability of the *tert*-butyl end-capped tetrayne derivative ^tBu–$(C{\equiv}C)_4$–^tBu, a minor amount of the sample (2 %) underwent graphitization when exposed to an increased pressure of 60 kbar over extended periods of time. In contrast to these experiments at high pressures, Hlavaty and coworkers [81, 82] explored the reactivity of different hexatriyne derivatives that lacked sterically demanding end-capping groups in organic solutions at or below room temperature. Thus, solutions of the unsubstituted hexatriyne as well as the 1-iodohexatriyne derivative started to rapidly react at 0°C and room temperature, respectively. Interestingly, these nondirected, low-temperature reactions of the molecular precursors furnished a mix of amorphous, carbonaceous materials that, upon closer inspection, featured a few instances of multi-walled carbon nanotubes with diameters of 10–20 nm, lengths of up to 200 nm and attached "onion-like" carbon nanoparticles [81]. The investigations with the corresponding dilithium and dipotassium salts of the parent hexatriyne showed that the latter were even more reactive as a carbonization already occurred in the course of their preparation. The authors showed that the mixed carbonaceous material obtained after hydrolysis contained few examples of multi-walled carbon nanotubes [82]. An ill-defined carbonaceous material with a graphitic microstructure was also obtained when a TIPS-hexatriyne derivative was deprotected in situ [83]. While these studies represent the first accounts of the deliberate use of oligoynes as reactive carbon precursors, it should be noted that the interpretation of some of the analytical data was ambiguous, the representative nature of the observed carbon nanostructures remained an open question and no explanation for their selective formation was put forward.

In contrast to this spontaneous solution-phase carbonization of simple molecular precursors, Zhao and coworkers [84, 85] investigated the solid-state polymerization of different fullerene-substituted tetrayne derivatives. The precursor molecules were obtained through oxidative coupling of iodoaryl acetylenes with trimethylsilylacetylene under Hay coupling conditions. The iodoarene was thereafter subjected to a Sonogashira coupling reaction with triisopropylsilylacetylene. Subsequently, selective removal of the trimethylsilyl group followed by an oxidative homocoupling under Hay conditions yielded a symmetric tetrayne. Desilylation of the TIPS-substituted tetrayne allowed for a functionalization with C_{60} fullerene moieties in an ethynylation reaction using an excess of C_{60} and lithium hexamethyldisilazide (Figure 10.5).

Thin films of the molecular precursors were prepared by drop-casting or spin-coating of toluene solutions on a mica surface, and their reaction was thereafter induced by a mild thermal treatment at a temperature of 160°C. According to the investigations by means of UV/Vis spectroscopy, the tetrayne moieties were consumed in the course

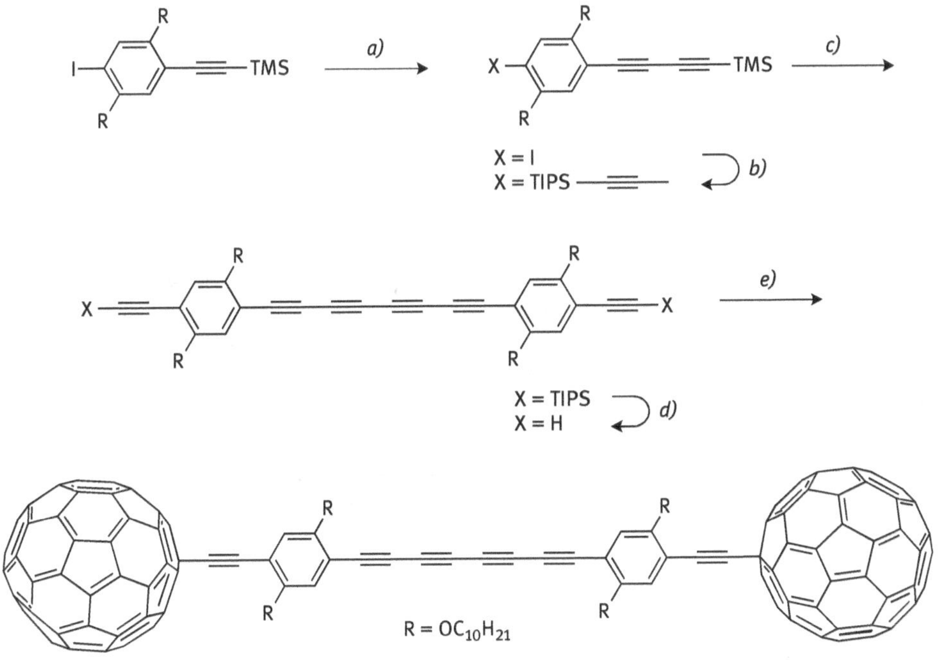

Figure 10.5: Reagents and conditions: (a) (i) K_2CO_3, MeOH/THF, (ii) TMS–C≡C–H, CuCl/TMEDA, O_2, acetone; (b) TIPS–C≡C–H, $PdCl_2(PPh_3)_2$, CuI, Et_3N, THF, 60°C; (c) (i) K_2CO_3, MeOH/THF, (ii) TMS–C≡C–H, CuCl/TMEDA, O_2, acetone, (d) TBAF; and (e) (i) C_{60}, LHMDS, THF and (ii) TFA. Synthetic scheme illustrating route toward fullerene-substituted tetrayne derivatives [84].

of the thermal treatment, as their characteristic absorption pattern disappeared, and the formation of carbon materials with different morphologies was observed by atomic force microscopy (AFM). Notably, an amorphous film of one of the investigated tetrayne derivatives gave rise to homogeneously distributed carbon nanospheres, with a uniform diameter below 20 nm after the thermal treatment (Figure 10.6) [84]. Although not discussed by the authors, this formation of a well-defined pattern of uniform carbon nanospheres can probably be rationalized by a thermodynamically driven dewetting of the precursor thin film from the substrate under the carbonization conditions.

Further approaches have been reported, wherein carbon materials were prepared from acetylene-containing molecular precursors by pyrolytic treatment at elevated temperatures, and some of these have been reviewed recently [86]. However, such high-temperature methods do not exploit the potential of oligoyne derivatives as molecular precursors for the preparation of carbon materials under mild conditions, which has certainly been illustrated by the preceding examples. In all of the discussed cases, the conversion of the inherently reactive molecular precursors lacked a defined means of supramolecular control that guided the carbonization process toward a predetermined morphology. With the exception of the conversion of one of the tetrayne derivatives reported by Zhao and coworkers, these carbonization processes therefore furnished product mixtures with minor amounts of distinct carbon nanostructures. Obviously, a preorganization of the molecular precursors prior to and potentially throughout the carbonization is necessary to allow for the preparation of defined carbon nanostructures with control over their nanoscopic morphology.

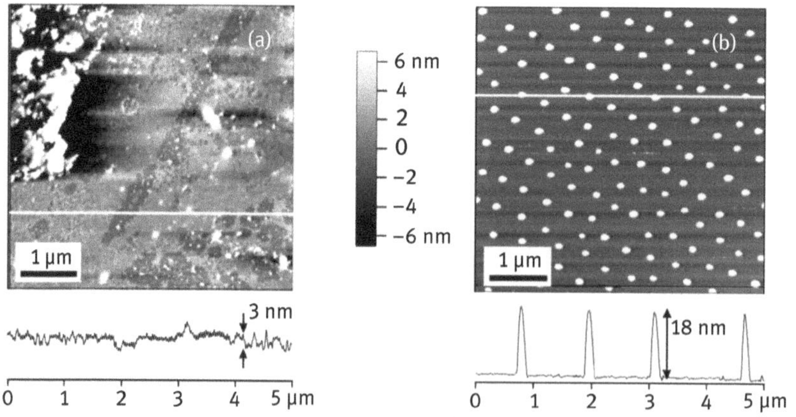

Figure 10.6: The carbonization of a fullerene-substituted tetrayne furnished homogeneously distributed carbon nanospheres of uniform size. (a) AFM image of the amorphous film before carbonization and (b) after thermal annealing at 160°C for 1 h [84].

10.5 Topochemical Polymerization of Oligoynes

The lack of control over the morphology of the obtained carbonaceous materials prevalent in the preceding examples of undirected conversions of oligoynes may, for example, be overcome by conducting the carbonization in single-crystalline specimen, in analogy to the well-understood topochemical polymerization of diacetylene derivatives [87–89]. In the context of an extensive investigation of the topochemical conversion of various linear and cyclic acetylene derivatives, Baughman and Yee [90] were the first to report an attempt to topochemically polymerize a octatetrayne bis(carbamate). However, a detailed analysis of the product structure was impeded by the loss of crystalline order in the course of the polymerization, and only the comparison with reactions of other acetylene derivatives led to the conclusion that the polymerization most likely proceeded in the sense of a 1,4-polyaddition with unclear regioselectivity. Nakanishi and coworkers [91] investigated the reactivity of tetrayne, pentayne and hexayne derivatives with different terminal substituents in terms of their solid-state topochemical polymerization. The polymerization of dithiophene- and diquinoline-substituted tetraynes was achieved by thermal annealing of single-crystalline specimen, and the recorded UV/Vis spectra displayed a broad absorption that the authors claimed to be consistent with the product of a controlled 1,2-polymerization [92]. While this notion has often been repeated in the pertinent literature, it should be noted that it had not been substantiated with analytical data and appears to be highly unlikely on the basis of packing and steric considerations. The same authors reported that the reactivity of an investigated bis(tetradecyl)-substituted hexayne was even higher, and its single crystals readily polymerized at ambient temperature in the dark while consecutively changing the color from green to black [93]. In the course of the reaction, the infrared (IR) bands corresponding to the acetylene moieties disappeared, and UV/Vis/near-IR spectroscopy showed several peaks with a maximum absorption at 776 nm. Monitoring of the UV absorption over the course of a couple of days showed that the absorption band increasingly broadened. The authors interpreted the corresponding, ill-defined solid-state ^{13}C nuclear magnetic resonance spectra in terms of a slow conversion of an initially formed tetrayne-substituted poly(diacetylene) into a butadiyne-bridged poly(diacetylene) ladder polymer. Moreover, the authors later speculated [91] that a subsequent cyclization would result in graphitic ribbons. However, the actual experimental evidence appears to be insufficient to support this conclusion and, again, the proposed pathways are not plausible on the basis of packing considerations.

Without doubt, topochemical reactions offer the ultimate means of control over the preorganization of the molecular precursors. However, the strict geometric requirements of topochemical reactions drastically limit the choice of side groups. Whether an acetylene-substituted poly(diacetylene) furnished by a topochemical polymerization of higher oligoyne derivatives can subsequently actually be carbonized awaits further exploration. Moreover, if one takes into consideration that the resulting graphene ribbons have a significantly higher carbon atom density compared to a corresponding one-dimensional stack of oligoynes with the same number of carbon atoms per side group, the carbonization must inevitably be accompanied by a significant lateral contraction of such an arrangement. It seems unlikely that the concomitant rearrangements on the molecular level can proceed under preservation of the crystal structure. Moreover, such carbonizations in single-crystalline specimen would only yield insoluble material with a limited variation of accessible morphologies.

Despite these constraints, notable examples have been reported, where the topochemical polymerization of diacetylene derivatives has been employed for the preparation of reactive, polymeric precursors for nanostructured carbons [86, 94]. In an elegant approach that relied on a supramolecular host–guest strategy, Goroff and coworkers [95] prepared a highly ordered poly(diiododiacetylene) by a spontaneous polymerization of diiodobutadiyne. Thus, the crystallization of diiodobutadiyne led to the formation of crystals in which the molecules were disordered along the main axis of the molecule and a subsequent controlled topochemical polymerization was not successful. However, exploiting the Lewis acidity of diiodobutadiyne allowed to crystallize the latter in an ordered host–guest crystal with *N, N'*-biscyanoalkyl oxalamides as Lewis-basic hosts (Figure 10.7).

After the polymerization, the host was successfully removed by sonication as well as rinsing with organic solvents. As a result, the polymer strands immediately aggregated in solution to furnish fibers with diameters of 10–50 nm [96]. The fibers of the obtained poly(diiododiacetylene)s can be formally regarded as a "carbon subiodide" and, accordingly, constituted a highly reactive precursor material that underwent an explosive conversion into carbonaceous materials by application of an external stimulus such as shock, pressure or irradiation [97]. The investigations by Goroff et al. [98] revealed that a mild carbonization could be achieved either by irradiation of the bulk fibers under release of iodine or by a dehalogenation of suspensions of the fibrous polymers through treatment with a Lewis base at room temperature. The carbonization of the fibrous polymers furnished a one-dimensionally oriented nanostructured

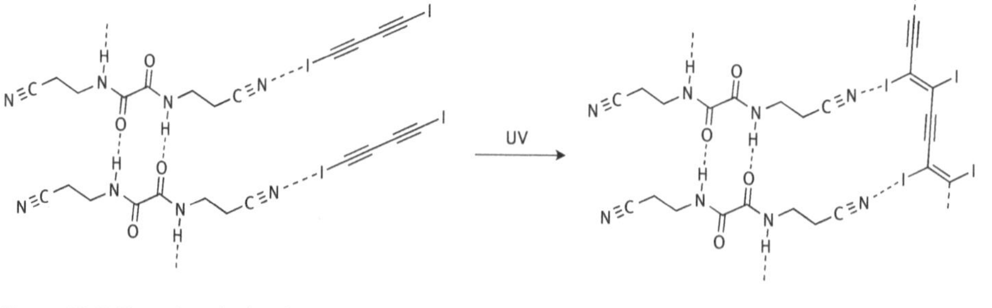

Figure 10.7: Topochemical polymerization of diiodobutadiyne in host–guest cocrystals with *N, N'*-biscyanoalkyl oxalamides triggered by UV irradiation [95].

carbon material rich in sp^2-hybridized carbons under very mild conditions. While constituting an unambiguous demonstration of the potential of reactive carbon-rich precursors toward the preparation of carbon nanostructures with defined morphology, the employed host–guest strategy cannot be straightforwardly transferred to other precursor molecules, rendering approaches based on supramolecular self-assembly worthwhile alternatives.

10.6 Self-Assembly and Carbonization of Functionalized Oligoynes

In order to allow for a defined preorganization of the reactive carbon precursor molecules before carbonization, an approach based on their self-assembly in solution or in the mesophase offers substantially more flexibility in comparison to a topochemical polymerization in single-crystalline specimen. In addition to the potential error correction before a covalent capture of the structures by cross-linking, the flexible nature of the self-assembled aggregates should be able to structurally adapt to the requirements of the carbonization process. The wealth of potentially accessible aggregates can be readily illustrated on the basis of the various examples for which the self-assembly and polymerization of diacetylene-containing molecules have already been successfully demonstrated. Among others, diacetylenes have been polymerized from mono- or multilayered structures [99–104], in molecular arrays organized on surfaces [105–107], as well as in self-assembled vesicles and tubular aggregates [108–114].

Expanding on the well-established strategy to covalently capture self-assembled aggregates by polymerization of incorporated diacetylene moieties, Morin and coworkers [115, 116] recently employed macrocycles with multiple diacetylene units and amide groups for the preparation of one-dimensional nanorods and nanotubes from organogels (Figure 10.8). The synthesis of the macrocycles featuring polymerizable diacetylene units was achieved by using 5-hexyn-1-ol as a commercially available starting material. Coupling with tetradecyne according to Cadiot–Chodkiewicz conditions and conversion of the hydroxyl group to an amine led to an amine-functionalized diacetylene. Subsequent reaction with diiodo-benzoic acid yielded the corresponding amide, which featured a diiodoarene group that was used as the key building block in three sequential Sonogashira couplings. These coupling reactions served to construct half of the desired macrocycle that eventually comprised two terminal alkyne groups. In a final coupling step under Eglinton–Galbraith conditions, two such half macrocycles were homocoupled to yield the full macrocycle containing two diacetylene and four acetylene moieties as well as further two diacetylenes in the attached side groups.

The obtained molecules were self-assembled into an organogel, and UV irradiation of the aggregated molecules led to the consumption of both the exocyclic and the macrocyclic butadiyne units, as proven by Raman spectroscopy. While indicating the

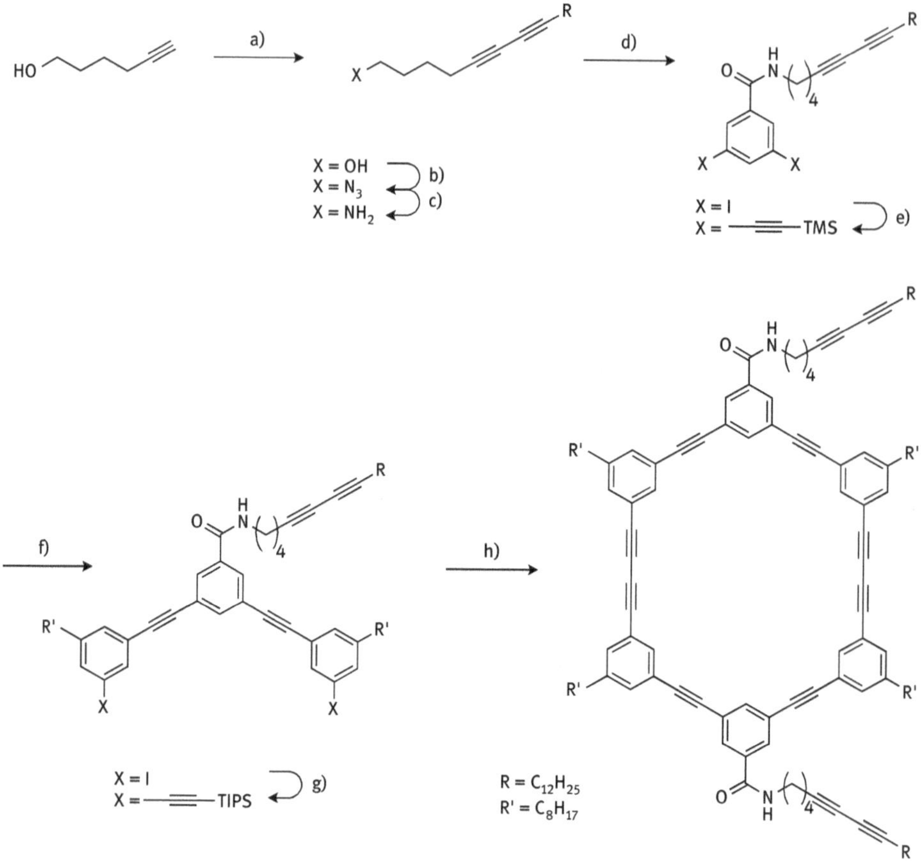

Figure 10.8: Reagents and conditions: (a) (i) Br_2, H_2O, (ii) 1-tetradecyne, CuCl, n-$PrNH_2$/n-$BuNH_2$, NH_2OH, Et_2O/H_2O; (b) (i) TsCl, Et_3N, DMAP, CH_2Cl_2 and (ii) NaN_3, DMF; (c) $SnCl_2$, MeOH; (d) $C_6H_3I_2COOH$, EDS, DMAP, CH_2Cl_2; (e) TMS–C≡C–H, $PdCl_2(PPh_3)_2$, CuI, Et_3N, THF; (f) (i) KOH, THF, MeOH and (ii) $C_6H_3I_2C_8H_{17}$, $PdCl_2(PPh_3)_2$, CuI, DIPEA, THF; (g) TIPS–C≡C–H, $PdCl_2(PPh_3)_2$, CuI, Et_3N, THF; and (h) (i) TBAF, THF and (ii) CuCl, $CuCl_2$, pyridine. Synthesis of macrocycles with multiple diacetylene units and amide groups as prepared by Morin and coworkers [115].

formation of multiple poly(diacetylene)s along the axis of the tubular aggregates, the exact nature of the obtained “nanorods” as well as a potential carbonization of the latter by mild thermal treatment remains to be demonstrated [86]. Following a similar approach, carbon-rich monomers featuring multiple diacetylene units were recently polymerized from a gel state and subsequently partly carbonized at temperatures as low as 55°C to furnish micrometer-sized flakes with a multilayered graphitic structure [117]. However, no control over the nanoscopic morphology was obtained in this case.

Demonstrating the usefulness of higher oligoynes as carbon precursors, Ding and Olesik [118] prepared carbon nanospheres in a wet-chemical approach by

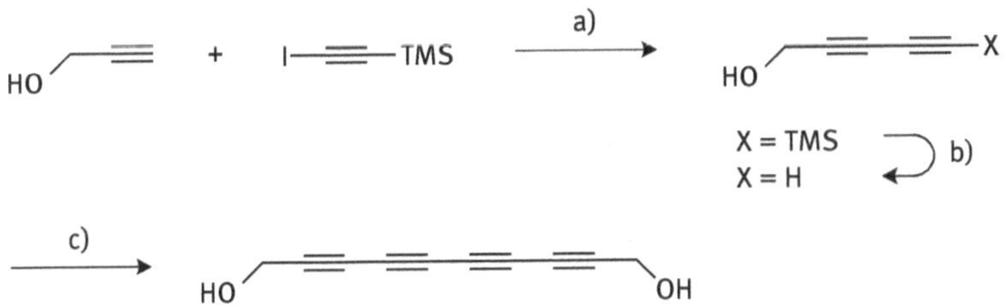

Figure 10.9: Reagents and conditions: (a) $(PPh_3)_2PdCl_2$, CuI, HCC-TMS, $(i\text{-}Pr)_2NH$; (b) KOH/H_2O, MeOH; and (c) $Cu(OAc)_2 \cdot H_2O$, air, TMEDA. Preparation of deca-2,4,6,8-tetrayne-1, 10-diol based on a sequence of Sonogashira and Hay reaction [118].

carbonization of a dispersion of deca-2,4,6,8-tetrayne-1,10-diol in a THF/water mixture. The tetrayne was obtained starting from commercially available propargyl alcohol and 1-iodo-2-(trimethylsilyl)acetylene that were cross-coupled in a Sonogashira-type reaction. Desilylation under alkaline conditions and subsequent homocoupling of the product under Hay conditions yielded the dihydroxyl functional tetrayne (Figure 10.9).

The carbonization of the precursor was carried out by heating the mixture to 70°C, and the addition of surfactants efficiently helped to control the size of the obtained water-soluble carbon nanospheres (Figure 10.10). From parallel observations on solid

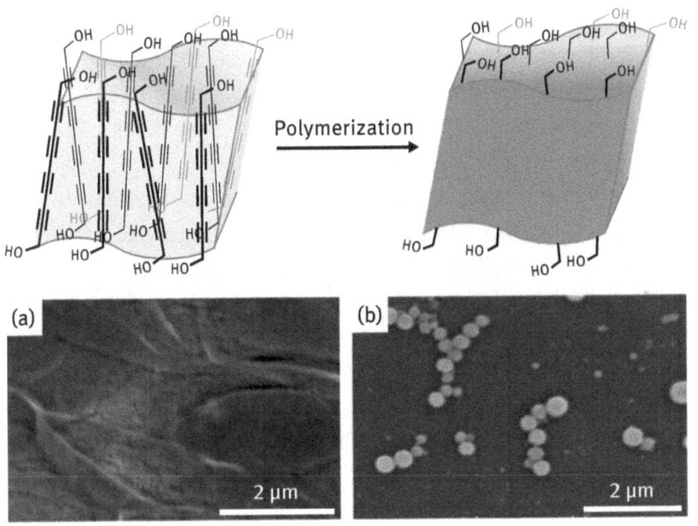

Figure 10.10: Schematic depiction of the wet-chemical preparation of carbon nanospheres. (a) A tetrayne-diol self-assembled into thin films in a THF/water mixture. (b) Droplets formed by heating and polymerization resulted in the formation of filled carbon spheres [13, 118].

thin films of the molecules, the authors suggested that the thermal treatment would first lead to a covalent cross-linking by a polymerization of the tetraynes, followed by further carbonization, although the validity of this analogy remained unclear. Most likely, the initially solid particles in the THF/water dispersion melt at the elevated temperature, resulting in an emulsion of liquid tetrayne droplets, of which the size can be efficiently controlled with the additional surfactant. In an extension of their approach, the same group investigated the polymerization of a dibutyl-substituted tetrayne derivative in an aqueous emulsion polymerization at 80°C with poly(vinyl alcohol) as the emulsifier [119]. Carbon nanospheres of various sizes were successfully prepared by these emulsion polymerizations of reactive precursor molecules, and the morphology of the nanostructures was preserved throughout further processing steps at elevated temperatures. While in these two examples, the oligoynes themselves have, hence, not been used to obtain a controlled nanostructure by means of supramolecular self-assembly, their phase segregation from a polar medium and the interfacial stabilization using surfactants was successfully employed as a soft template to obtain carbon nanostructures. In this context, their initial polymerization at still very benign temperatures served to covalently capture the template nanostructure that thus survived the final carbonization. Investigating the self-assembly and subsequent polymerization of another molecular precursor with a tetrayne moiety, Morin and coworkers [120] also obtained carbon nanospheres from an organogel. Whereas also this example thus involved a step in which the reactive molecular precursors were self-assembled, it is important to note that the nanoscopic morphology of the precursor phase was not preserved in the carbonized product.

In order to combine the aspect of supramolecular self-assembly into defined nanostructures with the use of similarly reactive molecular precursors, Frauenrath and coworkers [121, 122] have explored the phase segregation of suitable oligoyne amphiphiles in polar media. To this end, glycosylated hexayne derivatives were prepared that resembled typical glycolipid amphiphiles (Figure 10.11).

The preparation of glycosylated hexayne derivatives was achieved by employing a sequence of bromination and *sp*–*sp* Negishi coupling reactions (Figure 10.12). Thus, acetylated 4-pentyn-1-yl glucoside was first converted into the glycosylated triyne. To this end, 1,4-bis(trimethylsilyl)buta-1,3-diyne was used as an acetylenic building block and converted into a zinc acetylide via the monolithiated intermediate. The Pd-catalyzed coupling of this zinc acetylide with the bromoacetylene furnished the corresponding triyne. The latter was converted into the bromotriyne that was found to be highly reactive toward an unusual dimerization in the solid state and, therefore, had to be kept in solution [123]. It was coupled to 1-triisopropylsilyl-6-trimethylsilyl-hexa-1,3,5-triyne following the same procedure, yielding the glycosylated hexayne. In the final step, complete deacetylation of the glycosyl residue furnished the corresponding glycosylated hexayne amphiphile.

These carbon-rich amphiphiles reversibly self-assembled in aqueous solution to furnish a mixture of unilamellar and multilamellar bilayer vesicles with a broad

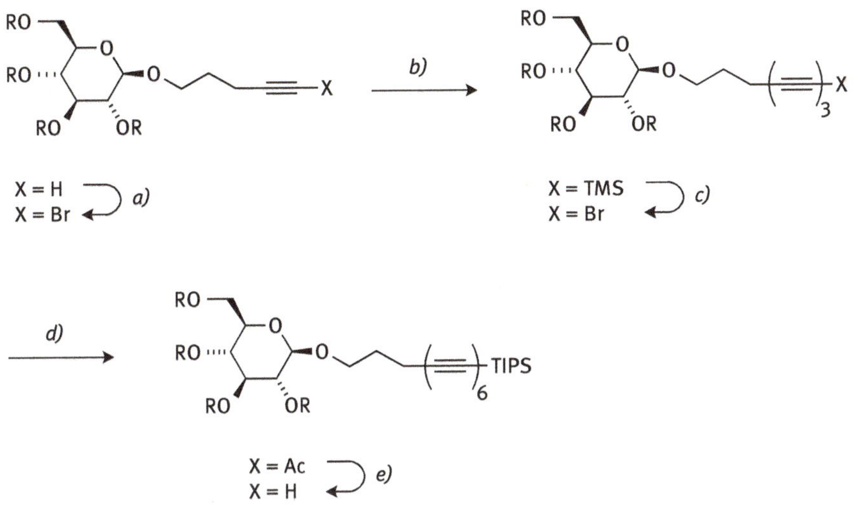

Figure 10.11: Reagents and conditions: (a) $AgNO_3$, NBS, MeCN; (b) 1,4-bis(trimethylsilyl) buta-1,3-diyne, MeLi · LiBr, $ZnCl_2$, $PdCl_2$(dppf) · DCM, THF/toluene; (c) AgF, NBS, MeCN; (d) 1-triisopropylsilyl-6-trimethylsilylhexa-1,3,5-triyne, MeLi · LiBr, $ZnCl_2$, $PdCl_2$(dppf) · DCM, THF/toluene; and (e) (i) NaOMe, 1,4-dioxane, MeOH and (ii) Amberlite IR-120 (H^+). Synthesis of reactive hexayne amphiphiles based on bromination and Negishi coupling reactions [121].

size distribution. Subsequent vesicle extrusion through polycarbonate membranes of controlled porosity furnished mostly unilamellar vesicles with a controlled size. In the aggregated state, irradiation by UV light at a temperature of only 1°C was sufficient as a mild external stimulus to induce the carbonization process. The hexayne segments were completely consumed according to UV/Vis and Raman spectroscopy,

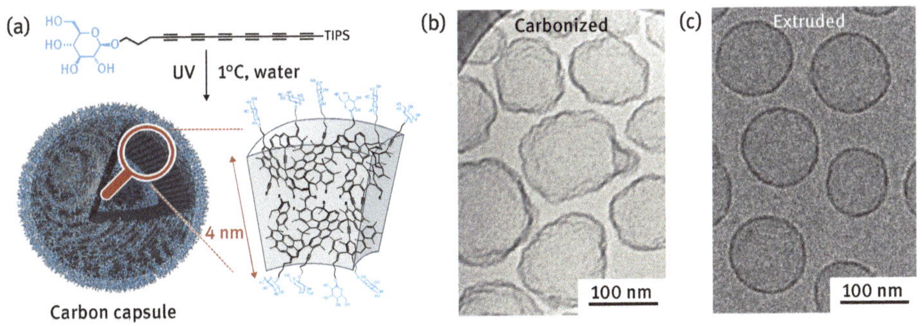

Figure 10.12: (a) Schematic representation of the wet-chemical preparation of carbon nanocapsules from glycosyl-functionalized amphiphiles containing a reactive carbon-only hexayne segment. (b) Unilamellar vesicles were obtained after self-assembly of the carbon-rich amphiphiles in water and vesicle extrusion through polycarbonate membranes. (c) The carbonization of these vesicles then proceeded under UV irradiation below room temperature, and the formed carbon nanocapsules comprised a graphite-like amorphous carbon with a hydrophilic carbohydrate shell [121].

and further characterization was carried out with the obtained carbon nanocapsules that provided conclusive evidence for an extensive carbonization. A thorough analysis of the nanocapsules suggested that the walls were constituted of "graphite-like amorphous carbon", which otherwise requires process temperatures above 600°C. At the same time, the carbohydrate shell of the vesicles was retained, as verified by a glucose-binding assay with Concanavalin A. Thus, combining a carbon-only reactive oligoyne segment with a hydrophilic head group served to furnish amphiphilic precursor molecules that allowed for the preparation of functional carbon nanocapsules under very mild conditions.

Along similar lines, choosing an oligoyne amphiphile with a smaller carboxylic acid ester head group allowed for the preparation of carbon nanostructures of a different morphology, that is, carbon nanosheets from a self-assembled monolayer of the precursors at the air–water interface [124].

The employed hexayne amphiphile was prepared following a similar sequence of bromination and Negishi cross-coupling as described before (Figure 10.13). The synthesis started from commercially available 1-hexynoic acid, and trityl phenol was introduced as a bulky protective group. The Negishi *sp*–*sp* cross-coupling toward the hexayne derivative was carried out with the zinc acetylide formed from 1-trimethylsilyloctadeca-1,3,5-triyne in order to introduce a lipophilic dodecyl terminus for improved film-forming properties. Subsequent transesterification furnished the amphiphilic hexayne methyl ester. The latter resembled simple fatty acid esters and, therefore, formed self-assembled monolayers upon spreading an organic solution at the air–water interface in a Langmuir trough (Figure 10.14).

The self-assembled monolayer featured a dense packing of the hexayne moieties, which consisted of a less than 7 Å thick carbon-only sublayer. Moreover, the investigations showed that multiple of the *sp*-hybridized carbon atoms of adjacent

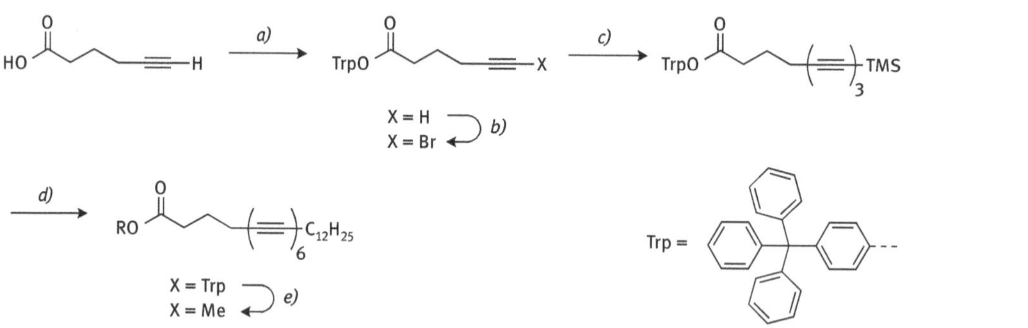

Figure 10.13: Reagents and conditions: (a) 4-tritylphenol, EDCI, DPTS, DCM/toluene; (b) $AgNO_3$, NBS, MeCN; (c) 1,4-bis(trimethylsilyl)buta-1,3-diyne, MeLi · LiBr, $ZnCl_2$, $PdCl_2$(dppf) · DCM, THF/toluene; (d) AgF, NBS, MeCN; *then* 1-trimethylsilyloctadeca-1,3,5-triyne, MeLi · LiBr, $ZnCl_2$, $PdCl_2$(dppf) · DCM, THF/toluene; and (e) NaOMe, DCM, MeOH. Preparation of methyl ester hexayne amphiphiles starting from 1-hexynoic acid [124].

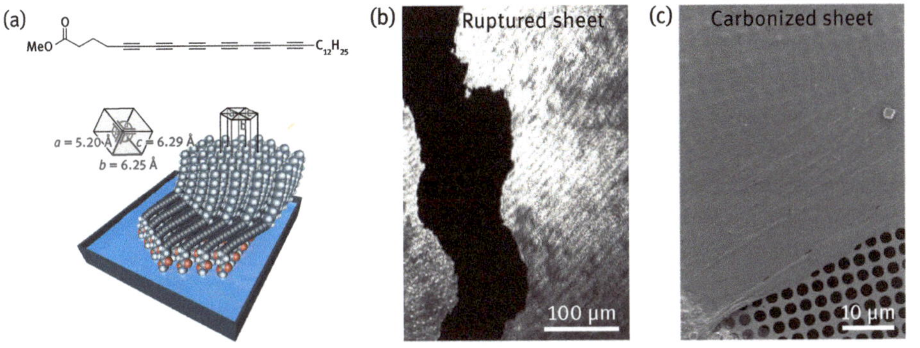

Figure 10.14: Preparation of carbon nanosheets from methyl ester hexayne amphiphiles: (a) structure of the molecular precursor and the compressed monolayer at the air–water interface; (b) Brewster-Angle microscopy image of a ruptured sheet (gray) on water as the sublayer (black); and (c) scanning electron microscopy image of a carbon nanosheet showing the edge of the carbon nanosheet [124].

hexayne moieties were at a reactive distance. UV irradiation at room temperature was sufficient to induce the conversion of the monolayer of aggregated molecules into a self-supporting carbon nanosheet with a thickness of 1.9 nm and lateral dimensions on the order of centimeters, that is, under preservation of the monolayer morphology and chemical functionalization. The carbon microstructure that resembled the one of graphene obtained by reduction of graphene oxide featured an sp^2-hybridized carbon content of 85 %. This is plausible for hexayne precursors in which two of the twelve acetylene carbons remain covalently linked to the substituents and must hence become tetrahedral defects that, due to the structure of the self-assembled precursors, will be homogeneously distributed throughout the carbon nanosheet.

10.7 Conclusions and Outlook

Molecules rich in *sp*-hybridized carbons represent an attractive platform to access carbon nanomaterials through cross-linking by UV irradiation or other mild stimuli. Although the preparation of (asymmetric) oligoynes remains challenging, it is possible to include a wide range of chemical functional groups into the precursor molecules. Hence, the presented approaches employing mild carbonization conditions constitute an interesting alternative to the largely established pyrolytic methods that employ temperatures above 600°C. Exploiting the high reactivity of the precursors allows to control the chemical functionality of the carbon structure through the predefined chemical functional groups in the precursor molecules. Along these lines, promising strategies to further improve the stability of oligoynes through host–guest interactions could be accompanied by control of the supramolecular arrangement of the oligoynes. It is important to point out that the self-assembly of the oligoynes into an ordered arrangement is required to achieve a control over the morphology of the carbon nanomaterials. Since the carbonization proceeds under mild conditions, a variety of interfaces can serve as soft templates, which further facilitates the access to carbon nanomaterials, since chemical etching for the removal of a hard template is no longer necessary. In this context, we conceive templating by, for example, covalent organic frameworks, nonspherical colloids and double emulsions to be highly attractive approaches for the preparation of novel carbon nanostructures and nanomaterials from oligoyne precursors.

Acknowledgments

B.S. acknowledges the support from the "EPFL Fellows" fellowship program cofunded by Marie Curie (FP7 grant agreement no. 291771) and from the Alexander von Humboldt foundation (Feodor Lynen fellowship). Moreover, funding from the Food and Nutrition Centre of EPFL, as well as the European Research Council (ERC grant 239831), is gratefully acknowledged.

References

[1] De Jong, KP, Geus JW. Carbon Nanofibers: Catalytic Synthesis and Applications, Catal Rev – Sci Eng 2000;42:481.
[2] Falcao EH; Wudl F. Carbon allotropes: beyond graphite and diamond, J Chem Technol Biotechnol 2007;82:524.
[3] Allen MJ, Tung VC, Kaner RB. Honeycomb Carbon: A Review of Graphene, Chem Rev 2010;110:132.
[4] Georgakilas V, Perman JA, Tucek J, Zboril R. Broad Family of Carbon Nanoallotropes: Classification, Chemistry, and Applications of Fullerenes, Carbon Dots, Nanotubes, Graphene, Nanodiamonds, and Combined Superstructures, Chem Rev 2015;115:4744.
[5] Chuenchom L, Kraehnert R; Smarsly BM. Recent progress in soft-templating of porous carbon materials, Soft Matter 2012;8:10801.
[6] Segura, JL, Martin N, Guldi DM. Fullerene for organic electronics, Chem Soc Rev 2005;34:31.
[7] Zhang Q, Huang J-Q, Qian W-Z, Zhang Y-Y, Wei F. The Road for Nanomaterials Industry: A Review of Carbon Nanotube Production, Post-Treatment, and Bulk Applications for Composites and Energy Storage Small 2013;9:1237.
[8] Burchell TD. Carbon materials for advanced technologies. Oxford, UK: Elsevier Science Ltd, 1999.
[9] Earnshaw A, Greenwood NN. Chemistry of the elements, 2nd ed. Oxford: Butterworth-Heinemann Ltd, 1997.
[10] Wei L, Kuo PK, Thomas RL, Anthony TR, Banholzer WF. Thermal conductivity of isotopically modified single crystal diamond Phys Rev Lett 1993;70:3764.
[11] Zhi L, Müllen K. A bottom-up approach from molecular nanographenes to unconventional carbon materials, J Mater Chem 2008;18:1472.
[12] Prasek J, Drbohlavova J, Chomoucka J, Hubalek J, Jasek O, Adam V, et al. Methods for carbon nanotubes synthesis—review, J Mater Chem 2011;21:15872.
[13] Hoheisel TN, Schrettl S, Szilluweit R, Frauenrath H. Nanostructured Carbonaceous Materials from Molecular Precursors, Angew Chem Int Ed 2010;49:6496.

[14] Schrettl S, Frauenrath H. Elements for a Rational Polymer Approach towards Carbon Nanostructures, Angew Chem Int Ed 2012;51:6569.
[15] Zhi L, Wu J, Li J, Kolb U, Müllen K. Carbonization of Disclike Molecules in Porous Alumina Membranes: Toward Carbon Nanotubes with Controlled Graphene-Layer Orientation, Angew Chem Int Ed 2005;44:2120.
[16] Werner JG, Hoheisel TN, Wiesner U. Synthesis and Characterization of Gyroidal Mesoporous Carbons and Carbon Monoliths with Tunable Ultralarge Pore Size, ACS Nano 2014;8:731.
[17] Liu J, Yang T, Wang D-W, Lu GQ, Zhao D, Qiao SZ. A facile soft-template synthesis of mesoporous polymeric and carbonaceous nanospheres, Nature Commun 2013;4:2798.
[18] Hirsch A, Englert JM, Hauke F. Wet Chemical Functionalization of Graphene, Acc Chem Res 2013;46:87.
[19] Eigler S, Hirsch A. Chemistry with Graphene and Graphene Oxide - Challenges for Synthetic Chemists, Angew Chem Int Ed 2014;53:7720.
[20] Chalifoux WA, Tykwinski RR. Synthesis of extended polyynes: Toward carbyne, C R Chim 2009;12:341.
[21] Sladkov AM, Kudryavtsev YP. Polyynes, Russ Chem Rev 1963;32:229.
[22] Armitage JB, Cook CL, Jones ER, Whiting MC. Researches on Acetylenic Compounds. Part XXX VI. The Synthesis of Symmetrical Conjugated Triacetylenic Compounds. J Chem Soc (resumed) 1952:2010.
[23] Klinger C, Vostrowsky O, Hirsch A. Synthesis of Alkylene-Bridged Diphenyl-Oligoynes, Eur J Org Chem 2006;2006:1508.
[24] Wang C, Batsanov AS, West K, Bryce MR. Synthesis and Crystal Structures of Isolable Terminal Aryl Hexatriyne and Octatetrayne Derivatives: $Ar{-}(C{\equiv}C)_nH$ ($n = 3, 4$), Org Lett 2008;10:3069.
[25] Kroto HW, Walton DR, Jones DE, Haddon RC. Polyynes and the Formation of Fullerenes [and Discussion], Philos Trans R Soc Lond Ser A 1993;343:103.
[26] Grösser T, Hirsch A. Dicyanopolyynes: Formation of New Rod-Shaped Molecules in a Carbon Plasma, Angew Chem Int Ed Engl 1993;32:1340.
[27] Cataldo F. Synthesis of polyynes in a submerged electric arc in organic solvents, Carbon 2004;42:129.
[28] Lagow RJ, Kampa JJ, Wei H-C, Battle SL, Genge JW, Laude DA, et al. Synthesis of Linear Acetylenic Carbon: The "*sp*" Carbon Allotrope, Science 1995;267:362.
[29] Nast R. Coordination chemistry of metal alkynyl compounds, Coord Chem Rev 1982;47:89.
[30] Mak TC, Zhao X-L, Wang Q-M, Guo G-C. Synthesis and structural characterization of silver(I) double and multiple salts containing the acetylenediide dianion, Coord Chem Rev 2007;251:2311.
[31] Cao Z, Xi B, Jodoin DS, Zhang L, Cummings SP, Gao Y, et al. Diruthenium–Polyyn-diyl–Diruthenium Wires: Electronic Coupling in the Long Distance Regime, J Am Chem Soc 2014;136:12174.
[32] Szafert S, Gladysz JA. Update 1 of: Carbon in One Dimension: Structural Analysis of the Higher Conjugated Polyynes, Chem Rev 2006;106:PR1.
[33] Johnson TR, Walton DRM. Silylation as a protective method in acetylene chemistry : Polyyne chain extensions using the reagents, $Et_3Si(CC)_mH$ ($m = 1,2,4$) in mixed oxidative couplings, Tetrahedron 1972;28:5221.
[34] Movsisyan LD, Kondratuk DV, Franz M, Thompson AL, Tykwinski RR, Anderson HL. Synthesis of Polyyne Rotaxanes, Org Lett 2012;14:3424.
[35] Weisbach N, Baranova Z, Gauthier S, Reibenspies JH, Gladysz JA. A new type of insulated molecular wire: a rotaxane derived from a metal-capped conjugated tetrayne, Chem Commun 2012;48:7562.

[36] Sahnoune H, Baranov. Z, Bhuvanesh N, Gladysz JA, Halet J-F. A Metal-Capped Conjugated Polyyne Threaded through a Phenanthroline-Based Macrocycle. Probing beyond the Mechanical Bond to Interactions in Interlocked Molecular Architectures, Organometallics 2013;32:6360.
[37] Schrettl S, Contal E, Hoheisel TN, Fritzsche M, Balog S, Szilluweit R, et al. Facile synthesis of oligoyne amphiphiles and their rotaxanes, Chem Sci 2015;6:564.
[38] Glaser C. Beiträge zur Kenntniss des Acetenylbenzols, Ber Dtsch Chem Ges 1869;2:422.
[39] Hay A. Communications- Oxidative Coupling of Acetylenes, J Org Chem 1960;25:1275.
[40] Hay AS. Oxidative Coupling of Acetylenes. II, J Org Chem 1962;27:3320.
[41] Eglinton G, Galbraith AR. Cyclic diynes, Chem Ind 1956:737.
[42] Eglinton G, Galbraith AR. Macrocyclic acetylenic compounds. Part I. Cyclotetradeca-1 :3-diyne and related compounds, J Chem Soc 1959:889.
[43] Sonogashira K, Tohda Y, Hagihara N. A convenient synthesis of acetylenes: catalytic substitutions of acetylenic hydrogen with bromoalkenes, iodoarenes and bromopyridines, Tetrahedron Lett 1975;16:4467.
[44] Chinchilla R, Nájera C. Recent advances in Sonogashira reactions, Chem Soc Rev 2011;40:5084.
[45] Chodkiewicz W, Cadiot P. Nouvelle synthèse de composés polyacétyleniques conjugués symétriques et dissymétriques, Compt Rend Acad Sci 1955;241:1055.
[46] Jevric M, Nielsen MB. Synthetic Strategies for Oligoynes, Asian J Org Chem 2015;4:286.
[47] Tykwinski RR. Carbyne: The Molecular Approach, Chem Rec 2015;15:1060.
[48] Negishi E, Okukado N, Lovich SF, Luo FT. A method for the preparation of terminal and internal conjugated diynes via palladium-catalyzed cross-coupling, J Org Chem 1984;49:2629.
[49] Negishi E-i, Hata M, Xu C. A Strictly "Pair"-Selective Synthesis of Conjugated Diynes via Pd-Catalyzed Cross Coupling of 1,3-Diynylzincs: A Superior Alternative to the Cadiot–Chodkiewicz Reaction, Org Lett 2000;2:3687.
[50] Métay E, Hu Q, Negishi E-i. Highly Efficient and Selective Synthesis of Conjugated Triynes and Higher Oligoynes of Biological and Materials Chemical Interest via Palladium-Catalyzed Alkynyl–Alkenyl Coupling, Org Lett 2006;8:5773.
[51] Hoheisel TN, Frauenrath H. A Convenient Negishi Protocol for the Synthesis of Glycosylated Oligo(ethynylene)s, Org Lett 2008;10:4525.
[52] Fritsch P. IV. Ueber die Darstellung von Diphenylacetaldehyd und eine neue Synthese von Tolanderivaten, Liebigs Ann Chem 1894;279:319.
[53] Buttenberg WP. Condensation des Dichloracetals mit Phenol und Toluol, Liebigs Ann Chem 1894;279:324.
[54] Wiechell H. Condensation des Dichloracetals mit Anisol und Phenetol, Liebigs Ann Chem 1894;279:337.
[55] Chalifoux WA, Tykwinski RR. Polyyne synthesis using carbene/carbenoid rearrangements, Chem Rec 2006;6:169.
[56] Siemsen P, Livingston RC, Diederich F. Acetylenic Coupling: A Powerful Tool in Molecular Construction, Angew Chem Int Ed 2000;39:2632.
[57] Allen SE, Walvoord RR, Padilla-Salinas R, Kozlowski MC. Aerobic Copper-Catalyzed Organic Reactions, Chem Rev 2013;113:6234.
[58] Sindhu KS, Thankachan AP, Sajitha PS, Anilkumar G. Recent developments and applications of the Cadiot–Chodkiewicz reaction, Org Biomol Chem 2015;13:6891.
[59] Marsden JA, Haley MM. Metal-Catalyzed Cross-Coupling Reactions. de Meijere, A. and Diederich, F, editors, Weinheim: Wiley-VCH Verlag GmbH, 2004:317.
[60] Alami M, Hamze A, Messaoudi S. Comprehensive Organic Synthesis II (Second Edition), Knochel P, editor. Amsterdam: Elsevier, 2014:528.

[61] Cai C, Vasella A. Oligosaccharide Analogues of Polysaccharides. Part 5. Studies on the cross-coupling of alkynes and haloalkynes, Helv Chim Acta 1995;78:2053.
[62] Shi W, Lei A. 1,3-Diyne chemistry: synthesis and derivations, Tetrahedron Lett 2014;55:2763.
[63] Negishi E-i, Anastasia L. Palladium-Catalyzed Alkynylation, Chem Rev 2003;103:1979.
[64] Hoheisel TN, Frauenrath H. Glycosylated Oligo(ethynylene)s via a Pd/Zn-Mediated Cross-Coupling Reaction, Chimia 2009;63:208.
[65] Tykwinski RR, Chalifoux W, Eisler S, Lucotti A, Tommasini M, Fazzi D, et al. Toward carbyne: Synthesis and stability of really long polyynes, Pure Appl Chem 2010;82:891.
[66] Tykwinski RR. Carbyne: The Molecular Approach, Chem Rec 2015;15:1060.
[67] Cook CL, Jones ER, Whiting MC. Researches on acetylenic compounds. Part XXXIX. General routes to aliphatic polyacetylenic hydrocarbons and glycols, J Chem Soc 1952:2883.
[68] Kendall J, McDonald R, Ferguson MJ, Tykwinski RR. Synthesis and Solid-State Structure of Perfluorophenyl End-Capped Polyynes, Org Lett 2008;10:2163.
[69] Jones ER, Lee HH, Whiting MC. Researches on acetylenic compounds. Part LXIV. The preparation of conjugated octa- and deca-acetylenic compounds, J Chem Soc 1960:3483.
[70] Chalifoux WA, McDonald R, Ferguson MJ, Tykwinski RR. *tert*-Butyl-End-Capped Polyynes: Crystallographic Evidence of Reduced Bond-Length Alternation, Angew Chem Int Ed 2009;48:7915.
[71] Eisler S, Slepkov AD, Elliott E, Luu T, McDonald R, Hegmann FA, et al. Polyynes as a Model for Carbyne: Synthesis, Physical Properties, and Nonlinear Optical Response, J Am Chem Soc 2005;127:2666.
[72] Gibtner T, Hampel F, Gisselbrecht J-P, Hirsch A. End-Cap Stabilized Oligoynes: Model Compounds for the Linear sp Carbon Allotrope Carbyne, Chem Eur J 2002;8:408.
[73] Zheng Q, Gladysz JA. A Synthetic Breakthrough into an Unanticipated Stability Regime: Readily Isolable Complexes in which C_{16}–C_{28}Polyynediyl Chains Span Two Platinum Atoms, J Am Chem Soc 2005;127:10508.
[74] Schermann G, Grösser T, Hampel F, Hirsch A. Dicyanopolyynes: A Homologuous Series of End-Capped Linear sp Carbon, Chem Eur J 1997;3:1105.
[75] Eastmond R, Johnson TR, Walton DRM. Silylation as a protective method for terminal alkynes in oxidative couplings : A general synthesis of the parent polyynes $H(CC)_nH$ (n = 4–10, 12), Tetrahedron 1972;28:4601.
[76] Chalifoux WA, Tykwinski RR. Synthesis of polyynes to model the *sp*-carbon allotrope carbyne, Nature Chem 2010;2:967.
[77] Rubin Y, Lin SS, Knobler CB, Anthony J, Boldi AM, Diederich F. Solution-spray flash vacuum pyrolysis: a new method for the synthesis of linear poliynes with odd numbers of C.tplbond.C bonds from substituted 3,4-dialkynyl-3-cyclobutene-1,2-diones, J Am Chem Soc 1991;113:6943.
[78] Tsuji M, Tsuji T, Kuboyama S, Yoon S-H, Korai Y, Tsujimoto T, et al. Formation of hydrogen-capped polyynes by laser ablation of graphite particles suspended in solution, Chem Phys Lett 2002;355:101.
[79] Cataldo F, Ursini O, Angelini G. Kinetics of polyynes formation with the submerged carbon arc, J Electroanal Chem 2007;602:82.
[80] Zhao C, Kitaura R, Hara H, Irle S, Shinohara H. Growth of Linear Carbon Chains inside Thin Double-Wall Carbon Nanotubes, J Phys Chem C 2011;115:13166.
[81] Hlavatý J, Kavan L, Kasahara N, Oya A. Polymerisation of 1-iodohexa-1,3,5-triyne and hexa-1,3,5-triyne: a new synthesis of carbon nanotubes at low temperatures, Chem Commun 2000:737.
[82] Hlavatý J, Kavan L, Okabe K, Oya A. Carbonaceous polymers and nanotubes from α,ω-dialkali hexatriynides, Carbon 2002;40:1147.

[83] Hlavatý J, Kavan L, Kubišta J. Carbonaceous materials from end-capped alkynes, Carbon 2002;40:345.
[84] Zhou N, Merschrod SE, Zhao Y. Preparation of Fullerene–Polyyne Nanospheres via Thermally Induced Solid-State Polymerization, J Am Chem Soc 2005;127:14154.
[85] Zhou N, Zhao Y. Conjugated Oligoyne-Bridged [60]Fullerene Molecular Dumbbells: Syntheses and Thermal and Morphological Properties, J Org Chem 2010;75:1498.
[86] Rondeau-Gagné S, Morin J-F. Preparation of carbon nanomaterials from molecular precursors, Chem Soc Rev 2014;43:85.
[87] Wegner G. Polymerisation von Derivaten des 2.4-Hexadiin-1.6-diols im kristallinen Zustand, Z Naturforsch B: Anorg Chem Org Chem 1969;24:824.
[88] Enkelmann V. Polydiacetylenes, Cantow H-J, editor. Berlin Heidelberg: Springer, 1984;63:91.
[89] Lauher JW, Fowler FW, Goroff NS. Single-Crystal-to-Single-Crystal Topochemical Polymerizations by Design, Acc Chem Res 2008;41:1215.
[90] Baughman RH, Yee KC. Solid-state polymerization of linear and cyclic acetylenes, J Polym Sci Macromol Rev 1978;13:219.
[91] Sarkar A, Okada S, Matsuzawa H, Matsuda H, Nakanishi H. Novel polydiacetylenes for optical materials: beyond the conventional polydiacetylenes, J Mater Chem 2000;10:819.
[92] Sarkar A, Okada S, Komatsu K, Nakanishi H, Matsuda H. Octatetraynes with Directly Linked Aromatic Sidegroups: Preparation and Polymerization, Macromolecules 1998;31:5624.
[93] Okada S, Hayamizu K, Matsuda H, Masaki A, Minami N, Nakanishi H. Solid-State Polymerization of 15,17,19,21,23,25-Tetracontahexayne, Macromolecules 1994;27:6259.
[94] Chernick ET, Tykwinski RR. Carbon-rich nanostructures: the conversion of acetylenes into materials, J Phys Org Chem 2013;26:742.
[95] Sun A, Lauher JW, Goroff NS. Preparation of Poly(diiododiacetylene), an Ordered Conjugated Polymer of Carbon and Iodine, Science 2006;312:1030.
[96] Luo L, Wilhelm C, Sun A, Grey CP, Lauher JW, Goroff NS. Poly(diiododiacetylene): Preparation, Isolation, and Full Characterization of a Very Simple Poly(diacetylene), J Am Chem Soc 2008;130:7702.
[97] Luo L, Wilhelm C, Young CN, Grey CP, Halada GP, Xiao K, et al. Characterization and Carbonization of Highly Oriented Poly(diiododiacetylene) Nanofibers, Macromolecules 2011;44:2626.
[98] Luo L, Resch D, Wilhelm C, Young CN, Halada GP, Gambino RJ, et al. Room-Temperature Carbonization of Poly(diiododiacetylene) by Reaction with Lewis Bases, J Am Chem Soc 2011;133:19274.
[99] Tieke B, Wegner G, Naegele D, Ringsdorf H. Polymerization of Tricosa-10,12-diynoic Acid in Multilayers, Angew Chem Int Ed Engl 1976;15:764.
[100] Batchelder DN, Evans SD, Freeman TL, Haeussling L, Ringsdorf H, Wolf H. Self-Assembled Monolayers containing Polydiacetylenes, J Am Chem Soc 1994;116:1050.
[101] Kuriyama K, Kikuchi H, Kajiyama T. Solid-State Polymerization Behaviors of Crystalline Diacetylene Monolayers on Hydrophilic Surfaces, Langmuir 1996;12:2283.
[102] Huggins KE, Son S, Stupp SI. Two-Dimensional Supramolecular Assemblies of a Polydiacetylene. 1. Synthesis, Structure, and Third-Order Nonlinear Optical Properties, Macromolecules 1997;30:5305.
[103] Britt DW, Hofmann UG, M.bius D, Hell SW. Influence of Substrate Properties on the Topochemical Polymerization of Diacetylene Monolayers, Langmuir 2001;17:3757.
[104] Day D, Ringsdorf H. Polymerization of diacetylene carbonic acid monolayers at the gas-water interface, J Polym Sci, Part C: Polym Lett 1978;16:205.
[105] Okawa Y, Aono M. Nanoscale control of chain polymerization, Nature 2001;409:683.
[106] Miura A, De Feyter S, Abdel-Mottaleb MM, Gesquière A, Grim PC, Moessner G, et al. Light- and STM-Tip-Induced Formation of One-Dimensional and Two-Dimensional Organic Nanostructures, Langmuir 2003;19:6474.

[107] Sullivan SP, Schnieders A, Mbugua SK, Beebe TP. Controlled Polymerization of Substituted Diacetylene Self-Organized Monolayers Confined in Molecule Corrals, Langmuir 2005;21:1322.
[108] Georger JH, Singh A, Price RR, Schnur JM, Yager P, Schoen PE. Helical and tubular microstructures formed by polymerizable phosphatidylcholines, J Am Chem Soc 1987;109:6169.
[109] Spector MS, Selinger JV, Singh A, Rodriguez JM, Price RR, Schnur JM. Controlling the Morphology of Chiral Lipid Tubules, Langmuir 1998;14:3493.
[110] Singh A, Wong EM, Schnur JM. Toward the Rational Control of Nanoscale Structures Using Chiral Self-Assembly: Diacetylenic Phosphocholines, Langmuir 2003;19:1888.
[111] Svenson S, Messersmith PB. Formation of Polymerizable Phospholipid Nanotubules and Their Transformation into a Network Gel, Langmuir 1999;15:4464.
[112] Cheng Q, Yamamoto M, Stevens RC. Amino Acid Terminated Polydiacetylene Lipid Microstructures: Morphology and Chromatic Transition, Langmuir 2000;16:5333.
[113] Frankel DA, O'Brien DF. Supramolecular assemblies of diacetylenic aldonamides, J Am Chem Soc 1991;113:7436.
[114] Fuhrhop JH, Blumtritt P, Lehmann C, Luger P. Supramolecular assemblies, a crystal structure, and a polymer of N-diacetylenic gluconamides, J Am Chem Soc 1991;113:7437.
[115] Rondeau-Gagné. S, Néabo JR, Desroches M, Larouche J, Brisson J, Morin J-F. Topochemical Polymerization of Phenylacetylene Macrocycles: A New Strategy for the Preparation of Organic Nanorods, J Am Chem Soc 2013;135:110.
[116] Rondeau-Gagné S, Néabo JR, Desroches M, Levesque I, Daigle M, Cantin K, et al. Rigid organic nanotubes obtained from phenylene-butadiynylene macrocycles, Chem Commun 2013;49:9546.
[117] Levesque I, Néabo JR, Rondeau-Gagné S, Vigier-Carrière C, Daigle M, Morin J-F. Layered graphitic materials from a molecular precursor, Chem Sci 2014;5:831.
[118] Ding L, Olesik SV. Synthesis of Polymer Nanospheres and Carbon Nanospheres Using the Monomer 1,8-Dihydroxymethyl-1,3,5,7- octatetrayne, Nano Lett 2004;4:2271.
[119] Ding L, Olesik SV. Carbon Microbeads Produced through Synthesis and Pyrolysis of Poly(1,8-dibutyl-1,3,5,7-octatetrayne), Chem Mater 2005;17:2353.
[120] Néabo JR, Vigier-Carrière C, Rondeau-Gagné S, Morin J-F. Room-temperature synthesis of soluble, fluorescent carbon nanoparticles from organogel precursors, Chem Commun 2012;48:10144.
[121] Szilluweit R, Hoheisel TN, Fritzsche M, Ketterer B, Fontcuberta i Morral A, Demurtas D, et al. Low-Temperature Preparation of Tailored Carbon Nanostructures in Water, Nano Lett 2012;12:2573.
[122] Schrettl, S.; Schulte, B.; Frauenrath, H. Templating for hierarchical structure control in carbon materials, Nanoscale 2016;8:18828.
[123] Hoheisel TN, Schrettl S, Marty R, Todorova TK, Corminboeuf C, Sienkiewicz A, et al. A multistep single-crystal-to-single-crystal bromodiacetylene dimerization, Nature Chem 2013;5:327.
[124] Schrettl S, Stefaniu C, Schwieger C, Pasche G, Oveisi E, Fontana Y, et al. Functional carbon nanosheets prepared from hexayne amphiphile monolayers at room temperature, Nature Chem 2014;6:468.

Index

FSC
www.fsc.org
MIX
Papier | Fördert gute Waldnutzung
FSC® C083411

Kolibri 360 GmbH
Ferdinand-Jühlke-Straße 7
99095 Erfurt, Deutschland
produktsicherheit@kolibri360.de